Red Hat Enterprise Linux 9 Administration

A comprehensive Linux system administration guide
for RHCSA certification exam candidates

Pablo Iranzo Gómez

Pedro Ibáñez Requena

Miguel Pérez Colino

Scott McCarty

BIRMINGHAM—MUMBAI

Red Hat Enterprise Linux 9 Administration

Group Product Manager: Mohd Riyan Khan
Publishing Product Manager: Shrilekha Malpani
Senior Editor: Romy Dias
Technical Editor: Arjun Varma
Copy Editor: Safis Editing
Project Coordinator: Ashwin Kharwa
Proofreader: Safis Editing
Indexer: Manju Arasan
Production Designer: Aparna Bhagat
Marketing Coordinator: Nimisha Dua

First published: November 2021

Second edition: October 2022

Production reference: 1211022

Published by Packt Publishing Ltd.
Livery Place
35 Livery Street
Birmingham
B3 2PB, UK.
ISBN 978-1-80324-880-6

www.packt.com

To all the people contributing to and being part of open source, building it one piece at a time, and pushing its boundaries to make huge things possible.

– Pablo Iranzo Gómez

To the ones that once crossed my path and made a difference.

– Pedro Ibáñez Requena

Contributors

About the authors

Pablo Iranzo Gómez is a software engineer who started his Linux exposure while studying physics. He was also involved in LUGs and some projects related to HPC clusters and system administration and consultancy.

Currently, he is a principal software engineer at Red Hat's TelCo 5G OpenShift management integration department, leveraging his experience in consulting, cloud technical account management, and OpenStack software maintenance in industries such as hotel, retail, airlines, government, Telco, 5G, Partner, and IT – covering system administration and automation, virtualization, PaaS, support, the cloud, and so on. He has a broad understanding of different views, needs, and risks across the industry.

Pablo was born in and lives in Valencia, Spain, with his family.

I want to thank my wife, Eva and my sons, Pau and Roi, for the support provided while working on this project, Javier for the moral support provided during this writing, and Miquel for always being there. Not forgetting about Miguel, who gave me this opportunity and helped get the prior version ready, and, of course, the whole Packt team that has been helping, guiding, and advising during the whole process.

Pedro Ibáñez Requena is a software engineer who started his Linux adventure playing with Linux distributions provided on CDs by magazines. He was also involved in LUGs while studying software engineering at Universitat de València.

Currently, he is a principal systems engineer in Red Hat's Telco 5G field engineering team, focusing on OpenShift, with experience in automation, security, high availability, and databases. His experience was built in industries such as retail, finance, insurance, and Telco. He has been involved in the design, configuration, and deployment of complex IT systems including data storage, networking, monitoring, and PaaS and SaaS systems where he has a broad understanding of different views, needs, and risks across the industry.

Pedro was born in Alcoy and lives in Valencia, Spain, with his family.

I want to thank my girls, Katja and Maya, for the endless days at home, playing, chatting, swimming, traveling, or discussing because, without them, I wouldn't be who I am today. To my parents, for supporting me to study what I wanted and for providing when I needed it. To Carles and Alex, because their help in this race was life-changing. To Pablo, who thought about me helping him with this book, and lastly, to all the people that make my day, the Telco 5G field engineering team, and the Packt team that made this possible.

Miguel Pérez Colino is an experienced technology enthusiast with a clear orientation toward IT, free/open source software, and open standards. His career has grown in the IT field with connections in every direction. From user support and teaching; through systems and network administration; to systems engineering, systems architecture, IT strategy definition, and now, as a senior manager for the Tanzu portfolio to bring Kubernetes closer to customers. He loves designing solutions and tools, making them useful and easy to use.

Scott McCarty is a principal product manager at Red Hat for the container subsystem team, which enables key product capabilities in OpenShift Container Platform and Red Hat Enterprise Linux. His focus areas include container runtimes, tools, and images. Scott is a social media start-up veteran, an e-commerce old-timer, and a weathered government research technologist, with experience across a variety of companies and organizations, from 7-person start-ups to 15,000-employee technology companies. This has culminated in a unique perspective on open source software development, delivery, and maintenance.

About the reviewers

Derek Thurston is a senior solutions architect at Red Hat and has been working with open source software since the mid-1990s. After graduating from a small merchant marine college in coastal Maine, Derek landed a job in the "IT department" at a naval architecture firm in Arlington, VA, where he started using Linux. Derek has worked with SGI and HP/UX systems and did his part to help save the world by converting COBOL programs to four-digit years in the late 1990s. Having worked with a variety of public and private sector entities, he has spent over 25 years helping customers find the right open source tools and technologies.

I would like to thank my wife, who inspires me every day, my son, who is always down for a cool adventure, and my two dogs, who are my steadfast officemates.

Luis Sampaio is a system engineer. He has managed servers since 2010. He has experience with Red Hat, Ubuntu, and Debian. He has some certifications such as RHCE and RHCSA.

Luis has diverse experiences in different private, hybrid cloud architectures with Amazon Web Services and Azure, as well as experience with DevOps tools such as Ansible and containers.

He likes to share his tech views and experiments with a community. Luis loves playing poker with friends and visiting new cities.

I would like to thank my family (my mother, Silvia, for all the support, my wonderful wife, Yolanda, and my terrific and smart son, Bruno Miguel) for unlimited patience and tolerance.

My wife is my best friend, and I love her dearly. Without her love and the support from her, I would never have been able to finish the book.

Table of Contents

Part 1 – Systems Administration – Software, User, Network, and Services Management

1

2

3

4

Tools for Regular Operations   103

5

Securing Systems with Users, Groups, and Permissions   137

6

Enabling Network Connectivity   165

7

Adding, Patching, and Managing Software   191

Part 2 – Security with SSH, SELinux, a Firewall, and System Permissions

8

9

Part 3 – Resource Administration – Storage, Boot Process, Tuning, and Containers

13

Flexible Storage Management with LVM   351

14

Advanced Storage Management with Stratis and VDO   379

15

Understanding the Boot Process   393

16

Kernel Tuning and Managing Performance Profiles with tuned   409

17

Managing Containers with Podman, Buildah, and Skopeo   433

Part 4 – Practical Exercises

18

Practice Exercises – 1   451

19

Preface

Linux is everywhere, from personal devices to the largest supercomputers, from the computer labs at universities to Wall Street or the International Space Station, and even Mars! **Red Hat Enterprise Linux (RHEL)** is the most used Linux distribution in enterprise environments and knowing how to use it is a key skill for anyone in technology. No matter whether you are completely into managing infrastructure or you are a developer interested in knowing more about the platform you want to deploy on, learning about Linux – and, more precisely, about RHEL – will help you be more effective and could even boost your career.

In this book, we cover the basic RHEL administration skills from a very practical perspective, providing examples and tips that we have learned from our experience in "the trenches." You will be able to follow it from beginning to end, being able to practice with each step while learning about how things are built and why they behave as they do.

We hope you enjoy this book, that you make the most of it, and that you end up, after reading it, with a strong foundation of RHEL administration skills. That's what we wrote it for.

Enjoy reading...and practicing!

Who this book is for

Anyone that aspires to build and work on IT infrastructures using Linux will benefit from this book as a reference for different useful tasks, tips, and best practices. It will help anyone seeking to pass the **Red Hat Certified Systems Administrator (RHCSA)** exam, although it will be no substitute for the official training, in which labs and specially crafted tests will be run during the whole process. The scope of the book is adjusted to the RHCSA, extending it with advice from real-world experience and many practical examples.

What this book covers

Chapter 1, Getting RHEL Up and Running, covers the installation of RHEL, from obtaining the software and the subscriptions to the installation of the system itself.

Chapter 2, RHEL 9 Advanced Installation Options, introduces advanced use cases for the installer, including deploying instances in the cloud and automating the installation.

Chapter 3, Basic Commands and Simple Shell Scripts, explains the daily commands that will be used during system administration, and how can they be automated via shell scripting.

Chapter 4, Tools for Regular Operations, shows the simple tools that are available in our system that can be used for regular daily operations, such as starting or enabling a system service or reviewing what is going on in the system through logs.

Chapter 5, Securing Systems with Users, Groups, and Permissions, covers how to manage users, groups, and permissions in any Linux system, with some specifics on RHEL.

Chapter 6, Enabling Network Connectivity, goes through the steps to connect a system to the network and the possible ways it can be configured.

Chapter 7, Adding, Patching, and Managing Software, reviews how the steps to add, remove, and update can be managed in our system, including examples for upgrades and rollbacks.

Chapter 8, Administering Systems Remotely, covers how to remotely connect to your system in order to be more effective. It includes using ssh connections to create keys and using a terminal multiplexer (tmux).

Chapter 9, Securing Network Connectivity with firewalld, instructs you on how the network firewall configuration works in RHEL and how to properly manage it, including managing zones, services, and ports.

Chapter 10, Keeping Your System Hardened with SELinux, covers usage and basic troubleshooting of SELinux.

Chapter 11, System Security Profiles with OpenSCAP, explains how to run security profiles with OpenSCAP and check compliance in RHEL with typical regulations.

Chapter 12, Managing Local Storage and Filesystems, covers filesystem creation, mount points, and general storage management.

Chapter 13, Flexible Storage Management with LVM, explains how **Logical Volume Manager** (**LVM**) empowers more flexible storage management by being able to add disks and extend logical volumes.

Chapter 14, Advanced Storage Management with Stratis and VDO, introduces **Virtual Data Optimizer** (**VDO**) and how it can be used in our system to deduplicate storage, as well as using Stratis to manage storage more easily.

Chapter 15, Understanding the Boot Process, explains how the system boots and the details that make it important.

Chapter 16, Kernel Tuning and Managing Performance Profiles with tuned, explains how kernel tunning works and how to use tuned for predefined profile usage.

Chapter 17, Managing Containers with Podman, Buildah, and Skopeo, covers containers and tools for managing and building them.

Chapter 18, Practice Exercises – 1, allows you to test your acquired knowledge.

Chapter 19, Practice Exercises – 2, gives more complex testing of your acquired knowledge.

To get the most out of this book

All software requirements will be indicated in the chapters. Note that this book assumes that you have access to a physical or virtual machine, or have access to the internet to create a cloud account in order to perform the operations that the book will guide you through.

Software/hardware covered in the book	Operating system requirements
Red Hat Enterprise Linux 9	You will need to install RHEL. It can be installed on a virtual machine on Linux, macOS, or Windows.

If you are using the digital version of this book, we advise you to type the code yourself or access the code from the book's GitHub repository (a link is available in the next section). Doing so will help you avoid any potential errors related to the copying and pasting of code.

Download the example code files

You can download the example code files for this book from GitHub at `https://github.com/PacktPublishing/Red-Hat-Enterprise-Linux-RHEL-9-Administration`. If there's an update to the code, it will be updated in the GitHub repository.

We also have other code bundles from our rich catalog of books and videos available at `https://github.com/PacktPublishing/`. Check them out!

Download the color images

We also provide a PDF file that has color images of the screenshots and diagrams used in this book. You can download it here: `https://packt.link/NcDqa`.

Conventions used

There are a number of text conventions used throughout this book.

`Code in text`: Indicates code words in text, database table names, folder names, filenames, file extensions, pathnames, dummy URLs, user input, and Twitter handles. Here is an example: "Mount the downloaded `RHEL9.iso` disk image file as another disk in your system."

A block of code is set as follows:

```
#!/bin/bash
echo "Hello world"
```

When we wish to draw your attention to a particular part of a code block, the relevant lines or items are set in bold:

```
[default]
branch = main
repo = myrepo
username = bender
protocol = https
```

Any command-line input or output is written as follows:

```
$ mkdir scripts
$ cd scripts
```

Bold: Indicates a new term, an important word, or words that you see on screen. For instance, words in menus or dialog boxes appear in **bold**. Here is an example: "Select **System info** from the **Administration** panel."

> **Tips or Important Notes**
> Appear like this.

Get in touch

Feedback from our readers is always welcome.

General feedback: If you have questions about any aspect of this book, email us at customercare@packtpub.com and mention the book title in the subject of your message.

Errata: Although we have taken every care to ensure the accuracy of our content, mistakes do happen. If you have found a mistake in this book, we would be grateful if you would report this to us. Please visit www.packtpub.com/support/errata and fill in the form.

Piracy: If you come across any illegal copies of our works in any form on the internet, we would be grateful if you would provide us with the location address or website name. Please contact us at copyright@packt.com with a link to the material.

If you are interested in becoming an author: If there is a topic that you have expertise in and you are interested in either writing or contributing to a book, please visit authors.packtpub.com.

Share your thoughts

Once you've read *Red Hat Enterprise Linux 9 Administration*, we'd love to hear your thoughts! Scan the QR code below to go straight to the Amazon review page for this book and share your feedback.

https://packt.link/r/1803248807

Your review is important to us and the tech community and will help us make sure we're delivering excellent quality content.

Download a free PDF copy of this book

Thanks for purchasing this book!

Do you like to read on the go but are unable to carry your print books everywhere? Is your eBook purchase not compatible with the device of your choice?

Don't worry, now with every Packt book you get a DRM-free PDF version of that book at no cost.

Read anywhere, any place, on any device. Search, copy, and paste code from your favorite technical books directly into your application.

The perks don't stop there, you can get exclusive access to discounts, newsletters, and great free content in your inbox daily

Follow these simple steps to get the benefits:

1. Scan the QR code or visit the link below

https://packt.link/free-ebook/9781803248806

2. Submit your proof of purchase
3. That's it! We'll send your free PDF and other benefits to your email directly

Part 1 –
Systems Administration –
Software, User, Network, and
Services Management

Deploying and configuring systems and keeping them up to date is the base task that every system administrator performs in their day-to-day work. In this part, the core parts of doing so are explored in a restructured way so that you can follow the tasks one by one and learn, practice, and understand them properly.

The following chapters are included in this part:

- *Chapter 1, Getting RHEL Up and Running*
- *Chapter 2, RHEL 9 Advanced Installation Options*
- *Chapter 3, Basic Commands and Simple Shell Scripts*
- *Chapter 4, Tools for Regular Operations*
- *Chapter 5, Securing Systems with Users, Groups, and Permissions*
- *Chapter 6, Enabling Network Connectivity*
- *Chapter 7, Adding, Patching, and Managing Software*

1
Getting RHEL Up and Running

The first step to start working with **Red Hat Enterprise Linux** (RHEL) is to have it running. Whether on your own laptop as the main system, on a virtual machine, or on a physical server, its installation is necessary in order to get your hands on the system you want to learn to use. It is highly encouraged that you get yourself a physical or virtual machine to use the system while reading this book.

In this chapter, you will deploy your own RHEL 9 system to be able to follow all the examples mentioned in this book, as well as discover more about Linux.

The topics to be covered in this chapter are as follows:

- Obtaining RHEL software and a subscription
- Installing RHEL 9

Technical requirements

The best way to get started is by having an **RHEL 9** virtual machine to work with. You can do it on your main computer as a virtual machine or using a physical machine. In the following section, we will review both options and you will be able to run your own RHEL 9 system.

> **Tip**
> A virtual machine is a way to emulate a complete computer. To be able to create this emulated computer on your own laptop, if you are using macOS or Windows, you will need to install virtualization software such as VirtualBox, for example. If you are running Linux, it is already prepared for virtualization, and you will only need to add the `virt-manager` package.

Obtaining RHEL and a subscription

To be able to deploy RHEL, you will need a **Red Hat subscription** to obtain the images to be used, as well as access to repositories with software and updates. You can obtain, free of charge, a developer subscription from the developers' portal site of Red Hat using the following link: `https://developers.redhat.com/`. You then need to follow these steps:

1. Log in or create an account at `https://developers.redhat.com/`.

2. Click on the **Log in** button:

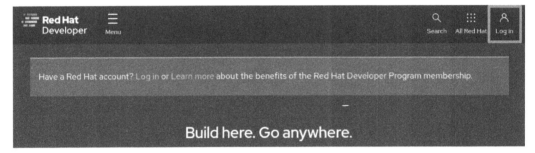

Figure 1.1 – The developers.redhat.com home page, indicating where to click to log in

3. Once on the login page, use your account or, if you do not have one, create one by clicking on **Register** in the top-right corner or on the **Register for a Red Hat account** button directly in the registration box, as follows:

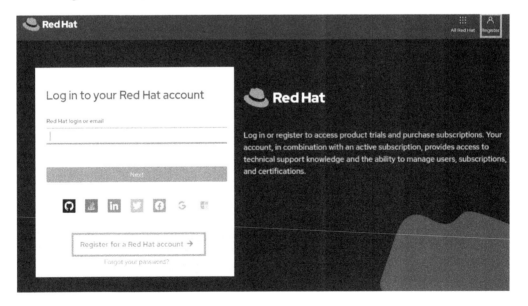

Figure 1.2 – Red Hat login page (common to all Red Hat resources)

You can choose to use your credentials in several services (in other words, *Google*, *GitHub*, or *Twitter*) if you prefer to do so.

4.   Once you have logged in, go to the **Products** section in the top bar. You can find the **Red Hat Enterprise Linux** section in the navigation bar before the content:

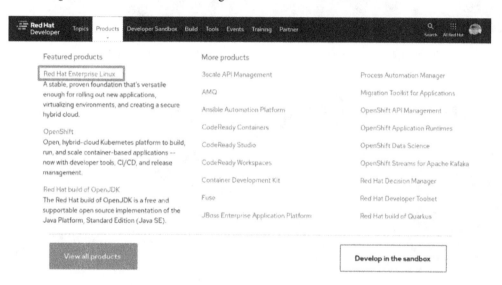

Figure 1.3 – Accessing the Linux page at developers.redhat.com

Click on **Download RHEL at no-cost**, which appears as a red button on the next page:

Figure 1.4 – Accessing the RHEL download page at developers.redhat.com

Then, the ISO image for the **x86_64 (8 GB)** architecture will start downloading:

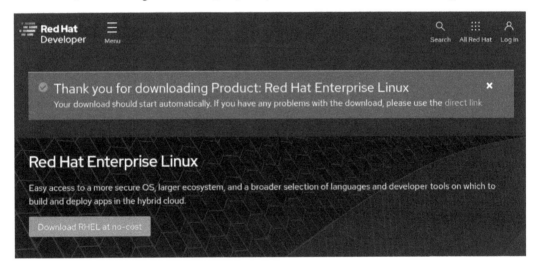

Figure 1.5 – Download dialog for RHEL 9

The ISO image is a file that contains an exact copy of the contents of a full DVD (even when we are not using a DVD). This file will later be used to install our machines, whether dumping it to a USB drive for *bare metal* installations, unpacking it for network installations, or attaching it for virtual machine installations (or using out-of-band capabilities in servers such as IPMI, iLO, or iDRAC)

> Tip
> To verify the ISO image, and ensure that the one we have obtained is not corrupted or altered, a mechanism called **checksum** can be used. Checksums are a way to review a file and provide a set of letters and numbers that can be used to verify that the file is precisely the same one as in the origin. Red Hat provides a list of sha256 checksums for doing so in the downloads section of the customer portal (access.redhat.com). An article describing the process is available here: https://access.redhat.com/solutions/8367.

We have the software, in this case, the ISO image, to install RHEL 9 on any computer. These are the same bits that are used in production machines worldwide, and that you can use yourself for learning purposes with your developer subscription. Now, it is time to give them a go in the next section.

Installing RHEL 9

In this section, we will follow the typical installation process to have RHEL installed on a machine. We will follow the default steps, reviewing the options available for each one.

Preparation for a physical server installation

A physical server requires some initial setup before beginning with installation. Common steps include configuring the disks in the *internal array*, connecting it to the networks, preparing the switches for any *interface aggregation* that is expected (teaming, bonding), preparing access to external *disk arrays* (in other words, *fiber channel arrays*), setting up out-of-band capabilities, and securing the **BIOS** configuration.

We will not get into the details of these preparations, except for the boot sequence. The server will require to boot (start loading the system) from an external device such as a *USB thumb drive* or *optical disk* (whether physical or emulated through the out-of-band capabilities).

Creating a bootable USB thumb drive from a machine with Linux or macOS is as simple as doing a "disk dump" with the dd application. Perform the following steps:

1. Find your USB device in the system, usually /dev/sdb in Linux, or /dev/disk2 in macOS (in macOS, this command requires special privileges; please run it as sudo dmesg | grep removable):

```
$ dmesg | grep removable
[66931.429805] sd 0:0:0:0: [sdb] Attached SCSI removable
disk
```

> **Important Note**
> Please verify the disk name very carefully, as the procedure for using "disk dump" will completely overwrite the disk target.

Check whether the USB is mounted and if so, dismount it (for macOS users, please use diskutil list to ascertain whether the device is mounted):

```
$ lsblk /dev/sdb
NAME    MAJ:MIN RM  SIZE RO TYPE MOUNTPOINT
sdb       8:0    1  3,8G  0 disk
```

```
├─sdb1    8:1    1  1,8G   0 part /run/media/miguel/USB
├─sdb2    8:2    1 10,9M   0 part
└─sdb3    8:3    1 22,9M   0 part
```

In this case, only partition one of the sdb disk, referred to as sdb1, is mounted. We will need to *unmount* all the partitions mounted. In this example, this is straightforward as there is only one. To do so, we can run the following command:

```
$ sudo umount /dev/sdb1
```

> **Important Note**
>
> By using **super-user do** (**sudo**) for administrative tasks such as unmounting devices, we can open an administrator shell (root in Linux and Unix-like systems) or run the command using sudo, which provides administrative privileges to the current user. When running commands with sudo, the user will be requested to enter their password (not the admin password, but the user's own password) to proceed with the execution (this default behavior may be overridden in the sudoers configuration file).

Dump the image! (Warning, this will erase the selected disk!):

```
$ sudo dd if=rhel-baseos-9.0-beta-x86_64-dvd.iso of=/dev/
sdb bs=512k
```

> **Tip**
>
> Alternative graphical tools are available for creating a boot device that can help select both the image and the target device. In Fedora Linux (the development branch used to create Red Hat Enterprise Linux, and a workstation for many engineers and developers), the **Fedora Media Writer** tool can be used. For other environments, the **UNetbootin** tool could also serve to create your boot media.

Now, with the USB thumb drive, we can install any physical machine, from a tiny laptop to a huge server. The next part involves making the physical machine boot from the **USB thumb drive**. The mechanism for doing that will depend on the server being used. However, it is becoming common to offer an option to select a boot device during bootup. The following is an example of how to select a temporary boot device for a laptop:

1. Interrupt the normal startup. In this case, the boot process shows that I can do that by pressing *Enter*:

Figure 1.6 – Example of a BIOS message to interrupt normal startup

2. Choose a temporary startup device, in this case, by pressing the *F12* key:

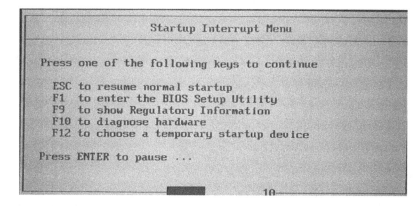

Figure 1.7 – Example of a BIOS menu for interrupted startup

3. Select the device to boot from. We want to boot from our USB thumb drive, which, in this case, is **USB HDD: ChipsBnk Flash Disk**:

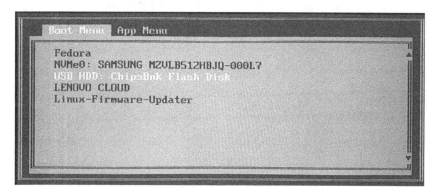

Figure 1.8 – Example of a BIOS menu to choose the USB HDD boot device

Let the system start the installer from the USB drive.

Once we know how to prepare a USB drive with an RHEL installer and how to make a physical machine boot from it, we can skip to the *Running an RHEL installation* section and proceed to install it. This can be pretty useful if we have a mini server (in other words, an Intel NUC), an old computer, or a laptop to be used as the machine for following along with this book.

Next, we will look at how to prepare a virtual machine in your installation, in case you are considering following this book with your current main laptop (or workstation) but you still want to keep a separate machine to work with.

Preparation for a virtual server installation

A **virtual server** works as if having some virtualization software that emulates a real machine in your current system. In a Linux workstation, installing `virt-manager` will add all the under-the-hood components required to run (for your information, these components are **KVM**, **Libvirt**, **Qemu**, and **virsh**, among others). Other no-cost virtualization software recommended for Windows or macOS systems includes **Oracle VirtualBox** and **VMware Workstation Player**.

The examples in this section will be executed using `virt-manager`, but are easily applicable to any other virtualization software, whether in a laptop or in the largest deployments.

The preliminary steps have been described previously and require obtaining the **Red Hat Enterprise Linux ISO** image, which, in this case, will be `rhel-baseos-9.0-beta-0-x86_64-dvd.iso`. Once downloaded and, if possible, having checked its integrity (as mentioned in the last tip of the *Obtaining RHEL software and a subscription* section), let's prepare to deploy a virtual machine:

1. Start your virtualization software, in this case, `virt-manager`:

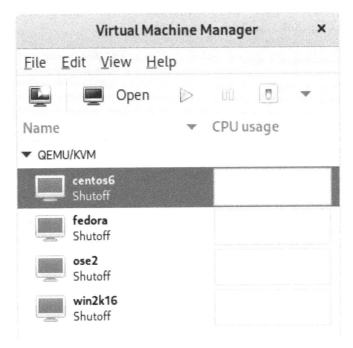

Figure 1.9 – The virtual manager main menu

2. Create a new virtual machine by going to **File** and then clicking on **New Virtual Machine**. Select **Local install media (ISO image or CDROM)**:

Figure 1.10 – Virtual manager – New Virtual Machine menu

3.   Select the *ISO image*. With this, the virtual machine will be configured with a **virtual DVD/
CDROM drive** and already prepared to boot from it. This is standard behavior. However, when
using different virtualization software, you may want to perform a check:

Figure 1.11 – The virtual manager menu to select an ISO image as an installation medium

4. Assign memory and CPU to the virtual machine we are creating (note: a virtual machine is usually referred to as a **VM**). For **Red Hat Enterprise Linux 9** (also referred to as **RHEL 9**), 1.5 GB of memory is the minimum, while 1.5 GB per logical CPU is recommended. We will use the minimum settings (1.5 GB memory, 1 CPU core):

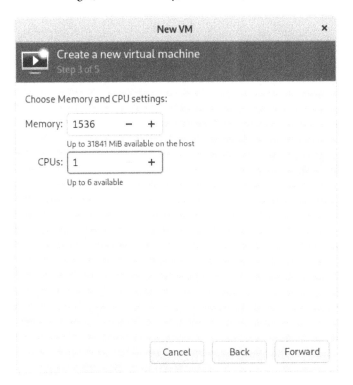

Figure 1.12 – The virtual manager menu for selecting memory and CPU

It is time to assign at least one disk to the virtual machine. In this case, we will assign a single disk with the minimum disk space, 10 GB, but in future chapters, we will be able to assign more disks to test other functionalities:

Figure 1.13 – The virtual manager menu to create a new disk and add it to the virtual machine

5. Our virtual machine has all that we need to get started: a boot device, memory, CPU, and disk space. In this last step, a network interface is added, so now we even have a network. Let's review the data and launch it:

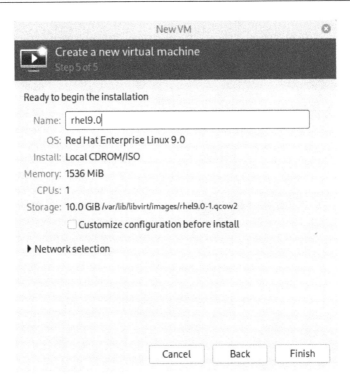

Figure 1.14 – The virtual manager menu for selecting the name of the virtual machine and the network

After taking these steps, we have a fully functional virtual machine available. Now, it is time to complete the process by installing the RHEL operating system on it. See how to do this in the next section.

Running an RHEL installation

Once we have prepared our virtual or physical server for installation, it's time to proceed with it. We will know whether all the previous steps were performed correctly if we arrive at the following screen:

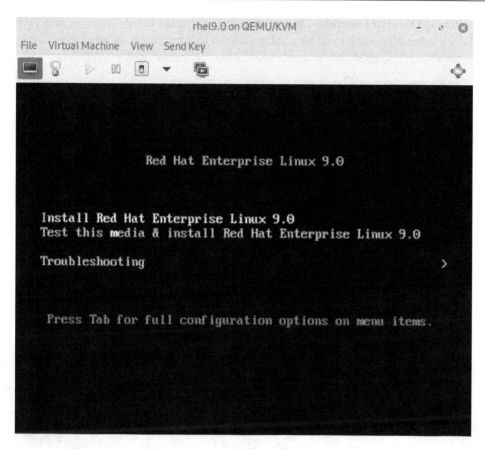

Figure 1.15 – Initial boot screen for RHEL 9 installation with Install selected

We are offered three options (*the selected option is in white*):

- **Install Red Hat Enterprise Linux 9.0**: This option will boot and run the installer.

- **Test this media & install Red Hat Enterprise Linux 9.0**: This option will check the image being used to ensure that it is not corrupt and that the installation can proceed with certainty. It is recommended to use this one for the first time using a just-downloaded ISO image or just-created media, such as a USB thumb drive or DVD (in a virtual machine, it takes approximately 1 minute to run the check).

- **Troubleshooting**: This option will help you review other options in case there are problems with installation, with a running system, or with hardware. Let's take a quick look at the available options on this menu:

  - **Install Red Hat Enterprise Linux 9.0 in basic graphics mode**: This option is useful for systems with an old graphics card and/or an unsupported one. It can help to get the system installed in case an issue with visualization is identified.

- **Rescue a Red Hat Enterprise Linux system**: This option can be used when we have a system with issues booting or when we want to access it to introspect it (in other words, review a possible compromised system). It will initiate a basic in-memory system to perform these tasks.

- **Run a memory test**: The system memory can be checked to prevent issues, as in the case of a brand-new server, for instance, where we want to ensure that its memory is running correctly, or a system suffering issues and panics that may indicate a memory-related issue.

- **Boot from local drive**: In case you booted from the install media, but you already have a system installed.

- **Return to main menu**: To go back to the previous menu.

Important Note

The RHEL boot menu will show several options. The one selected will show in white, with one single letter in a different color, in this case, "i" for install and "m" for test media. These are shortcuts. Pressing the key with that letter will take us directly to this menu item.

Let's proceed with **Test this media & install Red Hat Enterprise Linux 9.0** to let the installer review the ISO image we are using:

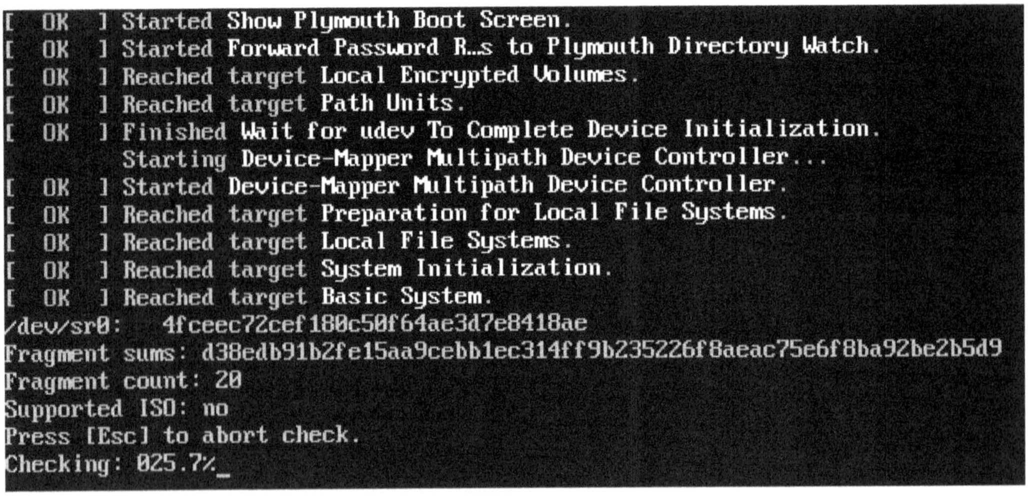

```
[  OK  ] Started Show Plymouth Boot Screen.
[  OK  ] Started Forward Password R...s to Plymouth Directory Watch.
[  OK  ] Reached target Local Encrypted Volumes.
[  OK  ] Reached target Path Units.
[  OK  ] Finished Wait for udev To Complete Device Initialization.
         Starting Device-Mapper Multipath Device Controller...
[  OK  ] Started Device-Mapper Multipath Device Controller.
[  OK  ] Reached target Preparation for Local File Systems.
[  OK  ] Reached target Local File Systems.
[  OK  ] Reached target System Initialization.
[  OK  ] Reached target Basic System.
/dev/sr0:    4fceec72cef180c50f64ae3d7e8418ae
Fragment sums: d38edb91b2fe15aa9cebb1ec314ff9b235226f8aeac75e6f8ba92be2b5d9
Fragment count: 20
Supported ISO: no
Press [Esc] to abort check.
Checking: 025.7%_
```

Figure 1.16 – RHEL 9 ISO image self-check

Important Note

If the media test fails, redownload the RHEL 9 ISO and try again.

Once completed, it will reach the first installation screen. The installer is called **Anaconda** (a joke, as it is written in a language called **Python**), and it follows a step-by-step approach. It is important to pay attention to the options we will select during installation. We will review them later in the *Automating deployments with Anaconda* section of the book.

Localization

The first step to installation is selecting the installation language. For this installation, we will select **English**, followed by **English (United States)**:

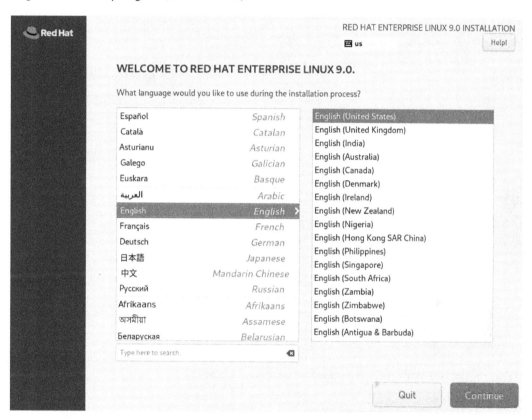

Figure 1.17 – RHEL 9 install menu – language

If you cannot easily find your language, you can type it in the box under the list to search for it. Once a language is selected, we can click the **Continue** button to proceed. This will take us to the **INSTALLATION SUMMARY** screen:

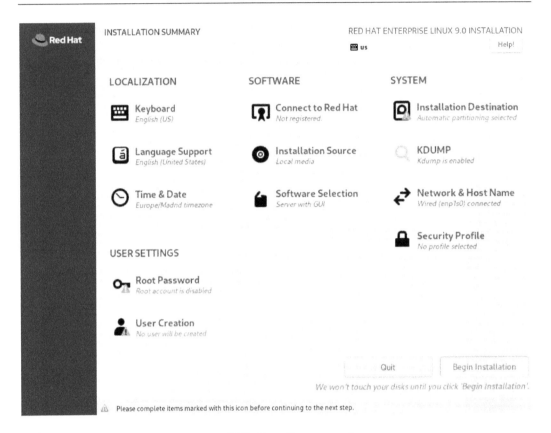

Figure 1.18 – RHEL 9 install menu – main page

On the **INSTALLATION SUMMARY** screen, all the configuration parts required are shown, with many of them (the ones without a warning sign and red text underneath) already preconfigured with defaults.

Let's review the **LOCALIZATION** settings, starting with **Keyboard**:

Figure 1.19 – RHEL 9 install – The Keyboard selection icon

We can review the **Keyboard** settings, which can help not just in changing the keyboard, but adding extra layouts in case we want to switch between them:

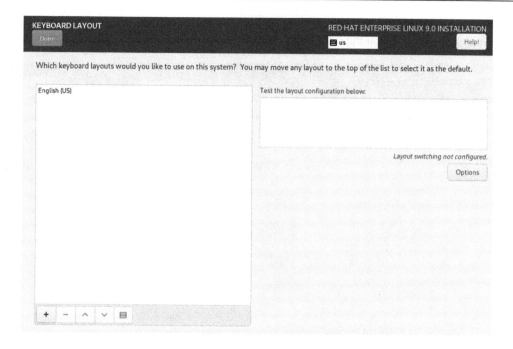

Figure 1.20 – RHEL 9 install – keyboard selection dialog

This can be done by clicking on the + button. Here is an example of adding the **Spanish; Castilian (Spanish)** layout. We search for spa until it appears, and then we select it, and then click **Add**, as follows:

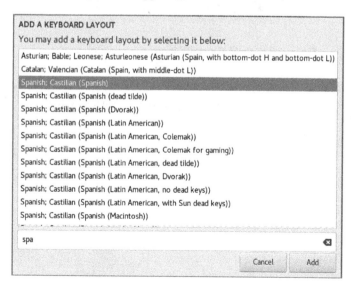

Figure 1.21 – RHEL 9 install – keyboard selection list

To make it the default option will require clicking on the ^ button underneath. In this case, we will keep it as a secondary option so that the supporting software gets installed. Once completed, click **Done**:

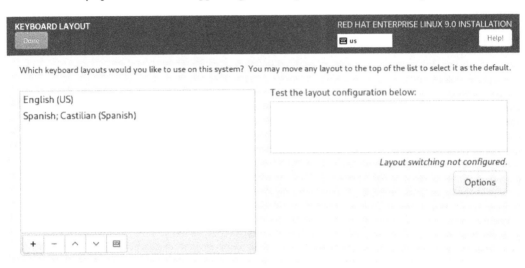

Figure 1.22 – RHEL 9 install – keyboard selection dialog with different keyboards

Now, we will move on to **Language Support**:

Figure 1.23 – RHEL 9 install – language selection icon

Here, we can also add our local language. In this example, I'll use **Español**, and then **Español (España)**. This will again include the software required to support the language that has been added:

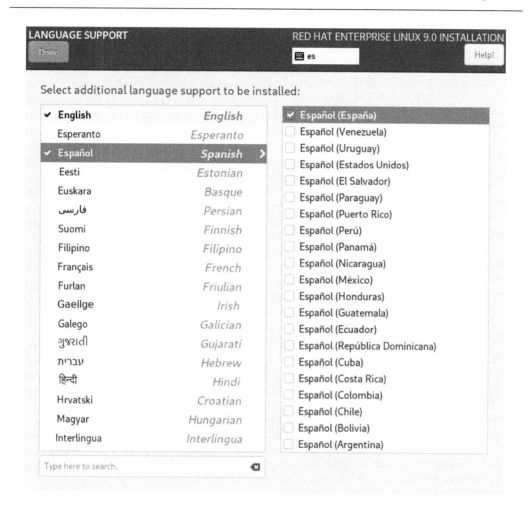

Figure 1.24 – RHEL 9 install – language selection dialog with different languages

We will proceed with both languages configured, although you may want to choose your own localized language.

Now, we will move on to **Time & Date**, which can be seen as follows:

Figure 1.25 – RHEL 9 install – Time and Date selection icon

The default configuration is set to the city of New York in the United States of America. You have two possibilities here:

- Use your local time zone. This is recommended when you want to have all the logs registered in that time zone (in other words, because you are only working in one time zone or because there are local teams for each time zone). In this example, we are selecting the *Spain, Madrid, and Europe* time zone:

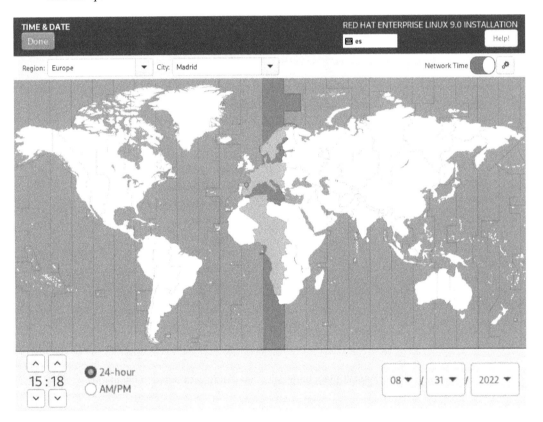

Figure 1.26 – RHEL 9 install – the Time & Date selection dialog – Madrid selected

- Use **Coordinated Universal Time** (**UTC**) to have the same time zone for all the servers around the globe. This can be selected under **Region | Etc**, and then **City | Coordinated Universal Time**:

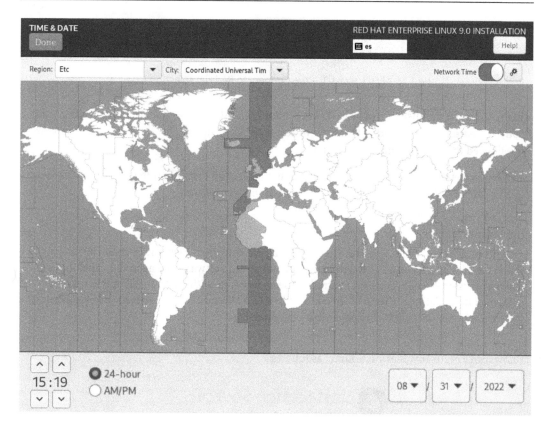

Figure 1.27 – RHEL 9 install – the Time & Date selection dialog – UTC selected

We will proceed with the localized time for Spain, Madrid, and Europe, although you may want to select your localized time zone.

> **Tip**
>
> As you can see on the screen, there is an option to select **Network Time** to have the machine's clock synchronized with other machines. This option can only be selected once the network is configured.

Software

With the **localization** configuration completed (or almost completed; we may come back for the network time later), we move on to the **software** section, or, more precisely, to **Connect to Red Hat** under it:

Figure 1.28 – RHEL 9 install – the Connect to Red Hat selection icon

In this section, we can use our own Red Hat account, like the one we created previously under `https://developers.redhat.com/`, to access the latest updates for the system. To configure it, we will need to configure the network first.

For this deployment, we will not configure this section now. We will review how to manage subscriptions and get updates in *Chapter 7, Adding, Patching, and Managing Software*.

> **Important Note**
>
> Systems management with Red Hat Satellite: For large deployments with more than 100 virtual servers, Red Hat offers **Red Hat Satellite**, with advanced software management capabilities (such as versioned content views, centralized security scans with OpenSCAP, and simplified patching and updating for RHEL). To connect to a Red Hat satellite, the activation key can be used, thereby simplifying the management of systems.

Let's now move on to **Installation Source**, as follows:

Figure 1.29 – RHEL 9 install – the Installation Source icon

This can be used for installation using remote sources. It is very useful when using the boot ISO image that only contains the installer. In this case, as we are using the full ISO image, it already contains all the software (also referred to as *packages*) needed to complete the installation.

The next step is **Software Selection**, as shown in the following screenshot:

Figure 1.30 – RHEL 9 install – the Software Selection icon

In this step, we can select a predefined set of packages to be installed on the system so that it can perform different tasks. While it can be very convenient to do so at this stage, we are going to adopt a more manual approach and select the **Minimal Install** profile to add software to the system later.

This approach also has the advantage of reducing the **attack surface** by installing just the minimum required packages in the system:

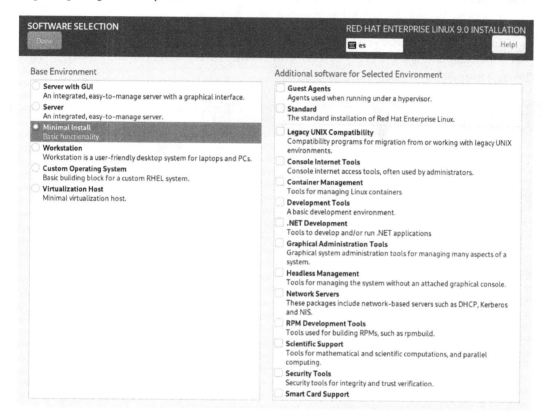

Figure 1.31 – RHEL 9 install – the Software Selection menu; Minimal Install selected

System

Once the set of packages has been selected, let's move on to the system configuration section. We will start with the destination of the installation, where we can choose the disk or disks to be used to install and configure them:

Figure 1.32 – RHEL 9 install – the Installation Destination icon
with a warning sign as this step is not complete

This task is very important as it will define not just the way the system is deployed on the disk, but also how the disk is distributed and with which tools. Even when in this section, we won't use the advanced options. We will take some time to review the main options.

This is the default **Device Selection** screen, with only one local standard disk discovered, no **Specialized & Network Disks** options, and ready to run the **Automatic** partitioning. This can be seen in the following screenshot:

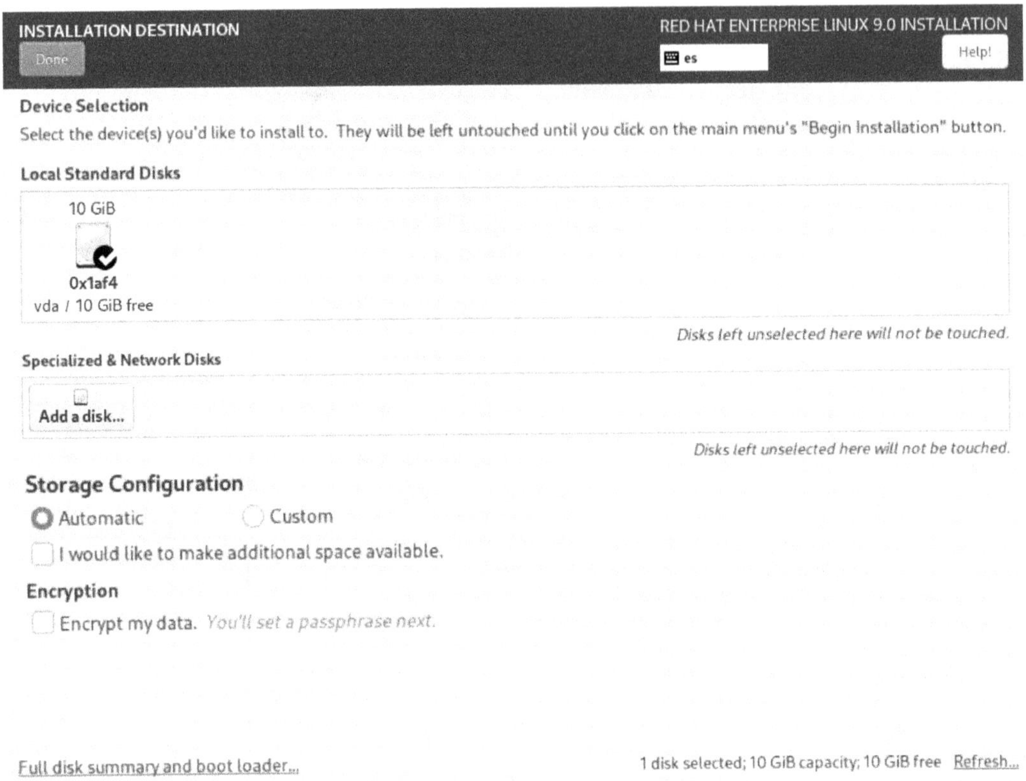

Figure 1.33 – RHEL 9 install – the INSTALLATION DESTINATION menu with automatic partitioning selected

Clicking **Done** in this section will complete the minimal set of data required to continue with the installation.

Let's review the sections.

Local Standard Disks shows a set of disks to be used by the installer. It may be the case that we have several disks, and we only want to use a specific disk:

Local Standard Disks
10 GiB
0x1af4
vda / 10 GiB free
2 GiB
0x1af4
vdb / 2 GiB free
4 GiB
0x1af4
vdc / 4 GiB free
Disks left unselected here will not be touched.

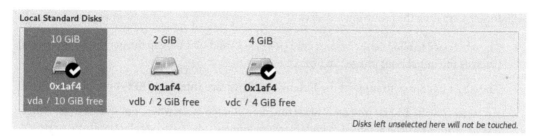

Figure 1.34 – RHEL 9 install – the INSTALLATION DESTINATION menu with several local disks selected

This is an example of having three available disks and using only the first and third ones.

In our case, we only have one disk, and it is already selected:

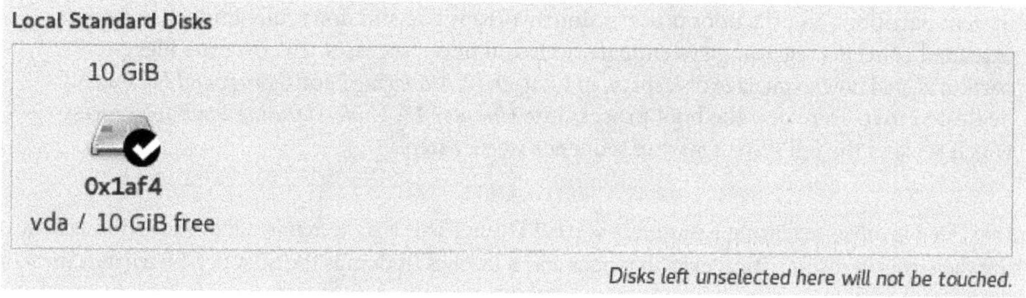

Local Standard Disks
10 GiB
0x1af4
vda / 10 GiB free
Disks left unselected here will not be touched.

Figure 1.35 – RHEL 9 install – the INSTALLATION DESTINATION menu with a single local disk selected

It would be easy to use full-disk encryption by selecting **Encrypt my data**, which is highly recommended for laptop installations or for installing in environments with low levels of trust:

Encryption
Encrypt my data. You'll set a passphrase next.

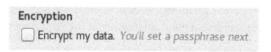

Figure 1.36 – RHEL 9 install – the INSTALLATION DESTINATION menu with the data encryption option (not selected)

For this example, we will not encrypt our drive.

The **Automatic** install option will distribute the disk space automatically:

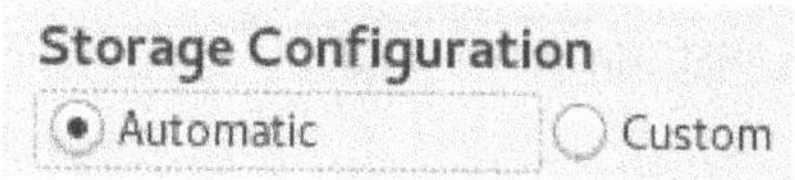

Storage Configuration
Automatic
Custom

Figure 1.37 – RHEL 9 install – the INSTALLATION DESTINATION menu; Storage Configuration (Automatic)

It will do so by creating the following resources:

- /boot: Space to allocate the system core (kernel) and files to help during the boot process (such as the initial boot image, initrd).

- /boot/efi: Space to support the **Extensible Firmware Interface** (**EFI**) boot process.

- /: the root filesystem. This is the main storage space where the system lives. Other disks/partitions will be assigned to folders (when doing so, they will be called mountpoints).

- /home: Space where the user will store personal files.

Let's select this option and then click **Done**.

> **Tip**
>
> System partitions and the boot process: don't worry if you still don't fully understand some extended concepts regarding system partitions and boot processes. To cover the filesystems, partitions, and how to manage disk space, in *Chapter 12, Managing Local Storage and Filesystems*, dedicated to it. To review the boot process, the *Chapter 15, Understanding the Boot Process*, which reviews the full system startup sequence step by step.

The next step involves reviewing **Kdump**, or **Kernel Dump**. This is a mechanism that allows the system to save the status in case a critical event happens and it crashes (it dumps the memory, hence its name):

Figure 1.38 – RHEL 9 install – the Kdump configuration icon

In order to work, it will reserve some memory for itself where it will stay, waiting to act if the system crashes. The default configuration does a good calculation of the requirements:

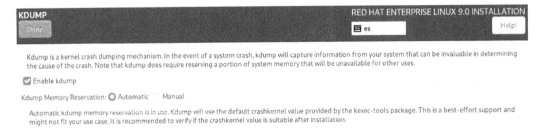

Figure 1.39 – RHEL 9 install – the Kdump configuration menu

Clicking **Done** will take us to the next step, **Network & Host Name**, which appears as follows:

Figure 1.40 – RHEL 9 install – the Network & Host Name configuration icon

This section will help to have the system connected to a network. In the case of a virtual machine, access to external networks will be handled by the **virtualization software**. It is very common that the default configuration uses **network address translation (NAT)** and the **Dynamic Host Configuration Protocol (DHCP)**, which will provide a network configuration to the virtual machine and access to external networks.

Once on the configuration page, we can see how many network interfaces are assigned to our machine. In this case, there is only one, as follows:

Figure 1.41 – RHEL 9 install – the NETWORK & HOST NAME configuration menu

First, we can enable the interface by clicking on the **ON/OFF** toggle on the right. To turn it off, it looks like this:

Figure 1.42 – RHEL 9 install – the NETWORK & HOST NAME configuration toggle (OFF)

And to turn it on, it should look like this:

Figure 1.43 – RHEL 9 install – the NETWORK & HOST NAME configuration toggle (ON)

We will see that the interface now has a configuration (**IP Address**, **Default Route**, and **DNS**):

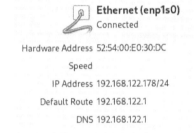

Figure 1.44 – RHEL 9 install – the NETWORK & HOST NAME configuration information details

To make this change permanent, we will click the **Configure** button at the bottom-right corner of the screen to edit the interface configuration:

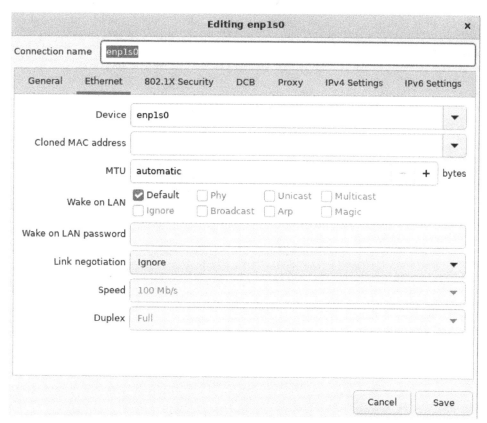

Figure 1.45 – RHEL 9 install – the NETWORK & HOST NAME
configuration, interface configuration, and Ethernet tab

Clicking on the **General** tab will present the main options. We will select **Connect automatically with priority** and leave the value as **-999**, just like this:

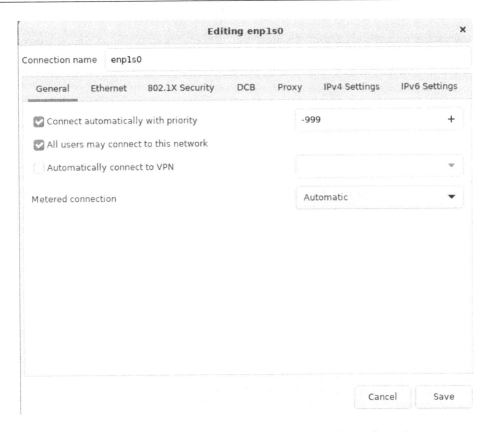

Figure 1.46 – RHEL 9 install – the NETWORK & HOST NAME configuration,
interface configuration, and the General tab

Clicking **Save** will make the changes permanent and have this network interface enabled by default.

Now, it's time to give a name to our virtual server. We will go to the **Host Name** section on the main page and type the name we want for it. We can use `rhel-instance.example.com`, and then click **Apply**:

Figure 1.47 – RHEL 9 install – the NETWORK & HOST NAME configuration with Host Name detail

> **Tip**
>
> The example.com domain is used for demonstration purposes and is safe to be used on any occasion, knowing that it will not collide or cause any trouble to other systems or domains.

The networking page will look like this:

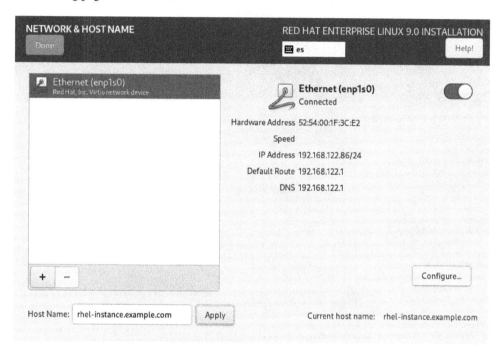

Figure 1.48 – RHEL 9 install – the NETWORK & HOST NAME
configuration menu with configuration complete

Clicking **Done** will take us back to the main installer page with a system connected to a network and prepared to connect once the installation is complete.

In *Chapter 6, Enabling Network Connectivity*, will describe in more detail the options available to configure the network in an RHEL system.

> **Important Note**
>
> Now that the system is connected to the network, we can go back to **Time & Date** and enable **Network Time** (which is done automatically by the installer), as well as go to **Connect to Red Hat** to subscribe the system to Red Hat's **content distribution network** (CDN). The subscription of the system to the CDN will be explained in detail in *Chapter 7, Adding, Patching, and Managing Software*.

It is now time to review the final system option, security profiles, by going to **Security Policy** as follows:

Figure 1.49 – RHEL9 install – the Security Policy configuration icon

In it, we will see a list of security profiles that can be enabled by default in our system:

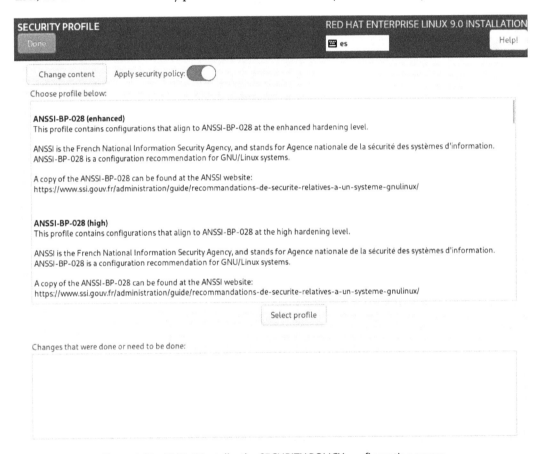

Figure 1.50 – RHEL 9 install – the SECURITY POLICY configuration menu

The security profiles have requirements that we are not covering in this installation (such as having separate /var or /tmp partitions). We can click on **Apply security policy** to turn it off, and then on **Done**:

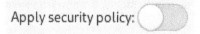

Figure 1.51 – RHEL 9 install – Security policy configuration toggle (off)

More on this topic will be covered in *Chapter 11*, *System Security Profiles with OpenSCAP*.

User settings

The main administrator user in a Unix or Linux system is called `root`.

We can enable a `root` user by clicking in the **Root Password** section, although this is not necessary and, in security-restricted environments, you are advised not to do so. We will do so in this chapter in order to learn how to do it and explain the cases covered:

Figure 1.52 – RHEL 9 install – the Root Password configuration icon (warning as it is not set)

After clicking on **Root Password**, we are presented with a dialog to type it:

ROOT PASSWORD		RED HAT ENTERPRISE LINUX 9.0 INSTALLATION

Done   ⌨ es   Help!

The root account is used for administering the system. Enter a password for the root user.

Root Password: `●●●●●●●●●●●●●●|`   👁

_____ **Strong**

Confirm: 👁

☐ Lock root account

☐ Allow root SSH login with password

Figure 1.53 – RHEL 9 install – the Root Password configuration menu

It is recommended that the password has the following:

- More than 10 characters (and a minimum of six)
- Lowercase and uppercase letters
- Numbers
- Special characters (such as $, @, %, and &)

If the password does not meet those requirements, it will warn us and force us to click **Done** twice to use a weak password.

It is now time to create a user for the system by clicking on **User Creation**:

Figure 1.54 – RHEL 9 install – the User Creation configuration icon (with a warning as it is not complete)

This will take us to a section to input user data:

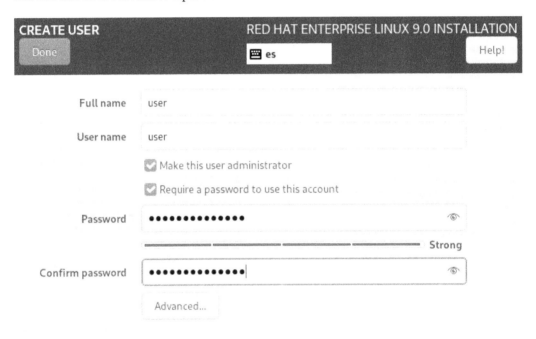

Figure 1.55 – RHEL 9 install – the User Creation configuration menu

The same password rules will apply here as in the previous section.

Clicking on **Make this user administrator** will enable the performance of administrative tasks (and also, no need to configure the `root` password).

Tip

As a good practice, do not use the same password for the root account and for the user account.

Chapter 5, Securing Systems with Users, Groups, and Permissions, includes a section on how to use and manage administrative privileges for users with the `sudo` tool.

Click on **Done** to return to the main installer screen. The installer is ready to proceed with the installation. The main page will look like this:

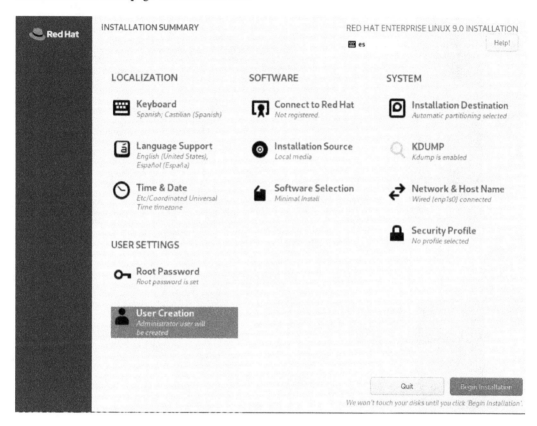

Figure 1.56 – RHEL 9 install – the main menu once completed

Clicking on **Begin Installation** will launch the installation process:

Important Note

If any of the steps required to start the installation are omitted, the **Begin Installation** button will be grayed out and, therefore, not available to be clicked.

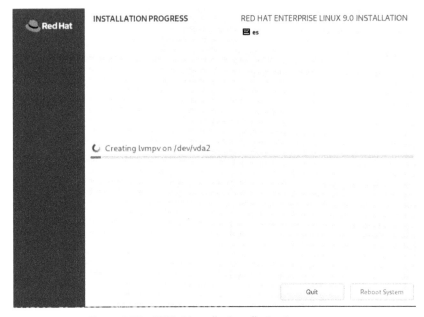

Figure 1.57 – RHEL 9 install – installation in progress

Once the installation is complete, we can click on **Reboot System** and it will be ready to use:

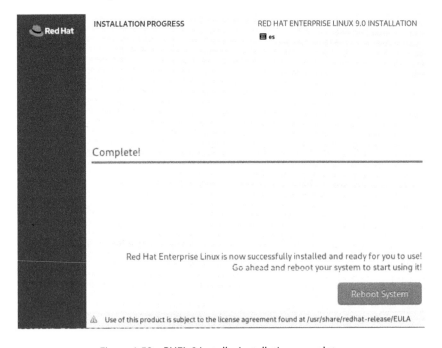

Figure 1.58 – RHEL 9 install – installation complete

It is important to remember to detach the ISO image from the virtual machine (or remove the USB thumb drive from the server) and check that the boot order is properly configured in the system.

Your first Red Hat Enterprise Linux 9 system is now ready! Congratulations.

As you can see, it is easy to install RHEL in a virtual or physical machine and have it ready to be used for any service we want to run on it. In the cloud, the process is very different, as machines are instantiated from images to run. In the next chapter, we will review how to run RHEL in a virtual machine instance in the cloud.

Summary

The *Red Hat Certified System Administrator* exam is entirely practical and based on real-world experience. The best way to prepare for it is by practicing as much as possible, which is why this book begins by providing access to RHEL 9 and offering alternatives on how to deploy your own virtual machine.

Different scenarios are covered regarding installation. These are the most common ones and include using a physical machine, a virtual machine, or a cloud instance. In this chapter, we focused on using a virtual machine or a physical one.

When using physical hardware, we will be focusing on the fact that many people like to reuse old hardware, buy second-hand or cheap mini servers, or even use their laptops as the primary installation for their Linux experience.

In the case of virtual machines, we are thinking about the people that want to keep all their work on the same laptop, but without messing with their current operating system (which may not even be Linux). This could also work well with the previous option by having virtual machines on your own mini server.

After this chapter, you are ready to proceed with the rest of the book, having at least an instance of RHEL 9 available to work with and practice on.

In the next chapter, we will review a number of advanced options, such as using the cloud for RHEL instances, automating the installation, and best practices.

Let's get started!

2

RHEL 9 Advanced Installation Options

In the previous chapter, we learned how to install **Red Hat Enterprise Linux**, or **RHEL**, on a physical or **virtual machine** (**VM**) so that we can use it while we're reading this book. In this chapter, we will review how to use RHEL instances in the cloud and the main differences that appear when doing so.

You will also learn not just how to deploy a system, but the best choices to do so, and be able to perform the deployment in an automated fashion.

To complete the installation, a section on best practices has been included so that you can start avoiding long-term issues from day one.

These are the topics that will be covered in this chapter:

- Automating RHEL deployments with Anaconda
- Deploying RHEL on the cloud
- Installation best practices

Technical requirements

In this chapter, we will review the automated installation process using **Anaconda**. For that, you will need to use the *RHEL 9 deployment* we created in the previous chapter.

We will also create cloud instances, for which you will need to create an account in the cloud environment of your choice. We will be using **Google Cloud Platform** (**GCP**).

Automating RHEL deployments with Anaconda

Once you have finished your first deployment of RHEL locally, you can log in as `root` on the machine and list the files that the `root` user has in their folder, as follows:

```
[root@rhel-instance ~]# ls /root/
anaconda-ks.cfg
```

You will find the `anaconda-ks.cfg` file. This is an important file, called a kickstart file, and it contains the responses given to the installer, **Anaconda**, during the installation process. Let's review the content of this file.

> **Important Note**
> In cloud images, there is no `anaconda-ks.cfg` file.

This file can be reused to install other systems with the same options as the ones we used for this installation. Let's review the options that we added during our previous installation.

Lines starting with # are comments and have no effect on the installation process.

The comment specifying the version that is being used is shown here:

```
#version=RHEL9
```

Then, a type of installation was performed. It can be `graphical` or `text` (for headless systems, it is common to use the second one). Here, you can see that it is a `graphical` installation:

```
# Use graphical install
graphical
```

The software source for installing application packages, or any other package, is specified with the `repo` entry. As we were using the **International Organization for Standardization** (**ISO**) image, it was accessed (mounted, in Linux parlance) as if it were a *CD-ROM*. The code is illustrated here:

```
repo --name="AppStream" --baseurl=file:///run/install/sources/
mount-0000-cdrom/AppStream
```

Sections are specified with the `%` symbol. In this case, we will enter the `packages` section with a list of packages to be installed and use the `%end` special tag to close them. There are two selections: a group of packages that are defined by it starting with the `@^` symbol (in this case, `minimal-environment`), and the name of a package that doesn't require any prefix (in this case, the package is `kexec-tools`, which is responsible for installing the `kdump` capability we explained previously).

The code is illustrated in the following snippet:

```
%packages
@^minimal-environment
kexec-tools
%end
```

We continue to define options without a specific section. In this case, we have the keyboard layouts and system language support. As you can see in the following code snippet, we added the *English (United States) US American keyboard* (marked as us) and the *Spanish, Spain* one (marked as es):

```
# Keyboard layouts
keyboard --xlayouts='us','es'
```

For the system language, we also added English US American (en_US), and Spanish, Spain (es_ES). There are several ways to manage, store, and represent text in operating systems. The most common one nowadays is **Unicode (or Universal Coded Character Set) Transformation Format-8 – 8 bit (UTF-8)**, which enables us to have many character sets under one single standard. That's why the system language has .UTF-8 appended to it, as shown in the following code snippet:

```
# System language
lang en_US.UTF-8 --addsupport=es_ES.UTF-8
```

> **Tip**
>
> UTF-8 is a character encoding that extends the capabilities of previous ones in order to support Chinese, Cyrillic, or Arabic (among many others) in the same text (such as the one representing a web page or a console). UTF-8 was presented in 1993 and is used by 95.9% of the **World Wide Web's (WWW's)** pages. Previous character sets supported US English or Latin characters only, such as the **American Standard Code for Information Interchange**, or **ASCII**, published in 1963. To learn more about character encodings and their evolution, check out the *Wikipedia* pages for both UTF-8 and ASCII.

Now, it's time to configure the network interface. In this case, we only have one, named enp1s0, as shown in the following code snippet. The configuration uses **Internet Protocol version 4 (IPv4)** with the **Dynamic Host Configuration Protocol (DHCP)** and IPv6, both of which are activated at boot. The hostname is configured as rhel9.example.com:

```
# Network information
network  --bootproto=dhcp --device=enp1s0 --ipv6=auto
--activate
network  --hostname=rhel9.example.com
```

Now, we need to define the installation media. In this case, we used an emulated CD-ROM/DVD using the ISO image file we downloaded, as shown here:

```
# Use CDROM installation media
cdrom
```

The option for `firstboot` is enabled by default. In this case, as the installation does not include a *graphical interface*, it won't be run but will be added to the kickstart file. We can safely remove it, like so:

```
# Run the Setup Agent on first boot
firstboot --enable
```

Now, let's configure the disks. First, to be safe, we will instruct the installer to ignore all disks except for the target one; in this case, vda, as shown here:

```
ignoredisk --only-use=vda
```

> **Important Note**
>
> The disk's name will vary, depending on the platform you are running on. Typically, it will be vda, xda, or sda. In this example, we show the vda disk that was defined by the installer, Anaconda, as used in the previous chapter.

Now, we must install the bootloader to enable the system to boot. We will do so in the **Master Boot Record** or **MBR** of the main disk, vda, and we will instruct it to use the `crashkernel` option, which enables the kdump mechanism (this dumps memory in case of a system crash). The code is illustrated in the following snippet:

```
# System bootloader configuration
bootloader --append="crashkernel=auto" --location=mbr --boot-
drive=vda
```

Now, we must partition the disk. In this case, it will be fully automated, as indicated here:

```
autopart
```

Space to be used by the system must be declared. We will clear the whole disk for this example, like so:

```
# Partition clearing information
clearpart --none --initlabel
```

Let's set the time zone to Madrid, Europe, as follows:

```
# System timezone
timezone Europe/Madrid --isUtc
```

Now, we will set the root password and create a user (note that the encrypted password was redacted for security purposes). Here's the code we need to execute:

```
# Root password
rootpw --iscrypted $xxxxxxxxxxxxxxxxxxxxxxxxxxxxxxxxxxxxxxxxxxx
xxxxxxxxxxxxxxxxxxxxxxxxxxxxxxxxxxxxxxxxxxxxxxxxxxxxxxxxxxxxxxx
user --groups=wheel --name=user
--password=$xxxxxxxxxxxxxxxxxxx
xxxxxxxxxxxxxxxxxxxxxxxxxxxxxxxxxxxxxxxxxxxxxxxxxxxxxxxxxxxxxxx
xx
xxxxxxxxxxxxxxxxxxxxx --iscrypted --gecos="user"
```

> **Tip**
>
> The generated Anaconda file from the previous chapter contains an example of the encrypted password hash. If we want to change it, a new encrypted password hash, to be included here, can be generated if we run the `python -c 'import crypt,getpass;pw=getpass.getpass();print(crypt.crypt(pw) if (pw==getpass.getpass("Confirm: ")) else exit())'` command.

Now, we need a special section where we can configure `kdump` so that we can reserve memory automatically. Here's how we can create this:

```
%addon com_redhat_kdump --enable --reserve-mb='auto'
%end
```

We also need a special section specifying the password policy that will be used for installation, so we'll run the following code to create this:

```
%anaconda
pwpolicy root --minlen=6 --minquality=1 --notstrict --nochanges
--notempty
pwpolicy user --minlen=6 --minquality=1 --notstrict --nochanges
--emptyok
pwpolicy luks --minlen=6 --minquality=1 --notstrict --nochanges
--notempty
%end
```

And with this, our kickstart file to reinstall our system is complete.

To use it, we will need to pass the kickstart option to the installer. To do so, we edit the kernel parameters. Let's see how this is done.

We start by pressing *Tab*, during boot, while the line `Install Red Hat Enterprise Linux 9.0` is selected. The boot line, starting with `vmlinuz`, will appear at the bottom of the screen, as indicated in the following screenshot:

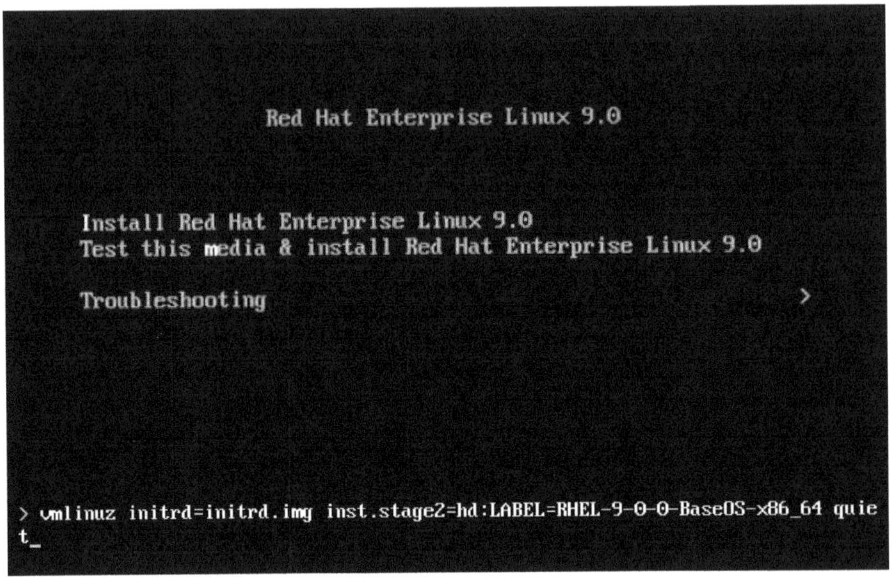

Figure 2.1 – RHEL 9 installer: Editing the boot line

Let's remove the `quiet` option and add one that lets the installer know where the kickstart is, as follows:

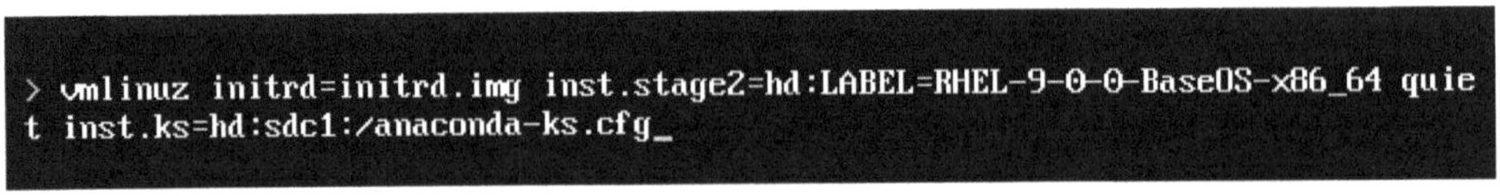

Figure 2.2 – RHEL 9 installer: Adding the kickstart option to the boot line

The option we've added looks like this:

```
inst.ks=hd:sdc1:/anaconda-ks.cfg
```

There are three parts to it that we can take a look at here:

- hd: The kickstart will be in a disk, such as a second **Universal Serial Bus (USB)** drive.
- sdc1: The device that hosts the file.
- /anaconda-ks.cfg: The path to the kickstart file in the device.

With this, we can reproduce the full installation we have done.

> **Tip**
>
> The *Red Hat Enterprise Linux 9 Customizing Anaconda* guide provides detailed options you can follow if you wish to create your own Anaconda kickstart file or further customize this one. It can be accessed here: https://access.redhat.com/documentation/en-us/ red_hat_enterprise_linux/9/html/customizing_anaconda/index. Note that the kickstart file itself is parsed by Anaconda but should be fine to have ordered in any suitable way—just remember that the scripts configured will be evaluated in the order they appear in the kickstart file.

As you have seen, it is very easy to create a kickstart file and automate the deployment of RHEL.

Now, let's move and look at a different way to make an RHEL 9 instance available: in the cloud.

Deploying RHEL on the cloud

Deploying RHEL on the cloud has some differences from the previous deployments we've done. Let's look at what these differences are here:

- We won't use an ISO image or Anaconda to perform a deployment, but a preconfigured image, usually prepared and made available by the cloud provider. The image can be later customized and adapted to our needs.
- We will not be able to choose the configuration details of our system (such as selecting a time zone, for example) during installation time, but will be able to after.
- An automated mechanism will be in place to change settings, such as adding a user and their credentials to access the system or configure the network:
  - The most extended and well-known mechanism used by cloud providers to do so is cloud-init.
  - Some of the images that are delivered by the cloud provider include the cloud-init software.

- Systems are usually accessed remotely using the **Secure Shell** (**SSH**) protocol and SSH keys that are generated by the user in the cloud provider (please check out *Chapter 8, Administering Systems Remotely*, for more details on how to access a system).

> **Important Note**
>
> When it comes to creating RHEL images, it's possible to create our own for the cloud or virtualization. To do so, we can use the RHEL Linux 8 Image Builder tool (`https://developers.redhat.com/blog/2019/05/08/red-hat-enterprise-linux-8-image-builder-building-custom-system-images/`). However, it is not part of **Red Hat Certified System Administrator** (**RHCSA**), so will not be covered in this book. Instead, we will follow the approach of taking the default image and customizing it. Note that the article targets RHEL 8, and we're now running RHEL 9, but the process should be very similar. You can read more information about `cloud-init` at the official guide at `https://access.redhat.com/documentation/en-us/red_hat_enterprise_linux/9/html/configuring_and_managing_cloud-init_for_rhel_9/introduction-to-cloud-init_cloud-content`.

Cloud providers propose an initial *getting started* offer where you try their services at no cost. It's a good way to get started with RHEL and cloud services.

In this book, we'll be using **Google Cloud Platform** as an example, so other clouds will not be covered. We will provide a brief example of how an RHEL 9 instance can be created and modified in this cloud environment. To do so, we will use Google Cloud Platform (it provides, as of December 2020, an initial credit that could last the whole duration required to complete this book). To follow this chapter, you will need to complete the following steps:

1. If you do not have a Google account, you will need to create one (if you use Gmail and/or an Android phone, you will have one already).

2. Log in to your Google account at `https://accounts.google.com` (or check whether you have already logged in). You will be required to sign in for a free trial, at which point you will have to provide a credit card number.

3. Go to `https://cloud.google.com/free` and claim your free credits.

4. Go to the cloud console at `https://console.cloud.google.com`.

5. Go to the **Projects** menu, which is shown here as **No organization** on the top bar, to show the projects for the new account:

Dashboard

Figure 2.3 – RHEL 9 in Google Cloud Platform: Organization menu access

6. Click on **NEW PROJECT**, as highlighted in the following screenshot:

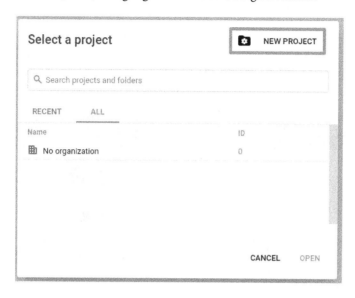

Figure 2.4 – RHEL 9 in Google Cloud Platform: Organization menu

7. Name it `Enterprise Linux` and click **CREATE**, as shown in the following screenshot:

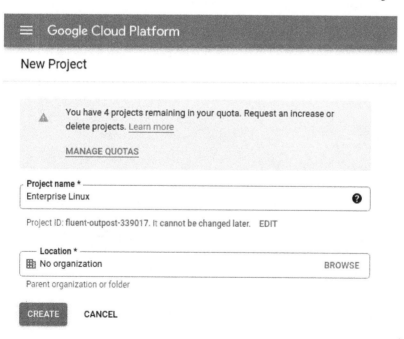

Figure 2.5 – RHEL 9 in Google Cloud Platform: Organization menu – creating a new project

> **Important Note**
> Depending on how your Google account is configured, you may need to enable billing after this step.

8. Go to the top-left menu (also called hamburger menu, with three horizontal lines next to it), click on **Compute Engine**, and then click on **VM instances**, as illustrated in the following screenshot:

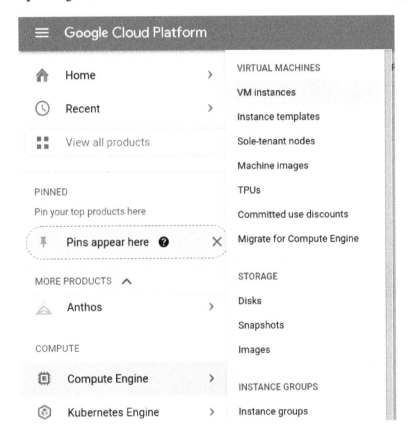

Figure 2.6 – RHEL 9 in Google Cloud Platform: Accessing the VM instances menu

9. Once **Compute Engine** is ready (this may take a few minutes), click on **Create**, as illustrated in the following screenshot (note that it might be required to enable the **application programming interface (API)** before having the **Create** button available):

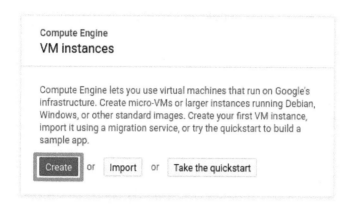

Figure 2.7 – RHEL 9 in Google Cloud Platform: Creating a new VM instance

10. We will name the instance `rhel-instance`, as illustrated here:

Figure 2.8 – RHEL 9 in Google Cloud Platform: Naming a new VM instance

11. Select the most convenient region (or leave the one already provided). You can see how to do this here:

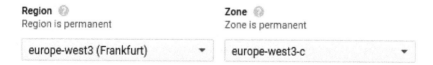

Figure 2.9 – RHEL 9 in Google Cloud Platform: Creating a new VM instance, region, and zone

12. Set the machine family and type to **General-purpose | e2-medium**, as illustrated here:

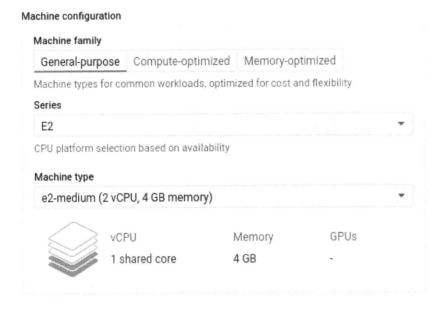

Figure 2.10 – RHEL 9 in Google Cloud Platform: Creating a new VM instance, type, and size

13. Click **Change** next to **Boot disk**, as illustrated here:

Figure 2.11 – RHEL 9 in Google Cloud Platform: Changing the boot disk

14. Change **Operating system** to **Red Hat Enterprise Linux** and **Version** to **Red Hat Enterprise Linux 9**, as illustrated in the following screenshot. Then, click **SELECT**:

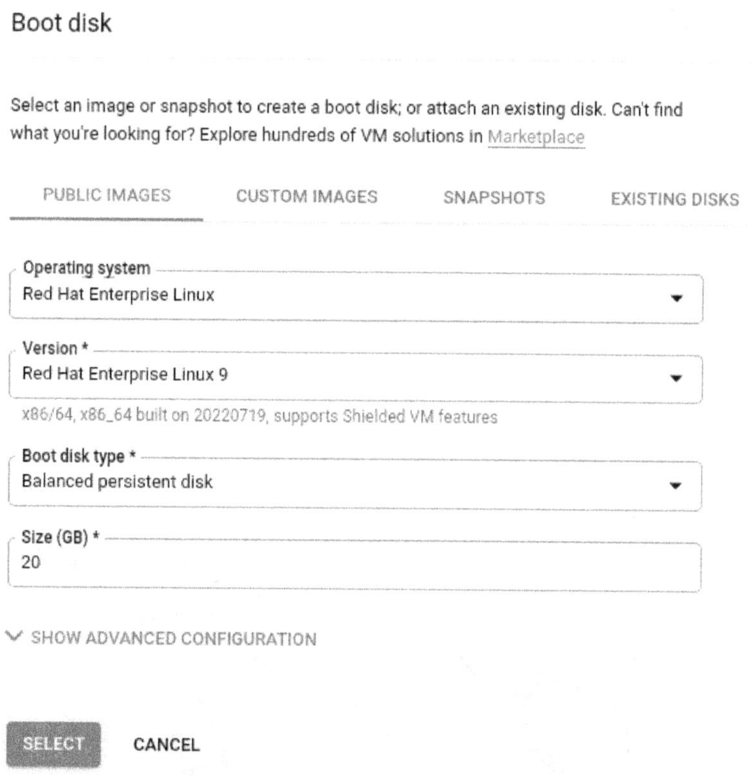
Boot disk
Select an image or snapshot to create a boot disk; or attach an existing disk. Can't find what you're looking for? Explore hundreds of VM solutions in Marketplace
PUBLIC IMAGES
CUSTOM IMAGES
SNAPSHOTS
EXISTING DISKS
Operating system
Red Hat Enterprise Linux
Version *
Red Hat Enterprise Linux 9
x86/64, x86_64 built on 20220719, supports Shielded VM features
Boot disk type *
Balanced persistent disk
Size (GB) *
20
SHOW ADVANCED CONFIGURATION
SELECT
CANCEL

Figure 2.12 – RHEL 9 in Google Cloud Platform: Creating a new VM instance, image selection, and disk size

15. Click **Create** and wait for the instance to be created. When it has been created, it will appear in the VM instance list shown here:

Filter
Enter property name or value
Status
Name
Zone
Recommendations
In use by
Internal IP
External IP
Connect
rhel-instance
us-east1-b
Save $33 / mo
10.0.0.5 (nic0)
SSH

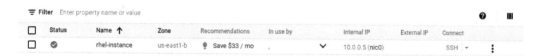

Figure 2.13 – RHEL 9 in Google Cloud: VM instance list

16. Later, we will learn how to connect via SSH. Now, click on the triangle next to **SSH**, under **Connect**, and select **Open in browser window**, as follows:

Figure 2.14 – RHEL 9 in Google Cloud Platform: VM instance – access console

17. With that, your fresh RHEL 9 instance will be deployed, as shown in the following screenshot:

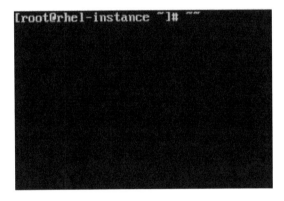

Figure 2.15 – RHEL 9 in Google Cloud Platform: VM instance – console

It takes some time to get set up in the cloud, configure your account, and find the **SSH** key (which will be shown in *Chapter 8, Administering Systems Remotely*), but once it's all set up, it's easy to get a new instance up and running.

To become an administrator, you only need to run the following command:

```
[miguel@rhel-instance ~]$ sudo -i
[root@rhel-instance ~]#
```

Now, you can check the time configuration with `timedatectl` and change it, as follows:

```
[root@rhel-instance ~]# timedatectl
              Local time: Sat 2021-12-12 17:13:29 UTC
          Universal time: Sat 2021-12-12 17:13:29 UTC
                RTC time: Sat 2021-12-12 17:13:29
               Time zone: UTC (UTC, +0000)
System clock synchronized: yes
             NTP service: active
         RTC in local TZ: no
[root@rhel-instance ~]# timedatectl set-timezone Europe/Madrid
[root@rhel-instance ~]# timedatectl
              Local time: Sat 2021-12-12 18:20:32 CET
          Universal time: Sat 2021-12-12 17:20:32 UTC
                RTC time: Sat 2021-12-12 17:20:32
               Time zone: Europe/Madrid (CET, +0100)
System clock synchronized: yes
             NTP service: active
         RTC in local TZ: no
```

You can also change the language configuration with `localectl`, like so:

```
[root@rhel-instance ~]# localectl
   System Locale: LANG=en_US.UTF-8
       VC Keymap: us
      X11 Layout: n/a
```

To change `locale` or language support, you will need to install its *language package* first, as follows:

```
[root@rhel-instance ~]# yum install glibc-langpack-es -y
... [output omitted] ...
[root@rhel-instance ~]# localectl set-locale es_ES.utf8
[root@rhel-instance ~]# localectl
   System Locale: LANG=es_ES.utf8
       VC Keymap: us
      X11 Layout: n/a
```

You now have a machine configured that you can use throughout this book. These locale changes are not needed to proceed—they're only needed to create a machine with the same configuration as in the previous chapter.

Now that we know how to automatically redeploy VMs using Anaconda and how to get instances in the cloud, let's move on and look at some of the best practices to be taken into account when performing installations.

Installation best practices

RHEL installations have many options you can choose from, and what you choose should be tailored for your specific use case. However, some common recommendations apply. Let's look at the most common types.

The first type is **blueprints**. Here's how you can use them:

- Standardize the core installation and create a blueprint for it:

  - This blueprint will be minimal enough to serve as the base for all other blueprints and deployments.

- Build a set of blueprints for common cases when needed:

  - Try to use an automation platform to build extended cases (that is, Ansible).

  - Try to make the cases modular (that is, an app server; database blueprints can be combined into one single machine).

  - Be aware of the requirements you must apply to your templated blueprints and adapt to the environments you will use.

The second type is **software**. Here are some guidelines regarding this:

- The less software that's installed, the smaller the attack surface. Try to keep servers with the minimal set of packages required on them for them to run and operate (that is, try not to add a **graphical user interface** (**GUI**) to your servers).

- Standardize the installed tools where possible to be able to react quickly in case of emergency.

- Package your third-party applications so that you have healthy life cycle management (whether with **RPM Package Manager** (**RPM**) or in containers).

- Establish a patching schedule.

The third type is **networking**. Here are some recommendations in terms of this:

- In VMs, try not to overuse the number of network interfaces.

- In physical machines, use interface teaming/bonding whenever possible. Segment networks using **virtual local area networks (VLANs)**.

The fourth type is **storage**. Here are some useful suggestions on how best to use this:

- For servers, use **Logical Volume Manager (LVM)** where possible (usually, everything but /boot or /boot/efi).

- If you think you will need to reduce your filesystems, use ext4; otherwise, go for the default of xfs.

- Partition the disk carefully by doing the following:

  - Keep the default boot partition with its default size. If you change it, enlarge it (you may need space there during upgrades).

  - The default swap partition is the safest bet unless the third-party software has specific requirements.

  - For long-lived systems, have at least separate partitions for / (root) /var, /usr, /tmp, and /home, and consider even a separate one for /var/log and /opt (for ephemeral cloud instances or short-lived systems, this does not apply).

The fifth type is **security**. Follow these guidelines:

- Do not disable **Security-Enhanced Linux (SELinux)**. It has been improved a lot in the latest versions and it's unlikely to interfere with your system (if required, set it in permissive mode instead of fully disabling it).

- Do not disable the firewall. Automate port opening with the service deployment.

- Redirect logs to a central location whenever possible.

- Standardize the security tools and configuration that you want to install to check system integrity and audit (that is, **Advanced Intrusion Detection Environment (AIDE)**, logwatch, fapolicyd, **Integrity Measurement Architecture (IMA)**, and auditd).

- Review software install (*RPM*) **GNU Privacy Guard (GPG)** keys, as well as ISO images, to ensure integrity.

- Try to avoid using passwords (especially for your root account) and use strong ones where needed.

- Review your systems with *OpenSCAP* to check on security (if needed, create your own hardware **Security Content Automation Protocol (SCAP)** profile with help from your security team).

Finally, we will look at the **miscellanea** type, as follows:

- Keep system time synchronized.

- Review `logrotate` policies to avoid "*disk full*" errors due to logs.

Following these best practices will help you avoid issues and make the installed base more manageable. With that, you know how to deploy RHEL on a system in a structured, repeatable manner while providing services to other teams in a fast and resilient fashion.

Summary

In the previous chapter, we mentioned how to prepare a machine that we can work with throughout this book. An alternative to that is using cloud instances, with which we could be consuming VM instances from the public cloud, which may simplify our consumption and provide us with enough free credit to prepare for RHCSA. Also, once the self-training process is complete, the machines can be still used to provide your own public services (such as deploying a blog).

Understanding the need to standardize your environments and the impact of doing so is also important when you're working with Linux as a professional. It is key to start with a good set of practices (automating installations, keeping track of installed software, reducing the attack surface, and so on) from the beginning.

Now that you've completed this chapter, you are ready to continue with the rest of this book, since you now have an instance of RHEL 9 available to work and practice with. In the next chapter, we will review the basics of the system to make ourselves comfortable and gain confidence in using the system.

3
Basic Commands and Simple Shell Scripts

Once you have your first **Red Hat Enterprise Linux (RHEL)** system running, you want to start using it, practicing, and getting comfortable with it. In this chapter, we will review the basics of logging into the system, navigating through it, and getting to know the basics in terms of its administration.

The set of commands and practices described in this chapter will be used on many occasions when managing systems, so it is important to study them with care.

The following topics will be covered in this chapter:

- Logging in as a user and managing multi-user environments
- Changing users with the su command
- Understanding users, groups, and basic permissions
- Using the command line, environment variables, and navigating through the filesystem
- Understanding I/O redirection on the command line
- Filtering output with grep and sed
- Listing, creating, copying, and moving files, directories, links, and hard links
- Using tar and gzip
- Creating basic shell scripts
- Using system documentation resources

Logging in as a user and managing multi-user environments

Login is the process during which a user identifies themselves in the system – usually, by providing a **username** and **password**, a couple of pieces of information often referred to as *credentials*.

The system can be accessed in many ways. The initial case for this, which we are covering here, is how a user accesses it when they install a physical machine (such as a laptop) or via the virtualization software interface. In this case, we are accessing the system through a *console*.

During installation, the user was created with an assigned password, and no graphical interface was installed. We will access the system in this case via its *text console*. The first thing we are going to do is to log in to the system using it. Once we start the machine and the boot process is completed, we will enter, by default, the multi-user text mode environment in which we are being requested to provide our login:

```
Red Hat Enterprise Linux 9.0 (Plow)
Kernel 5.14.0-70.13.1.el9_0.x86_64 on an x86_64

rhel-instance login: _
```

Figure 3.1 – The login process and username request

The blinking cursor will let us know that we are ready to enter our username, in this case, user, and then press *Enter*. A line requesting the password will appear:

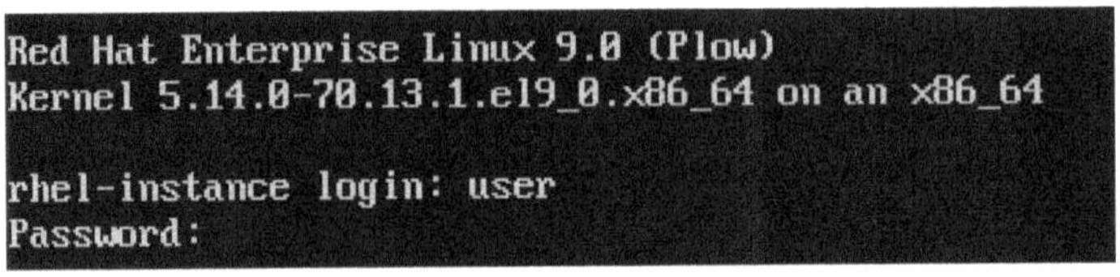

```
Red Hat Enterprise Linux 9.0 (Plow)
Kernel 5.14.0-70.13.1.el9_0.x86_64 on an x86_64

rhel-instance login: user
Password:
```

Figure 3.2 – The login process and password request

We may now type the user's password to complete the login and, by pressing *Enter* on your keyboard, start a session. Note that no characters will be displayed on the screen when typing the password to avoid eavesdropping. The following screenshot shows the session running:

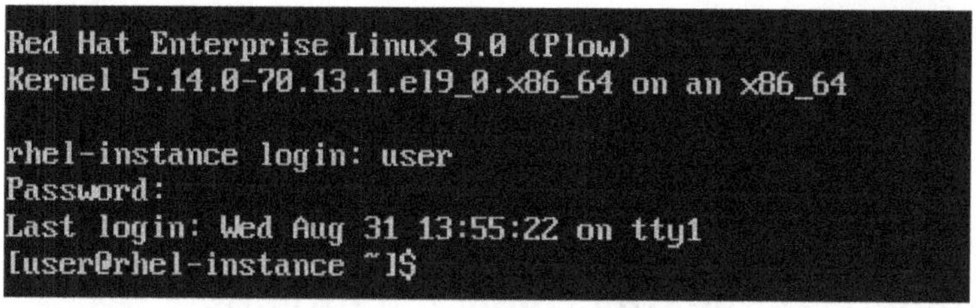

```
Red Hat Enterprise Linux 9.0 (Plow)
Kernel 5.14.0-70.13.1.el9_0.x86_64 on an x86_64

rhel-instance login: user
Password:
Last login: Wed Aug 31 13:55:22 on tty1
[user@rhel-instance ~]$
```

Figure 3.3 – The completed login process and the session running

Now, we are fully logged in to the system with the *credentials* for a user called user. This will define what we can do in the system, which files we can access, and even how much disk space we are assigned.

The console can have more than one session. To make that possible, we have different terminals through which we can log in. The default terminal can be reached by simultaneously pressing the *Ctrl + Alt + F1* keys. In our case, nothing will happen, as we are already in that terminal. We could move to the second terminal by pressing *Ctrl + Alt + F2*, to the third one by pressing *Ctrl + Alt + F3*, and so on for the rest of the terminals (by default, six are allocated). This way, we can run different commands in different terminals.

Using the root account

Regular users will not be able to make changes to the system, such as creating new users or adding new software to the whole system. To do so, we need a user with administrative privileges and for that, the default user is root. This user always exists in the system and its identifier – the **User Id** (**UID**) – has the value 0.

In the previous installation, we configured the root password, making the account accessible through the console. To use it by logging in to the system, we only need to type the user root into one of the terminals shown right next to the login, then hit *Enter*, and then provide its password, which won't be displayed. This way, we will access the system as the administrator, root:

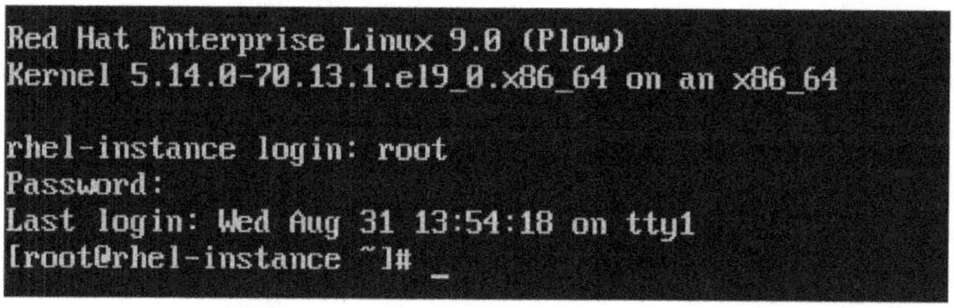

```
Red Hat Enterprise Linux 9.0 (Plow)
Kernel 5.14.0-70.13.1.el9_0.x86_64 on an x86_64

rhel-instance login: root
Password:
Last login: Wed Aug 31 13:54:18 on tty1
[root@rhel-instance ~]#
```

Figure 3.4 – The completed login process for root

> **Important Note**
>
> Above the login prompt, there is a message suggesting how the activation of the web console (cockpit) can be done – the cockpit is a set of tools that enables web management for the system. The cockpit is covered in *Chapter 4, Tools for Regular Operations*.

Using and understanding the command prompt

The command line that appears once we have logged in and are waiting for our commands to be typed and run is called the **command prompt**.

In its default configuration, it will show the *username* and *hostname* between brackets to let us know with which user we are working. Next, we see the path, in this case, ~, which is the shortcut for the **user's home directory** (in other words, /home/user for user, and /root for root).

The last part and, probably the most important one, is the symbol before the prompt:

- The $ symbol is used for regular users with no administrative privileges.

- The # symbol is used for root or once a user has acquired administrative privileges.

> **Important Note**
>
> Be careful when using a prompt with the # sign, as you will be running as an administrator and the system will likely not stop you from damaging it.

Once we have identified ourselves within the system, we are logged in and have a running session. It is time to learn how to change from one user to the other in the following section.

Changing users with the su command

As we have entered a **multi-user system**, it is logical to think that we will be able to change between users. Even when this can be done easily by opening a session for each, sometimes we want to act as several users within one session.

To do so, we can use the su tool. The name of the tool is usually referred to as **Substitute User**.

Let's use that last session, in which we logged in as root, and turn ourselves into user.

Before doing so, we can always ask which user we are logged in as by running the whoami command:

```
[@rhel-instance ~]# whoami
root
```

Now, we can make the change from `root` to `user`:

```
[root@rhel-instance ~]# su user
[user@rhel-instance root]$ whoami
user
```

Now, we have a session as `user`. We can finish this session by using the `exit` command:

```
[user@rhel-instance root]$ exit
exit
[root@rhel-instance ~]# whoami
root
```

As you may have seen, when we are logged in as `root`, we can act as any user without knowing its password. But how can we impersonate `root`? We can do so by running the `su` command and specifying the `root` user. In this case, the `root` user's password will be requested:

```
[user@rhel-instance ~]$ su root
Password:
[@rhel-instance user]# whoami
root
```

As `root` is the user with the ID 0 and the most important one, when running `su` without specifying the user we want to turn into, it will default to turning us into `root`:

```
[user@rhel-instance ~]$ su
Password:
[root@rhel-instance user]# whoami
root
```

Each user can define several options in their own environment, such as, for example, their preferred editor. If we want to fully impersonate the other user and take their preferences (or **environment variables**, as they are referred to on many occasions), we can do so by adding - after the `su` command:

```
[user@rhel-instance ~]$ su -
Password:
Last login: mar feb 15 04:57:29 CET 2022 on pts/0
[root@rhel-instance ~]#
```

We can also switch from `root` to `user`:

```
[root@rhel-instance ~]# su - user
Last login: Tue Feb 15 04:53:02 CET 2022 from 192.168.122.1 on
pts/0
[user@rhel-instance ~]$
```

As you can observe, it behaves as if a new login was done, but within the same session. Now, let's move on to managing the permissions for the different users in the system, as addressed in the following section.

Understanding users, groups, and basic permissions

Multi-user environments are defined by being able to handle more than one user simultaneously. But to be able to administer the system resources, two capabilities help with the tasks:

- **Groups**: Can aggregate users and provide permissions for them in blocks.

  Each user has a *primary group*.

  By default, a group is created for each user and assigned to it as a primary with the same name as the username.

- **Permissions**: Assigned to files, determining which users and groups can access each file.

  Standard Linux (and UNIX or POSIX) permissions include *user*, *group*, and *others* (`ugo`).

The whole system comes with a set of permissions assigned by default to each file and directory. Be careful when changing them.

There is a certain principle in UNIX that Linux has inherited: *everything is a file*. Even when there may be some corner cases to this principle, it remains true on almost any occasion. It means that a disk is represented as a file in the system (in other words, such as `/dev/sdb` mentioned in the installation), a process can be represented as a file (under `/proc`), and many other components in the system are also represented as files.

This means that, when assigning permissions to files, we can also assign permissions to many other components and capabilities implemented by them by virtue of the fact that, in Linux, everything is represented as a file.

> **Tip**
> The **Portable Operating System Interface** (**POSIX**) is a family of standards specified by the IEEE Computer Society: `https://en.wikipedia.org/wiki/POSIX`.

Users

Users are a way of providing security limits to people as well as programs running in a system. There are three types of users:

- **Regular users**: Assigned to individuals to perform their job. They have restrictions applied to them.

- **The superuser**: Also referred to as `root`. This is the main administrative account in the system and has full access to it.

- **System users**: These are user accounts usually assigned to running processes or "daemons" to limit their reach within the system. System users are not intended for logging in to the system.

Users have a number called the UID that the system uses to internally identify each one of them.

We previously used the `whoami` command to reveal which user we were working with, but to get more information here, we will use the `id` command:

```
[user@rhel-instance ~]$ id
uid=1000(user) gid=1000(user) groups=1000(user),10(wheel) conte
xt=unconfined_u:unconfined_r:unconfined_t:s0-s0:c0.c1023
```

We can also check the information related to other user accounts in the system, even to get info about `root`:

```
[user@rhel-instance ~]$ id root
uid=0(root) gid=0(root) groups=0(root)
```

Now, let's take a look at the information we have received for `user` by running `id`:

- `uid=1000(user)`: The UID is the numeric identifier of the user in the system. In this case, it is `1000`. Identifiers of 1,000 and above are used in RHEL for regular users, whereas 999 and below are reserved for system use.

- `gid=1000(user)`: The group ID is the numeric identifier for the principal group assigned to the user.

- `groups=1000(user),10(wheel)`: These are the groups that the user belongs to, in this case, `user` with a **Group ID (GID)** of `1000` and `wheel` with a GID of `10`. The `wheel` user group is a special one. It is used in RHEL and many other systems as the group for users that can become administrators by using the `sudo` tool (which we will explain later).

- `context=unconfined_u:unconfined_r:unconfined_t:s0-s0:c0.c1023`: This is the SELinux context for the user. It will define several restrictions in the system by using **SELinux** (which will be explained in greater depth in *Chapter 10, Keeping Your System Hardened with SELinux*).

ID-related data is stored in the system in the /etc/passwd file. Please note that this file is very sensitive and is better managed by using the tools related to it. If we want to edit it, we will do so by using vipw, a tool that will ensure (among other things) that only one admin is editing the file at any one time. The /etc/passwd file contains the info of each user per line. This is the line for user:

```
user:x:1000:1000:user:/home/user:/bin/bash
```

Each field is separated by a colon, :, in each line. Let's review what they mean:

- user: The username assigned to the user.
- x: The field for the encrypted password. In this case, it shows as x because it has moved to /etc/shadow, which is not directly accessible to regular users, to make the system more secure.
- 1000 (the first one): The UID value.
- 1000 (the second one): The GID value.
- user: A description of the account.
- /home/user: The home directory assigned to the user. This will be the default directory (or folder, if you prefer) that the user will work on and where their preferences will be stored.
- /bin/bash: The **command interpreter** for the user. Bash is the default interpreter in RHEL. Other alternatives, such as tcsh, zsh, or fish are available to be installed in RHEL.

Groups

Groups are a way of assigning certain permissions to a subset of users in a dynamic way. As an example, let's imagine a scenario where we have a finance team. We can create the finance group and provide permission to access, read, and write the /srv/finance directory. When the finance team has a new hire, to provide them with access to that folder, we only need to add the user assigned to this person to the finance group (this also works if someone leaves the team – we will only have to remove their account from the finance group).

Groups have a number called the **GID** that the system uses to identify them internally.

The data for groups is stored in the system in the /etc/group file. To edit this file in a way that ensures consistency and avoids corruption, we have to use the vigr tool. The file contains one group per line with different fields separated by a colon (:). Let's take a look at the line for the wheel group:

```
wheel:x:10:user
```

Let's review what each field means:

- wheel: This is the name of the group. In this case, this group is special, as it is configured by default to be used as the one to provide admin privileges to regular users.

- x: This is the group password field. It's currently obsolete and should always contain x. It is maintained for compatibility purposes.

- 10: This is the GID value for the group itself.

- user: This is the list of the users belonging to that group (separated by commas, such as user1, user2, and user3).

The types of groups are as follows:

- **The primary group**: This is the group assigned to the files newly created by the user.

- **A private group**: This is a specific group with the same name as the user that is created for each user. When adding a new user account, a private group will be automatically created for it. Commonly, the primary group and private group are the same.

- **A supplementary group**: This is another group usually created for specific purposes. By way of an example, we can see the wheel group for enabling admin privileges to users or the cdrom group for providing access to CDs and DVDs devices in the system.

File permissions

To review **file permissions**, we are going to log in to the system as root. We will use the ls command to list files and we will review the permissions associated with them. We will learn more about how to change permissions in *Chapter 5, Securing Systems with Users, Groups, and Permissions*.

Once logged in to the system as root, we can run the ls command:

```
[root@rhel-instance ~]# ls
anaconda-ks.cfg
```

This shows the files present in the *root user home directory*, represented by ~. In this case, it shows the *kickstart* file created by *Anaconda* that we reviewed in the previous chapter.

We can get the long version of the list by appending the -l option to ls:

```
[root@rhel-instance ~]# ls -l
total 4
-rw-------. 1 root root 1106 Feb 18 19:45 anaconda-ks.cfg
```

We see the following in the output:

- total 4: This is the total space in kilobytes occupied in the disk by the files (note that we are using 4 KB blocks, so every file under that size will occupy a minimum of 4 KB).

- -rw-------.: These are the permissions assigned to the file.

The structure of the permissions can be seen in the following diagram:

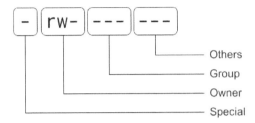

Figure 3.5 – The permissions structure for Linux

Let's discuss the preceding structure:

- The first character is for any *special permissions* that the file may have. If it is a regular file and has no special permission (as in this case), it will appear as -.

- Directories will appear with d. Consider that in Linux, everything is a file, and directories are a file with special permissions.

- Links, usually symbolic links, will appear with l. These act as a shortcut to a file from a different directory.

- Special permissions to run a file as a different user or group, called **setuid** or **setgid**, will appear as s.

- A special permission so that only the owner can remove or rename a file, called a **sticky bit**, will appear as t.

The next three characters, rw-, are the permissions for the *owner*:

- The first one, r, is the read permission assigned.

- The second one, w, is the write permission assigned.

- The third one, x, not present and shown as -, is the executable permission. Note that executable permission for directories means being able to enter them.

The next three characters, ---, are for the *group* permissions and work the same way as the owner permission. In this case, no group access is granted.

The final three characters, ---, are the permissions for *others*, which means users or groups do not show as the ones assigned to the file:

- 1: This indicates the number of **links** (hard links) to this file. This is intended, among other things, so that we do not delete a file used in another folder.

- root (the first instance): This indicates the owner of the file.

- `root` (the second instance): This indicates the group assigned to the file.
- `1393`: This indicates the size in bytes.
- `Dec 7 16:45`: This indicates the date and time that the file was last modified.
- `anaconda-ks.cfg`: This indicates the filename.

When we list a directory (referred to in other systems as a *folder*), the output will show the contents of the directory itself. We can list the info for the directory itself with the `-d` option. Let's now take a look at `/etc`, the directory that stores the system-wide configuration:

```
[root@rhel-instance ~]# ls -l -d /etc
drwxr-xr-x. 103 root root 8192 Feb 18 18:03 /etc
```

As you can see, it's quite easy to obtain information about files and directories in the system. Let's now learn more about the command line and how to navigate the filesystem easily in the following section.

Using the command line, environment variables, and navigating through the filesystem

As we have seen before, once we *log in* to the system, we have access to the command line. It's important to navigate the command line and the filesystem well to feel comfortable in the environment and make the most of it.

Command line and environment variables

The command line is provided by a program also known as an *interpreter* or **shell**. It will behave differently depending on which shell we use, but in this section, we will cover the most widespread shell used in Linux and the one provided by default in RHEL – **bash**.

A simple trick to know which shell you are using is to run the following command:

```
[root@rhel-instance ~]# echo $SHELL
/bin/bash
```

The `echo` command will show the content of whatever we give to it on the screen. Some content needs to be *substituted* or *interpreted*, such as environment variables. The content to be substituted starts with the $ symbol. In this case, we are telling the system to use the `echo` function on (and therefore, echo) the content of the SHELL variable. Let's use it for other variables:

```
[root@rhel-instance ~]# echo $USER
root
```

```
[root@rhel-instance ~]# echo $HOME
/root
```

These are environment variables that can be customized for every user. Let's now check these for a different user:

```
[root@rhel-instance ~]# su - user
Last login: Wed Feb 16 19:03:32 CET 2022 from 192.168.122.1 on
pts/0
[user@rhel-instance ~]$ echo $USER
user
[user@rhel-instance ~]$  echo $HOME
/home/user
```

As you can see, you can always refer to $USER and it will be substituted with the current user, or to $HOME and it will be substituted by the directory dedicated to the user, also known as the home directory.

These are some of the most common and important environment variables:

Variable	Value for user	Usage
EDITOR	(unset)	This establishes the default text editor for the user.
HOME	/home/user	This is the home directory for the current user.
HOSTNAME	rhel9.example.com	This is the hostname of the system we are logged in to.
LANG	en_US.UTF-8	This is the language configured for the current user. In this case, US American English with the UTF-8 extensions.
PATH	/home/user/.local /bin:/home/user/b in:/usr/local/bin:/ usr/bin:/usr/local/ sbin:/usr/sbin	This is a colon-separated list of directories that will be searched in order to run the command we are typing.
PS1	'[\u@\h \W]\$'	This defines what information will be shown at the "prompt", in this case, the user (\u), the @ hostname (\h), the working directory (\W), and the prompt symbol - $ for user or # for root (\$).
PWD	/home/user	This is the path to the working directory, the directory we are currently in.
SHELL	/bin/bash	This indicates the shell in user.
USER	user	This is the username of the current user, in this case, "user".

Table 3.1 – Common environment variables

The ~/.bashrc file is the one that should be edited to change these values for the current user.

> **Important Note**
> Files starting with a dot (.) are hidden files – to see them, you can use the ls -a command.

Navigating the filesystem

Now, it's time to move into the **directory tree** of the system. In Linux and Unix (macOS is a Unix-like system), there are no drive letters, but a single directory tree that starts with the *root directory*, represented by /. The rest of the content of the system will hang from that folder and any other accessible disk or device will be assigned an accessible directory.

> **Important Note**
> The *root directory* and the *home directory* for the *root user* are two different things. The *root user* is assigned the home directory, /root, by default, whereas the *root directory* is the mother of all directories in the system and is represented by /.`

We can see which directory we are in by running the pwd command:

```
[user@rhel-instance ~]$ pwd
/home/user
```

We can change the directory by using the cd command:

```
[user@rhel-instance ~]$ cd /var/tmp
[user@rhel-instance tmp]$ pwd
/var/tmp
```

As you already know, there is a **shortcut** for the home directory of the current user, ~. We can use this shortcut to go to it:

```
[user@rhel-instance tmp]$ cd ~
[user@rhel-instance ~]$ pwd
/home/user
```

Some shortcuts for directories include the following:

- ~: This is the home of the current user.
- .: This is the current directory.
- ..: This is the parent directory.
- -: This is the directory used previously.

More details on managing files and directories in Linux and RHEL are available in the *Listing, creating, copying, and moving files, directories, links, and hard links* section.

Bash auto-complete

Shortcuts are a faster way to reach commonly used directories or relative references to the current working directory. However, bash includes some capabilities to reach other directories in a fast way, which is called **auto-completion**. It relies on the *Tab* key (the one with two opposing arrows at the very left of your keyboard, right above *Caps Lock*).

When reaching a folder or a file, we can hit *Tab* to complete its name. For example, if we want to go to the /etc/NetworkManager folder, we type the following:

```
[user@rhel-instance ~]$ cd /et
```

Then, when we hit the *Tab* key, this will auto-complete it to /etc/, even adding the final forward slash (/), as it is a directory:

```
[user@rhel-instance ~]$ cd /etc/
```

Now, we type in the first letter of the directory we want to go to, NetworkManager, which is N:

```
[user@rhel-instance ~]$ cd /etc/N
```

Then, when we hit the *Tab* key, this will auto-complete it to /etc/NetworkManager/:

```
[user@rhel-instance ~]# cd /etc/NetworkManager/
```

Now, we can hit *Enter* and go there.

If we press *Tab* + *Tab* (pressing *Tab* twice during complete), this will show a list of available targets to complete – for example, see the following:

```
[user@rhel-instance ~]# cd /r
root/  run/
```

It can also be used to complete commands. We can type a letter, for example, h, hit *Tab + Tab*, and this will show all the commands starting with h:

```
[user@rhel-instance ~]# h
halt           hardlink      hash          h dparm        head
help           hexdump       history       hostid         hostname
hostnamectl    hwclock
```

This capability can be extended to help complete other parts of our commands by installing the bash-completion package:

```
[user@rhel-instance ~]# yum install bash-completion -y
```

> **Tip**
>
> A good practice when installing a system is to install bash-completion and also create or update the manuals database with the mandb command to get the latest manuals available using the man <command> command.

Previous commands

There is a way to recover the last commands run, which is referred to as **history**, in case you want to re-run them. Just press the *up arrow* key (the one with an arrow pointing up) and the previous commands will appear onscreen.

If there are too many commands in your history, you can search through them quickly by running the history command:

```
[user@rhel-instance ~]$ history
    1  su root
    2  su
    3  su -
    4  id
    5  id root
    6  grep user /etc/passwd
    7  echo $USER
    8   echo $HOME
    9  declare
   10   echo $SHELL
   11   echo EDITOR
```

```
12   echo $EDITOR
13   grep wheel /etc/gro
14   grep wheel /etc/group
15   cat /etc/group
16   grep nobody /etc/group /etc/passwd
```

You can run any of those commands again by using the ! command. Just run ! with the number of the command and it will run again:

```
[user@rhel-instance ~]$ !5
id root
uid=0(root) gid=0(root) groups=0(root)
```

> Tip
>
> The ! ! command will run the very last command again, no matter which number.

Now, it is time to enjoy your superfast command line. Let's learn more about the structure of directories in Linux, to know where to go to find things, in the following section.

The filesystem hierarchy

Linux has a standard, maintained by the *Linux Foundation*, that defines the **filesystem hierarchy** used in almost every Linux distribution, including *RHEL*. This standard is known as the **FHS**, or the **Filesystem Hierarchy Standard**. Let's review the most important folders in the standard and the system itself here:

Directory	Stands for	Use/Purpose
/	Root directory	The main directory of the filesystem hierarchy on which the other directories hang.
/boot	Bootable files	Files used by the system to boot. It usually has a dedicated partition for itself.
/dev	Devices	The directory that includes files representing the devices connected to it (whether disks, keyboards, or audio devices, they are represented here).
/etc	"Etcetera" or "editable text configuration"	Configuration files that apply system wide.
/home	Home	The directory to include home directories, where user personal settings and files are intended to be stored.
/media	External media	Intended to provide system-wide access to removable media.
/mnt	Mount	For temporarily mounted filesystems.
/opt	Optional	Optional application software packages. When building or installing your own software for more than one system, land it here.
/proc	Processes	A special virtual filesystem representing the processes running in the system.
/root	Root's home	The home folder for the "root" super-user. This is not present in /home in case it gets full so that root can always log in to the system.
/run	Run	This is the runtime data for processes. It is intended to be cleaned upon reboot.
/srv	Service	This is the content to be served externally, such as web pages of files in an FTP service.
/sys	System	This contains system information related to kernel features and connected devices.
/usr	User binaries and libs	This contains the read-only user data, including libraries, binaries, headers, sources, and other shared data.
/usr/bin	Binaries	This is the directory for the regular binaries used in the system.
/usr/lib	Libraries	This is the directory for holding the libraries used in the system.
/usr/local	Local binaries	This indicates local data specific to this host. When building local scripts or binaries for this specific system, host them here.
/usr/sbin	Super binaries	This is the directory for the binaries to be used by the superuser only.
/var	Variable content	This is the directory to be used to host content to be managed by different programs, from virtual machines to logs.

Table 3.2 – The most important folders in the filesystem

> **Tip**
> Previous versions of RHEL used to have /bin for the essential binaries and /usr/bin for the non-essential ones. Now, the content of both resides in /usr/bin. They also used /var/lock and /var/run for what is running in /run. In addition, they used to have /lib for the essential libraries and /usr/lib for the non-essential ones, which were consolidated into a single directory, /usr/lib. And last but not least, /sbin is the directory for the essential superuser binaries, and /usr/sbin is the directory for the non-essential ones, merged under /usr/sbin.

When partitioning, we may well be asking ourselves, where does the disk space go?

These are the allocation values for a *minimal* installation of RHEL 9 and the recommendations:

Directory	Used Space	Recommendation
/boot	179 M	Contains booting files and plays an important role when upgrading the system. 1GB is the minimum, while assigning 2 GB is recommended.
/etc	21 M	etc hardly ever grows a lot in size (rarely beyond 50 MB for a server). It is very convenient to leave this folder as part of the root partition and not create a partition for itself.
/home	16 K	home is assigned to user files. In a workstation, it holds all the working data of the user; however, in a server, it is usually kept with admin's temporary files. It is convenient to assign a partition to it.
/root	22 M	This is the home partition for the root user and again, hardly ever grows in size. It is convenient to keep it within the root partition.
/run	5.1 M	This is usually mounted as a temporary filesystem (tmpfs) that does not use disk space but memory. It is not recommended to make any changes to the default install.
/usr	1.3 GB	usr is the largest folder when installing a machine. It is 1.3 G in the minimal install, and reaches 4.6 GB for a full-blown workstation. Once the system is in production, the size of this directory won't change. On average, 10 GB is a very good size for it.
/var	65 M	var, although holding very little data initially, will hold most of the data available in the system once in production. For servers storing large amounts of data, it can be common to split this partition into several others, such as /var/log or /var/lib, or even /var/spool for email servers.

Table 3.3 – The allocation values for a minimal installation of RHEL 9

It's important to become familiar with the main directories in the system to make the best of them. It is recommended to navigate through the different system directories and look at what's in them to become comfortable with the structure. In the following section, we will look at how to perform redirections on the command line to learn more about command and file interaction.

Understanding I/O redirection on the command line

We have already run several commands to ascertain information about the system, such as listing files with `ls`, and we have got some information, or the **output**, from the running command, including, for example, filenames and file sizes. That information (or output) can be useful, and we want to be able to work with it, store it, and manage it properly.

When talking about command *output* and also **input**, there are three sources or targets for them that need to be understood:

- **STDOUT**: Also known as **Standard Output**, this is where commands will put their regular messages to provide information on what they are doing. In a terminal, on an interactive shell (as with the ones we are using so far), this output will show onscreen. This will be the main output managed by us.

- **STDERR**: Also known as **Standard Error**, this is where the commands will put their error messages to be processed. In our interactive shells, this output will also be shown on screen together with the standard output unless we specifically redirect it.

- **STDIN**: Also known as **Standard Input**, this is where the commands get data to be processed.

We will cover these in the following paragraphs to understand them better.

The way that the command input and output are used is defined by the following operators:

- `|`: A **pipe** operator is used to get the output from one command and make it the input of the next command. It channels data from one command to another.

- `>`: A **redirect** operator is used to redirect the entire output of a command into a file. If the file exists, it will be overwritten. Be careful with this, as you might clobber a file that could be important.

- `<`: **Reverse redirect** can be applied to use a file as the input for a command. Using it won't delete the file that is used as input.

- `>>`: A **redirect and add** operator is used to append the output of a command into a file. If the file does not exist, it will be created with the output provided to it.

- `2>`: A **redirect STDERR** operator will only redirect the output sent to the error message handler. (Note that no space should be included between 2 and > in order for this to work!)

- `1>`: A **redirect STDOUT** operator will only redirect the output sent to the standard output and not to the error message handler.

- `>&2`: A **redirect to STDERR** operator will redirect the output to the standard error handler.

- `>&1`: A **redirect to STDOUT** operator will redirect the output to the standard output handler.

To better understand these, we will go through several examples in this section and the following one.

Let's get a list of files and put it in a file. First, we list the files in `/var`, using the -m option to separate entries with commas:

```
[root@rhel-instance ~]# ls -m /var/
account, adm, cache, crash, db, empty, ftp, games, gopher,
kerberos, lib, local, lock, log, mail, nis, opt, preserve, run,
spool, tmp, yp
```

Now, we run the command again, redirecting the output to the `/root/var-files.txt` file:

```
[root@rhel-instance ~]# ls -m /var/ > /root/var-files.txt
[root@rhel-instance ~]#
```

As we can see, no output is shown on screen, but we will be able to find the new file in the current working directory – in this case, `/root`, the newly created file:

```
[root@rhel-instance ~]# ls /root
anaconda-ks.cfg  var-files.txt
```

To see the content of the file on screen, we use the cat command, intended to concatenate the output for several files, but regularly used for this purpose:

```
[root@rhel-instance ~]# cat var-files.txt
account adm, cache, crash, db, empty, ftp, games, gopher,
kerberos, lib, local, lock, log, mail, nis, opt, preserve, run,
spool, tmp, yp
```

We can also add to this file the content of `/var/lib`. First, we can list it:

```
[root@rhel-instance ~]# ls -m /var/lib/
alternatives, authselect, chrony, dbus, dhclient, dnf, games,
initramfs, logrotate, misc, NetworkManager, os-prober,
plymouth, polkit-1, portables, private, rhsm, rpm, rpm-state,
rsyslog, selinux, sss, systemd, tpm, tuned, unbound
```

Now, to append this content to the `/root/var-files.txt` file, we use the `>>` operator:

```
[root@rhel-instance ~]# ls -m /var/lib/ >> var-files.txt
[root@rhel-instance ~]# cat var-files.txt
account, adm, cache, crash, db, empty, ftp, games, gopher,
kerberos, lib, local, lock, log, mail, nis, opt, preserve, run,
spool, tmp, yp
alternatives, authselect, chrony, cni, containers, dnf, fprint,
games, initramfs, iscsi, kdump, logrotate, misc, mlocate,
NetworkManager, os-prober, PackageKit, plymouth, polkit-
1, private, rhsm, rpm, rpm-state, rsyslog, samba, selinux,
setroubleshoot, smartmontools, sss, systemd, tpm2-tss, udisks2,
unbound, xfsdump
```

The `/root/var-files.txt` file now contains both the comma-separated list for `/var` and for `/var/lib`.

Now, we can try to list a non-existing directory to see an error being printed:

```
[root@rhel-instance ~]# ls -m /non
ls: cannot access '/non': No such file or directory
```

The output we see is an error and is treated differently by the system than regular messages. We can try to redirect the output to a file:

```
[root@rhel-instance ~]# ls -m /non > non-listing.txt
ls: cannot access '/non': No such file or directory
[root@rhel-instance ~]# cat non-listing.txt
[root@rhel-instance ~]#
```

We can see that using the standard redirect, with a command providing an error message, will show the error message (via STDERR) onscreen and create an empty file. This is because the file contains the output of the common information messages that are shown via STDOUT. We can still capture the output of the error, redirecting STDERR, by using `2>`:

```
[root@rhel-instance ~]# ls -m/non 2> /root/error.txt
[root@rhel-instance ~]# cat /root/error.txt
ls: cannot access '/non': No such file or directory
```

Now, we can redirect the standard output and the error output independently.

Now, we want to count the number of files and directories in /var. For that, we are going to use the wc command, which stands for *word count,* with the -w option to only focus on counting words. To do so, we will redirect the output of ls to it by using a *pipe* represented by |:

```
[root@rhel-instance ~]# ls -m /var/ | wc -w
21
```

We can also use it to count the entries in /etc:

```
[root@rhel-instance ~]# ls -m /etc/ | wc -w
199
```

Pipes, |, are great for reusing the output of one command and sending it to another command to process that output. Now, we know more about using the more common operators to redirect input and output. There are several ways to process that output and we will see more examples in the following section.

Filtering output with grep and sed

The grep command is heavily used (and commonly mistyped) in system administration. It helps when finding a pattern in a line, whether in a file or via **standard input** (**STDIN**).

Let's do a recursive search of the files in /usr with find and put it in /root/usr-files.txt:

```
[root@rhel-instance ~]# find /usr/ > /root/usr-files.txt
[root@rhel-instance ~]# ls -lh usr-files.txt
-rw-r--r--. 1 root root 2,1M Feb 18 12:38 usr-files.txt
```

As you can see, it's a 2.1 MB file and it isn't easy to go through. There is a utility in the system called gzip and we want to know which files in /usr contain the gzip pattern. To do so, we can run the following command:

```
[root@rhel-instance ~]# grep gzip usr-files.txt
/usr/bin/gzip
/usr/lib64/python3.9/__pycache__/gzip.cpython-39.opt-2.pyc
/usr/lib64/python3.9/__pycache__/gzip.cpython-39.opt-1.pyc
/usr/lib64/python3.9/__pycache__/gzip.cpython-39.pyc
/usr/lib64/python3.9/gzip.py
/usr/share/licenses/gzip
/usr/share/licenses/gzip/COPYING
/usr/share/licenses/gzip/fdl-1.3.txt
/usr/share/man/man1/gzip.1.gz
```

```
/usr/share/doc/gzip
/usr/share/doc/gzip/AUTHORS
/usr/share/doc/gzip/ChangeLog
/usr/share/doc/gzip/NEWS
/usr/share/doc/gzip/README
/usr/share/doc/gzip/THANKS
/usr/share/doc/gzip/TODO
/usr/share/info/gzip.info.gz
/usr/share/mime/application/gzip.xml
```

As you can see, we have found all the files with gzip under the /usr directory by creating a file with all the content and searching through it with grep. Could we do the same without creating the file? We sure could – by using a *pipe*. We can redirect the output of find to grep and get the same output:

```
[root@rhel-instance ~]# find /usr/ | grep gzip
/usr/bin/gzip
/usr/lib64/python3.9/__pycache__/gzip.cpython-39.opt-2.pyc
/usr/lib64/python3.9/__pycache__/gzip.cpython-39.opt-1.pyc
/usr/lib64/python3.9/__pycache__/gzip.cpython-39.pyc
/usr/lib64/python3.9/gzip.py
/usr/share/licenses/gzip
/usr/share/licenses/gzip/COPYING
/usr/share/licenses/gzip/fdl-1.3.txt
/usr/share/man/man1/gzip.1.gz
/usr/share/doc/gzip
/usr/share/doc/gzip/AUTHORS
/usr/share/doc/gzip/ChangeLog
/usr/share/doc/gzip/NEWS
/usr/share/doc/gzip/README
/usr/share/doc/gzip/THANKS
/usr/share/doc/gzip/TODO
/usr/share/info/gzip.info.gz
/usr/share/mime/application/gzip.xml
```

In this command, the standard output from find was sent to grep to process it. We can even count the number of instances of files with wc, but this time, using the -l option to count the lines:

```
[root@rhel-instance ~]# find /usr/ | grep gzip | wc -l
18
```

We have now concatenated two pipes, one to filter the output and another to count it. We will find ourselves doing this kind of plumbing often when searching for and finding information in the system.

Some very common options for `grep` are as follows:

- `-i`: for **ignore-case**. This will match the pattern whether it's uppercase, lowercase, or a combination thereof.

- `-v`: for **invert match**. This will show all entries that do not match the pattern being searched for.

- `-r`: for **recursive**. We can tell `grep` to search for a pattern in all the files within a directory while going through all of them (if we have permission).

There is also a way to also filter columns in the output provided. Let's say we have a list of files in our home directory and we want to see its size. We run the following command:

```
[root@rhel-instance ~]# ls -l
total 1888
-rw-------. 1 root root    1393 Feb 18 19:45 anaconda-ks.cfg
-rw-r--r--. 1 root root      52 Feb 16 12:17 error.txt
-rw-r--r--. 1 root root       0 Feb 16 12:08 non-listing.txt
-rw-r--r--. 1 root root 1917837 Feb 16 12:40 usr-files.txt
-rw-r--r--. 1 root root     360 Feb 16 12:12 var-files.txt
```

Let's say we only want the size, which is the fifth column, of the content that has `files` in its name. We can use `awk` for that:

```
[root@rhel-instance ~]# ls -l | grep files | awk '{ print $5}'
1917837
360
```

The `awk` tool will help us to filter according to the correct column. It is very useful for finding identifiers in processes or for getting a specific list of data from a long output.

> **Tip**
> Consider that `awk` is super powerful for processing output and that we will use the minimal capability for it.

We could replace the separator with `-F` and get a list of available users in the system:

```
[root@rhel-instance ~]# awk -F: '{ print $1}' /etc/passwd
root
bin
```

```
daemon
adm
lp
sync
shutdown
halt
mail
operator
games
ftp
nobody
dbus
systemd-coredump
systemd-resolve
tss
polkitd
unbound
sssd
chrony
sshd
rngd
user
```

The `awk` and `grep` tools are very common processing tools in the life of a Linux sysadmin, and it is important to understand them well to manage the output provided by the system. We have applied the base knowledge to filter the output received by row and column. Let's now move on to how to manage files in a system so that we can better handle the stored output we have just generated.

Listing, creating, copying, and moving files, directories, links, and hard links

It is important to know how to **manage files and directories** in a system from the command line. It will serve as the basis for managing and copying important data, such as configuration files or data files.

Directories

Let's start by creating a directory to keep some working files. We can do so by running mkdir, short for "make directory":

```
[user@rhel-instance ~]$ mkdir mydir
[user@rhel-instance ~]$ ls -l
total 0
drwxrwxr-x. 2 user user 6 Feb 13 19:53 mydir
```

Folders can be deleted with the rmdir command, short for "remove directory":

```
[user@rhel-instance ~]$ ls -l
total 0
drwxrwxr-x. 2 user user 6 Feb 13 19:53 mydir
[user@rhel-instance ~]$ mkdir deleteme
[user@rhel-instance ~]$ ls -l
total 0
drwxrwxr-x. 2 user user 6 Feb 13 20:15 deleteme
drwxrwxr-x. 2 user user 6 Feb 13 19:53 mydir
[user@rhel-instance ~]$ rmdir deleteme
[user@rhel-instance ~]$ ls -l
total 0
drwxrwxr-x. 2 user user 6 Feb 13 19:53 mydir
```

However, rmdir will only delete empty directories:

```
[user@rhel-instance ~]$ ls /etc/ > ~/mydir/etc-files.txt
[user@rhel-instance ~]$ rmdir mydir
rmdir: failed to remove 'mydir': Directory not empty
```

How can we delete a directory and all the other files and directories it contains using the remove (rm) command? First, let's just create and remove a single file, var-files.txt:

```
[user@rhel-instance ~]$ ls /var/ > ~/var-files.txt
[user@rhel-instance ~]$ ls -l var-files.txt
-rw-rw-r--. 1 user user 109 Feb 13 15:31 var-files.txt
[user@rhel-instance ~]$ rm var-files.txt
[user@rhel-instance ~]$ ls -l var-files.txt
ls: cannot access 'var-files.txt': No such file or directory
```

To remove a full directory branch, including its contents, we may use the -r option, short for "recursive":

```
[user@rhel-instance ~]$ rm -r mydir/
[user@rhel-instance ~]$ ls -l
total 0
```

> **Important Note**
> Be very careful when using recursive mode to delete things, as there is neither a recovery command for it nor a trash bin to keep files that have been removed in the command line.

Let's take a look at the review table:

Command	Usage
mkdir	Creates directories
rmdir	Deletes empty directories
rm -r	Deletes directories that contain files or other directories recursively

Table 3.4 – The directory commands

Now that we know how to create and delete directories in a Linux system, let's start copying and moving content.

Copying and moving

Now, let's copy some files to play with them using the cp (which stands for **copy**) command. We may get some powerful awk examples copied to our home directory:

```
[user@rhel-instance ~]$ mkdir myawk
[user@rhel-instance ~]$ cp /usr/share/awk/* myawk/
[user@rhel-instance ~]$ ls myawk/ | wc -l
27
```

To copy more than one file at the same time, we have used **globbing** with the * sign. This works so that by specifying the files one by one, we can just type * for everything. We can also type the initial characters and then *, so let's try this by copying some more files using globbing:

```
[user@rhel-instance ~]$ mkdir mysystemd
[user@rhel-instance ~]$ cp /usr/share/doc/systemd/* mysystemd/
[user@rhel-instance ~]$ cd mysystemd/
[user@rhel-instance mysystemd]$ ls
```

```
20-yama-ptrace.conf  CODING_STYLE  DISTRO_PORTING  ENVIRONMENT.
md  GVARIANT-SERIALIZATION  HACKING  NEWS  README  TRANSIENT-
SETTINGS.md  TRANSLATORS  UIDS-GIDS.md
```

You will see that running `ls  TR*` only shows files that start with TR:

```
[user@rhel-instance mysystemd]$ ls TR*
TRANSIENT-SETTINGS.md  TRANSLATORS
```

It will work the same way with the file ending:

```
[user@rhel-instance mysystemd]$ ls *.md
ENVIRONMENT.md  TRANSIENT-SETTINGS.md  UIDS-GIDS.md
```

As you can see, it only shows files ending in `.md`.

We can copy a full branch of files and directories with the *recursive* option for cp, which is `-r`:

```
[user@rhel-instance mysystemd]$ cd ~
[user@rhel-instance ~]$ mkdir myauthselect
[user@rhel-instance ~]$ cp -r /usr/share/authselect/*
myauthselect
[user@rhel-instance ~]$ ls myauthselect/default/
minimal sssd winbind
```

The recursive option is very useful for copying complete branches. We could also move directories or files easily using the mv command. Let's put all our new directories together into a newly created directory called `docs`:

```
[user@rhel-instance ~]$ mkdir docs
[user@rhel-instance ~]$ mv my* docs/
[user@rhel-instance ~]$ ls docs/
myauthselect  myawk  mysystemd
```

You can see that with mv, you do not need to use the recursive option to manage a full branch of files and directories. It can also be used to rename files or directories:

```
[user@rhel-instance ~]$ cd docs/mysystemd/
[user@rhel-instance mysystemd]$ ls
20-yama-ptrace.conf  CODING_STYLE  DISTRO_PORTING  ENVIRONMENT.
md  GVARIANT-SERIALIZATION  HACKING  NEWS  README  TRANSIENT-
SETTINGS.md  TRANSLATORS  UIDS-GIDS.md
[user@rhel-instance mysystemd]$ ls -l NEWS
```

```
-rw-r--r--. 1 user user 451192 Feb 16 19:59 NEWS
[user@rhel-instance mysystemd]$ mv NEWS mynews
[user@rhel-instance mysystemd]$ ls -1 NEWS
ls: cannot access 'NEWS': No such file or directory
[user@rhel-instance mysystemd]$ ls -1 mynews
-rw-r--r--. 1 user user 451192 Feb 16 19:59 mynews
```

There is a special command for creating empty files, which is touch:

```
[user@rhel-instance mysystemd]$ cd ~
[user@rhel-instance ~]$ ls -1  docs/
total 4
drawer-x. 4 user user   35 Feb 16 20:08 myauthselect
drwxrwxr-x. 2 user user 4096 Feb 16 19:51 myawk
drwxrwxr-x. 2 user user  238 Feb 16 20:21 mysystemd
[user@rhel-instance ~]$ touch docs/mytouch
[user@rhel-instance ~]$ ls -1  docs/
total 4
drwxrwxr-x. 4 user user   35 Feb 16 20:08 myauthselect
drwxrwxr-x. 2 user user 4096 Feb 16 19:51 myawk
drwxrwxr-x. 2 user user  238 Feb 16 20:21 mysystemd
-rw-rw-r--. 1 user user    0 Feb 16 20:27 mytouch
```

When applied to an existing file or folder, it will update its access time to the current one:

```
[user@rhel-instance ~]$ touch docs/mysystemd
[user@rhel-instance ~]$ ls -1  docs/
total 4
drwxrwxr-x. 4 user user   35 Feb 16 20:08 myauthselect
drwxrwxr-x. 2 user user 4096 Feb 16 19:51 myawk
drwxrwxr-x. 2 user user  238 Feb 16 20:28 mysystemd
-rw-rw-r--. 1 user user    0 Feb 16 20:27 mytouch
```

Let's check the review table:

Command	Usage
cp	Copies a set of files in the same source directory
rm -r	Deletes a set of files in the same directory
cp -r	Copies a full directory branch, recursively, to a target directory
touch	Creates empty files or sets the access time to a file to the current time
mv	Renames a file or directory
mv	Moves a full directory branch recursively to a target directory

Table 3.5 – Copying and moving commands

Now we know how to copy, delete, rename, and move files and directories, even full directory branches. Let's now take a look at a different way to work with them – links.

Symbolic and hard links

We can have the same file in two places using **links**. There are two types of links:

- **Hard links**: There will be two entries (or more) to the same file in the filesystem. The content will be written once to disk. Hard links for the same file cannot be created in two different filesystems. Hard links cannot be created for directories.

- **Symbolic links**: A symbolic link is created pointing to a file or directory in any place in the system.

Both are created using the ln (which stands for link) utility.

Let's now create hard links:

```
[user@rhel-instance ~]$ cd docs/
[user@rhel-instance docs]$ ln mysystemd/README MYREADME
[user@rhel-instance docs]$ ls -l
total 20
drwxrwxr-x. 4 user user    35 Feb 16 20:08 myauthselect
drwxrwxr-x. 2 user user  4096 Feb 16 19:51 myawk
-rw-r--r--. 2 user user 13826 Feb 16 20:59 MYREADME
drwxrwxr-x. 2 user user   238 Feb 16 20:28 mysystemd
-rw-rw-r--. 1 user user     0 Feb 16 20:27 mytouch
[user@rhel-instance docs]$ ln MYREADME MYREADME2
[user@rhel-instance docs]$ ls -l
total 36
```

```
drwxrwxr-x. 4 user user    35 Feb 16 20:08 myauthselect
drwxrwxr-x. 2 user user  4096 Feb 16 19:51 myawk
-rw-r--r--. 3 user user 13831 Feb 16 20:32 MYREADME
-rw-r--r--. 3 user user 13831 Feb 16 20:32 MYREADME2
drwxrwxr-x. 2 user user   238 Feb 16 20:28 mysystemd
-rw-rw-r--. 1 user user     0 Feb 16 20:27 mytouch
drwxrwxr-x. 2 user user     6 Feb 16 20:35 test
```

Check the increasing number of references to the file (in bold in the previous example).

Now, let's create a symbolic link to a directory with ln -s (with s for symbolic):

```
[user@rhel-instance docs]$ ln -s mysystemd mysystemdlink
[user@rhel-instance docs]$ ls -l
total 36
drwxrwxr-x. 4 user user    35 Feb 16 20:08 myauthselect
drwxrwxr-x. 2 user user  4096 Feb 16 19:51 myawk
-rw-r--r--. 3 user user 13831 Feb 16 20:32 MYREADME
-rw-r--r--. 3 user user 13831 Feb 16 20:32 MYREADME2
drwxrwxr-x. 2 user user   238 Feb 16 20:28 mysystemd
lrwxrwxrwx. 1 user user     9 Feb 16 20:40 mysystemdlink ->
mysystemd
-rw-rw-r--. 1 user user     0 Feb 16 20:27 mytouch
drwxrwxr-x. 2 user user     6 Feb 16 20:35 test
```

Check how the symbolic link created is treated as a different type when listing it, as it starts with l for link (in bold in the previous example) instead of d for directory (also in bold in the previous example).

> **Tip**
> When in doubt about whether to use a hard link or a symbolic link, use the symbolic link as the default choice.

Let's check the review table:

Command	Usage
ln	Creates a hard link for a file in the same filesystem
ln -s	Creates a symbolic link to a file or directory that could cross different filesystems

Table 3.6 – Link commands

As you can see, creating links and symbolic links is super simple and can help us access the same file or directory from different locations. In the following section, we will cover how to pack and compress a set of files and directories.

Using tar and gzip

Sometimes, we want to pack a full directory (including files) into a single file for backup purposes or to simply share it more easily. The command that can help aggregate files into one is tar.

First, we need to install tar:

```
[root@rhel-instance ~]# yum install tar -y
```

We can try by creating (as root) a backup of the /etc directory branch:

```
[root@rhel-instance ~]# tar -cf etc-backup.tar /etc
tar: Removing leading '/' from member names
[root@rhel-instance ~]# ls -lh etc-backup.tar
-rw-r--r--. 1 root root 20M Feb 17 20:08 etc-backup.tar
```

Let's check the options used:

- -c: Short for create. tar can put files together but also unpack them.

- -f: Short for file. We specify that the next parameter will be working with a file.

We can try to unpack it:

```
[root@rhel-instance ~]# mkdir tmp
[root@rhel-instance ~]# cd tmp/
[root@rhel-instance tmp]# tar -xf ../etc-backup.tar
[root@rhel-instance tmp]# ls
etc
```

Let's check the new options used:

- -x: for extraction. It unpacks a tar file.

  Observe that we created a directory called tmp to work on and that we pointed to the parent directory of tmp by using the .. shortcut (which refers to the parent directory of the current working directory).

Let's use `gzip` to compress a file. We can copy `/etc/services` and compress it:

```
[root@rhel-instance tmp]# cp /etc/services .
[root@rhel-instance tmp]# ls -lh services
-rw-r--r--. 1 root root 677K Jun 23 2020 services
[root@rhel-instance tmp]# gzip services
[root@rhel-instance tmp]# ls -lh services.gz
-rw-r--r--. 1 root root 140K Feb 17 20:16 services.gz
```

Please note that when using `gzip`, this will compress the specified file, adding the `.gz` extension to it, and the original file will not be kept. Also, be aware that the newly created file is one fifth of the size of the original file.

To recover it, we can run `gunzip`:

```
-rw-r--r--. 1 root root 140K Feb 17 20:16 services.gz
[root@rhel-instance tmp]# gunzip services.gz
[root@rhel-instance tmp]# ls -lh services
-rw-r--r--. 1 root root 677K Feb 17 20:16 services
```

Now, we can combine the two of them, packing and compressing them:

```
[root@rhel-instance ~]# tar cf etc-backup.tar /etc/
tar: Removing leading '/' from member names
[root@rhel-instance ~]# ls -lh etc-backup.tar
-rw-r--r--. 1 root root 20M Feb 17 20:08 etc-backup.tar
[root@rhel-instance ~]# gzip etc-backup.tar
[root@rhel-instance ~]# ls etc-backup.tar.gz
etc-backup.tar.gz
[root@rhel-instance ~]# ls -lh etc-backup.tar.gz
-rw-r--r--. 1 root root 4,9M Feb 17 20:20 etc-backup.tar.
gz
```

This way, we pack and compress in two steps.

The `tar` command is smart enough to be able to perform packing and compression in a single step:

```
[root@rhel-instance ~]# rm -f etc-backup.tar.gz
[root@rhel-instance ~]# tar -czf etc-backup.tar.gz /etc/
tar: Removing leading '/' from member names
[root@rhel-instance ~]# ls -lh etc-backup.tar.gz
-rw-r--r--. 1 root root 4,9M Feb 17 20:22 etc-backup.tar.
gz
```

- -z: This compresses the newly created tar file with `gzip`. It is also usable for decompression.

  We may want to review that same option when decompressing:

  ```
  [root@rhel-instance ~]# cd tmp/
  [root@rhel-instance tmp]# rm -rf etc
  [root@rhel-instance tmp]# tar -xzf ../etc-backup.tar.gz
  [root@rhel-instance tmp]# ls
  etc
  ```

As you can see, it's very easy to pack and compress files using `tar` and `gzip`. There are other available compression methods with higher ratios, such as `bzip2` or `xz`, that you may want to try, too. Now, let's move on to combine all the commands that we have learned into a powerful way to automate – by creating shell scripts.

Creating basic shell scripts

As a system administrator (or sysadmin), there will be times when you want to run a series of commands more than once. You can do this manually by running each command every time. However, there is a more efficient way to do so – by creating a **shell script**.

A shell script is nothing more than a text file with a list of commands to be run and a reference to the shell that will interpret it.

In this book, we will not cover how to use a **text editor**; however, we will provide three recommendations for text editors in Linux that could be of help:

- **Nano**: This is probably the easiest text editor to use for beginners. Lean, simple, and straightforward, you may want to start by installing it and giving it a try.

- **vi** or **vim**: vi is the default text editor available in RHEL, even included in the minimal install and in many Linux distributions. Even if you are not going to use it every day, it's good to familiarize yourselves with the basics of it as it will be present in almost any Linux system you will use. **vim** stands for **vi-improved**.

- **Emacs**: This is probably the most advanced and complex text editor ever. It can do everything and beyond, including reading emails or helping with a bit of psychoanalysis via **Emacs Doctor**.

We can create our first shell script by editing a new file called `hello.sh` with the following line as its content:

```
echo ''hello world!''
```

Then, we can run it by using the `bash` command interpreter with the following line:

```
[root@rhel-instance ~]# bash hello.sh
hello world!
```

There is a different way to do this where we do not need to type `bash`. We can add an initial line referencing the interpreter, so the file content of `hello.sh` looks as follows:

```
#!/bin/bash
echo ''hello world!''
```

Now, we change the permissions to make it executable:

```
[root@rhel-instance ~]# ls -l hello.sh
-rw-r--r--. 1 root root 32 Feb 17 20:32 hello.sh
[root@rhel-instance ~]# chmod +x hello.sh
[root@rhel-instance ~]# ls -l hello.sh
-rwxr-xr-x. 1 root root 32 Feb 17 20:32 hello.sh
```

And we run it as follows:

```
[root@rhel-instance ~]# ./hello.sh
hello world!
```

We have created our first shell script. Congratulations!

> **Tip**
>
> The commands must be in the path to run in any working directory, as stated by the `$PATH` variable. If our command (or shell script) is not in one of the directories specified in the path, we will specify the running directory, in this case, using the `.` shortcut for the current directory and the `/` separator.

Let's use some variables. We can define a variable by simply providing its name and the value we want. Let's try replacing `world` with a variable. To use it, we prepend the `$` symbol to the name of the variable and it will be used. The script will look as follows:

```
#!/bin/bash
PLACE=''world''
echo ''hello $PLACE!''
```

We can run the script, obtaining the same output as before:

```
[root@rhel-instance ~]# ./hello.sh
hello world!
```

To have more clarity, when using the value of the variable, we will put the name of it between curly braces, { and }', and take this as a good practice.

The previous script will look as follows:

```
#!/bin/bash
PLACE=''world''
echo ''hello ${PLACE}!''
```

We now know how to create a basic script, but we may want to have greater control over it by using some programmatic capabilities, starting with loops. Let's go for it!

for loops

What if we want to run the same command over a list of places? That's what a `for` **loop** is used for. It can help iterate over a set of elements, such as a list or a counter, for example.

The `for` loop syntax is as follows:

- `for`: To specify the iteration
- do: To specify the action
- done: To close the loop

We can define a space-separated list to try it and iterate through it with our first `for` loop:

```
#!/bin/bash
PLACES_LIST="Madrid Boston Singapore World"
for PLACE in ${PLACES_LIST}; do
echo "hello ${PLACE}!"
done
```

Let's run it. The output will look as follows:

```
[root@rhel-instance ~]# ./hello.sh
hello Madrid!
hello Boston!
```

```
hello Singapore!
hello World!
```

Using the `for` loop can be very interesting when **reading the list from an external command**. We can do so by putting the external command between $ (and) .

> **Tip**
> Backticks, `'`, can also be used to run a command and get its output as a list, but we will stick to the previous expression for clarity.

One example of a usable external command can be `ls`. Let's create the `txtfiles.sh` script with the following content:

```
#!/bin/bash
for TXTFILE in $(ls *.txt); do
  echo ''TXT file ${TXTFILE} found! ''
done
```

Make it executable and run it:

```
[root@rhel-instance ~]# chmod +x txtfiles.sh
[root@rhel-instance ~]# ./txtfiles.sh
TXT file error.txt found!
TXT file non-listing.txt found!
TXT file usr-files.txt found!
TXT file var-files.txt found!
```

You see how we can now iterate over a set of files, including, for example, changing their names, finding and replacing content in them, or simply making a specific backup of a selection of files.

We've seen several ways in which to iterate a list with the `for` loop, which can be very useful when it comes to automating tasks. Now, let's move on to another programmatic capability in scripts – conditionals.

if conditionals

Sometimes, we may want to execute something different for one of the elements in a list, or only if a **condition** is happening. We can use the `if` conditional for this.

The `if` conditional syntax is `if;` to specify the condition.

Conditions are usually specified between brackets ([and]):

- then: To specify the action.

- fi: To close the loop.

- else: This is used as a then element when the condition is *not* matched.

Let's change our previous hello.sh script to say "hello to Madrid" in Spanish, as follows:

```
#!/bin/bash
PLACES_LIST="Madrid Boston Singapore World"
for PLACE in ${PLACES_LIST}; do
    if [ ${PLACE} = "Madrid" ]; then
        echo ''¡Hola ${PLACE}!''
    fi
done
```

Then, run it:

```
[root@rhel-instance ~]# ./hello.sh
¡Hola Madrid!
```

We have a problem; it only says hello to Madrid. What happens if we want to run the previous code on the ones not matching the condition? That's when we extend the conditional using else for the items that do not match.

And now, we have an example of a conditional using else:

```
#!/bin/bash
PLACES_LIST="Madrid Boston Singapore World"
for PLACE in ${PLACES_LIST}; do
    if [ ${PLACE} = "Madrid" ]; then
        echo ''¡Hola ${PLACE}!''
    else
        echo ''hello ${PLACE}!''
    fi
done
```

We can run it:

```
[root@rhel-instance ~]# ./hello.sh
¡Hola Madrid!
```

```
hello Boston!
hello Singapore!
hello World!
```

As you see, it's simple to use the conditionals in a script and provide a lot of control over the conditions under which a command is run. We now need to control when something may not be running correctly. That's what the exit codes (or error codes) are for. Let's go for it!

Exit codes

When a program is run, it provides an **exit code**, specifying whether it ran OK or whether there was an issue. That *exit code* is stored in a special variable called $?.

Let's take a look at it by running ls hello.sh:

```
[root@rhel-instance ~]# ls hello.sh
hello.sh
[root@rhel-instance ~]# echo $?
0
```

When the program runs OK, the *exit code* is zero, 0.

What happens when we try to list a file that doesn't exist (or run any other command incorrectly, or that is having issues)? Let's try listing a nonexistent file:

```
[root@rhel-instance ~]# ls nonexistentfile.txt
ls: cannot access 'nonexistentfile.txt': No such file or
directory
[root@rhel-instance ~]# echo $?
2
```

You see, the *exit code* is not zero. We will go to the documentation and check the number associated with it to understand the nature of the issue.

When running a command in a script, check for the exit code and act accordingly. Let's now review where to find further information on the commands, such as exit codes or other options, in the following section.

Using system documentation resources

The system includes resources to help you while working with it and guide you to improve your sysadmin skills. This is referred to as the **system documentation**. Let's check three different resources available by default in your RHEL installation: man pages, info pages, and other documents.

man pages

The most common resource used to obtain documentation is **manual pages**, also referred to by the command used to invoke them: man.

Almost any utility installed in the system has a man page to help you use it (in other words, specifying all the options for the tools and what they do). You can run man tar and check the output:

```
[root@rhel-instance ~]# man tar
TAR(1)                                           GNU TAR
Manual                                           TAR(1)
NAME
        tar - an archiving utility
SYNOPSIS
    Traditional usage
        tar {A|c|d|r|t|u|x}[GnSkUWOmpsMBiajJzZhPlRvwo] [ARG...]
    UNIX-style usage
        tar -A [OPTIONS] ARCHIVE ARCHIVE
        tar -c [-f ARCHIVE] [OPTIONS] [FILE...]
        tar -d [-f ARCHIVE] [OPTIONS] [FILE...]
```

You can see in it (navigate with the *arrow* keys, the *space bar*, or *Page Up* and *Page Down*) and exit it by hitting q (for *quit*).

There are sections within the man page on related topics. It is pretty simple to search those by using the apropos command. Let's see this for tar:

```
[root@rhel-instance ~]# apropos tar
dbus-run-session (1) - start a process as a new D-Bus session
dnf-needs-restarting (8) - DNF needs_restarting Plugin
dracut-pre-udev.service (8) - runs the dracut hooks before
udevd is started
gpgtar (1)            - Encrypt or sign files into an archive
gtar (1)             - an archiving utility
open (1)             - start a program on a new virtual
terminal (VT).
openvt (1)           - start a program on a new virtual
terminal (VT).
scsi_start (8)       - start one or more SCSI disks
setarch (8)          - change reported architecture in new
program environment and set personalit...
```

```
sg_reset (8)          - sends SCSI device, target, bus or host
reset; or checks reset state
sg_rtpg (8)           - send SCSI REPORT TARGET PORT GROUPS
command
sg_start (8)          - send SCSI START STOP UNIT command:
start, stop, load or eject medium
sg_stpg (8)           - send SCSI SET TARGET PORT GROUPS command
systemd-notify (1)    - Notify service manager about start-up
completion and other daemon status c...
systemd-rc-local-generator (8) - Compatibility generator for
starting /etc/rc.local and /usr/sbin...
systemd.target (5)    - Target unit configuration
tar (1)               - an archiving utility
tar (5)               - format of tape archive files
unicode_start (1)     - put keyboard and console in unicode mode
```

As you can see, it matches not only `tar` but also `start`. This isn't perfect, but it can provide helpful information related to `tar`, such as `gpgtar`.

`man` pages have sections. As you can see in the previous example, for `tar`, there are manual pages in two sections, one for the command-line utility (section `1`), and one for the archiving format (section `5`):

```
tar (1)               - an archiving utility
tar (5)               - format of tape archive files
```

We can access the page in section 5 to understand the format by running the following command:

```
[root@rhel-instance ~]# man 5 tar
```

Now, we can see the `tar format` page:

```
TAR(5)                           BSD File Formats
Manual                           TAR(5)
NAME
     tar — format of tape archive files
DESCRIPTION
     The tar archive format collects any number of files,
directories, and other file system objects (symbolic links,
device nodes, etc.) into a single stream of bytes.  The format
was ...
```

You can see that manual pages are a great resource for learning more about the typical commands being used. This is also a fantastic resource as far as the **Red Hat Certified System Administrator** exam is concerned. One recommendation is to review all man pages for the commands shown previously in this chapter, as well as for the forthcoming chapters. Consider man pages the main information resource in the system. Let's now review other information resources available.

info pages

info pages are usually more descriptive than man pages and are more interactive. They help more in getting started on a topic.

We can try to get info for the ls command by running the following:

```
[root@rhel-instance ~]# info ls
```

> **Important Note**
> If the info command is not available, you can install it first using the dnf install info -y command.

We can see the info page for it:

```
Next: dir invocation,   Up: Directory listing

10.1 'ls': List directory contents
====================================

The 'ls' program lists information about files (of any type,
including

directories).   Options and file arguments can be intermixed
arbitrarily,
```

info pages can *redirect to other topics, shown underlined,* and these can be followed by putting the cursor over them and hitting *Enter*.

As with man pages, press q to quit.

Please take some time to review the info pages for the main topics covered in this chapter (in several cases, info pages will not be available, but the ones that are will be very valuable).

What if we do not find a man or info page for a topic? Let's cover this in the following section.

Other documentation resources

For other documentation resources, you can go to the /usr/share/doc directory. There, you will find other documents that come with the tools installed in the system.

Let's see how many items we have:

```
[root@rhel-instance doc]# cd /usr/share/doc/
[root@rhel-instance doc]# ls | wc -l
333
```

You can see that there are 333 directories available under /usr/share/doc.

As a good example, let's enter the bash directory:

```
[root@rhel-instance doc]# cd bash/
```

Then, let's take a look at the INTRO file using less to read it (remember, you use q to quit):

```
[root@rhel-instance bash]# ls
bash.html  bashref.html  FAQ  INTRO  RBASH  README
[root@rhel-instance bash]# less INTRO
                    BASH - The Bourne-Again Shell
Bash is the shell, or command language interpreter, that will
appear in the GNU operating system.  Bash is an sh-compatible
shell that
incorporates useful features from the Korn shell (ksh) and C
shell
(csh).  It is intended to conform to the IEEE POSIX P1003.2/
ISO 9945.2 Shell and Tools standard.  It offers functional
improvements
```

This is a good read for a better understanding of bash. Now, you have a lot of documentation resources that you will be able to use during your daily tasks, as well as in the **RHCSA** exam.

Summary

In this chapter, we have learned about how to log in to a system with user and with root, understanding the basics of permissions and security. We are now also more comfortable using the command line with auto-complete, navigating through the directories and files, packing and unpacking them, redirecting command output and parsing it, and even automating processes with shell scripts. More importantly, we have a way to obtain information on what we are doing (or want to do) available in any RHEL system with the included documentation. These skills are the basis of the upcoming chapters. Don't hesitate to revisit this chapter if you feel stuck or if your progress is not as fast as you thought.

Now, it is time to extend your knowledge to encompass more advanced topics in the upcoming chapters. In the following chapter, you will be getting used to the *tools for regular operations*, in which you will review the most common actions taken when managing a system. Enjoy!

Tools for Regular Operations

At this point in this book, we've installed a system, and we've covered some of the scripts we can create to automate tasks, so we've reached the point where we can focus on the system itself.

Having a system properly configured requires not only installing it but understanding how to run tasks at specific times, keeping all the services running appropriately, and configuring time synchronization, service management, boot targets (runlevels), and scheduled tasks, all of which we will be covering in this chapter. We'll be additionally starting to mention **Cockpit** (web console), which will be covered during the whole book as a tool to perform administration tasks from a web interface.

In this chapter, you will learn how to check the statuses of services and how to start, stop, and troubleshoot them, as well as how to keep the system clock in sync for your server or your whole network.

The list of topics that will be covered is presented here:

- Managing system services with `systemd`
- Scheduling tasks with `cron` and `systemd`
- Learning about time synchronization with `chrony` and the **Network Time Protocol** (NTP)
- Checking for free resources—memory and disk (`free` and `df`)
- Finding logs, using `journald`, and reading log files, including log preservation and rotation

Technical requirements

It is possible for you to complete this chapter by using the **virtual machine** (VM) we created at the beginning of this book. Additionally, for testing the **NTP server**, it might be useful to create a second VM that will connect to the first one as a client, following the same procedure we used for the first one. Additionally, required packages will be indicated within the text.

Managing system services with systemd

In this section, you will learn how to manage **system services**, runtime targets, and all about the service status with systemd. You will also learn how to manage system boot targets and services that should start at system boot.

systemd (which you can learn a bit about at https://www.freedesktop.org/wiki/Software/systemd/) is defined as a **system daemon** that's used to manage the system. It came as a rework of how a system boots and starts, and it looks at the limitations related to the traditional way of doing it.

When we think about the system starting, we have the initial **kernel** and **ramdisk** load and execution, but right after that, services and scripts take control to make filesystems available. This helps prepare the services that provide the functionality we want from our system, such as the following:

- Hardware detection
- Additional filesystem activation
- Network initialization (wired, wireless, and so on)
- Network services (time sync, remote login, printers, network filesystems, and so on)
- User-space setup

However, most of the tools that existed before systemd came into play worked on this in a sequential way, causing the whole boot process (from boot to user login) to become lengthy and be subject to delays.

Traditionally, this also meant we had to wait for the required service to be fully available before the next one that depended on it could be started, increasing the total boot time.

Some approaches were attempted, such as using **Monit** or other tools that allow us to define dependencies, monitor processes, and even recover from failures, but in general, it was reusing an existing tool to perform other functions, trying to win the race regarding the fastest-booting system.

> **Important Note**
>
> systemd redesigned the process to focus on simplicity: start fewer processes and do more parallel execution. The idea itself sounds easy but requires redesigning a lot of what was taken for granted in the past, to focus on the needs of a new approach to improve operating system performance.
>
> This redesign, which has provided a lot of benefits, also came with a cost: it drastically changed the way systems booted, so there has been a lot of controversy on the adoption of systemd by different vendors, and even some efforts by the community to provide systemd-free variants.

Rationalizing how services start so that only those that are required are started is a good way to accomplish efficiency—for example, there is no need to start Bluetooth, printer, or network services when the system is disconnected, there is no Bluetooth hardware, or no one is printing. With fewer services waiting to start, the system boot is not delayed by those waits and focuses on the ones that really need attention.

On top of that, parallel execution allows us to have each service taking the time it needs to get ready but not make others wait, so in general, running services initialization in parallel allows us to maximize the usage of the **central processing unit (CPU)**, disk, and so on, and the wait times for each service are used by other services that are active.

systemd also pre-creates listening sockets before the actual daemon is started, so services that have requirements on other services can be started and be on a wait status until their dependencies are started. This is done without them losing any messages that are sent to them, so when the service is finally started, it will act on all pending actions.

Let's learn a bit more about systemd as it will be required for several operations we're going to describe in this chapter.

systemd comes with the concept of units, which are nothing but configuration files. These units can be categorized as different types, based on their file extension, as illustrated in the following screenshot:

Unit type	File extension	Description
Timer	`.timer`	A systemd timer. Described later in this chapter.
Socket	`.socket`	An inter-process communication socket.
Service	`.service`	Describes a system service.
Target	`.target`	A group of system units.
Automount	`.automount`	Defines a filesystem automount point.
Device	`.device`	Defines a device file recognized by the kernel.
Scope	`.scope`	An externally created process.
Slice	`.slice`	Defines a group of hierarchical units that manage system processes.
Path	`.path`	A file or directory in the system.
Mount	`.mount`	Defines a mountpoint in the filesystem.
Swap	`.swap`	A swap device or swap file definition.

Table 4.1 – systemd unit types description

> **Tip**
>
> Don't feel overwhelmed by the different systemd unit types. In general, the most common ones are **service**, **timer**, **socket**, and **target**.

Of course, these unit files are expected to be found in some specific folders, as shown here:

Folder	Description
/etc/systemd/system/	Created with the systemctl-enabled services. Takes preference over the runtime ones.
/usr/lib/systemd/system/	Units distributed with the packages we've installed on our system.
/run/systemd/system/	Units created at runtime. It takes preference over the ones defined by installed packages.

Table 4.2 – System folders containing systemd files

As we mentioned earlier about sockets, unit files for path, bus, and more are activated when a system's access to that path is performed, allowing services to be started when another one is requiring them. This adds more optimization for lowering system startup times.

With that, we have learned about systemd unit types. Now, let's focus on the file structure of unit files.

systemd unit file structure

Let's get our hands dirty with an example: a system has been deployed with sshd enabled, and we need to get it running once the network has been initialized in the **runlevels**, which provide connectivity.

As we mentioned previously, systemd uses unit files, and we can check the aforementioned folders or list them with systemctl list-unit-files. Remember that each file is a configuration file that defines what systemd should do; for example, /usr/lib/systemd/system/chronyd.service, as shown in the following screenshot:

Figure 4.1 – chronyd.service contents

This file defines not only the traditional program to start and the **process identifier** (**PID**) file, but the dependencies, conflicts, and soft dependencies, which provides enough information to systemd to decide on the right approach.

If you're familiar with INI files, this file uses that approach, in that it uses square brackets, [and], for sections and then pairs of key=value for the settings in each section.

Section names are case-sensitive, so they will not be interpreted correctly if the proper naming convention is not used.

Section directives are named like so:

- [Unit]
- [Install]

There are additional entries for each of the different types, as follows:

- [Service]
- [Socket]
- [Mount]
- [Automount]

- [Swap]
- [Path]
- [Timer]
- [Slice]

As you can see, we have specific sections for each type. If we execute man systemd.unit, it will give you examples, along with all the supported values, for the systemd version you're using, as illustrated in the following screenshot:

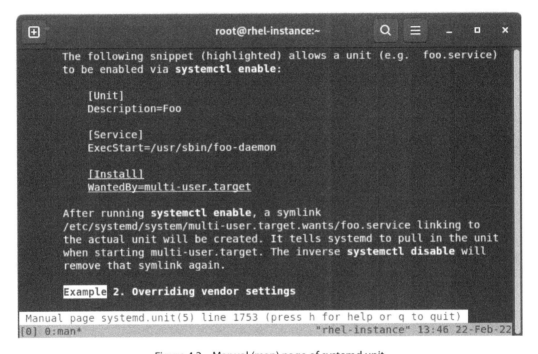

Figure 4.2 – Manual (man) page of systemd.unit

With that, we have reviewed the file structure of unit files. Now, let's use systemctl to actually manage the service's status.

Managing services to be started and stopped at boot

Services can be enabled or disabled; that is, the services will or won't be activated on system startup.

If you're familiar with the previous tools available in **Red Hat Enterprise Linux** (**RHEL**), it was common to use chkconfig to define the status of services based on their default rc.d/ settings.

The sshd service can be enabled via the following command:

```
#systemctl enable sshd
```

It can also be disabled via the following command:

```
#systemctl disable sshd
```

This results in creating or removing /etc/systemd/system/multi-user.target.wants/ sshd.service. Notice multi-user.target in the path, which is the equivalent of the runlevel we used to configure other approaches such as initscripts.

> **Tip**
> Although traditional usage of chkconfig (once installed) is provided for compatibility so that chkconfig sshd on/off or service start/stop/status/restart sshd is valid, it is better to get used to the systemctl approach described in this chapter.

The previous commands enable or disable the service at boot, but for executing an immediate action, we need to issue different commands.

To start the sshd service, use the following command:

```
#systemctl start sshd
```

To stop it, use the following command:

```
#systemctl stop sshd
```

Of course, we can also check the service's status. Here is an example of looking at systemd via systemctl status sshd:

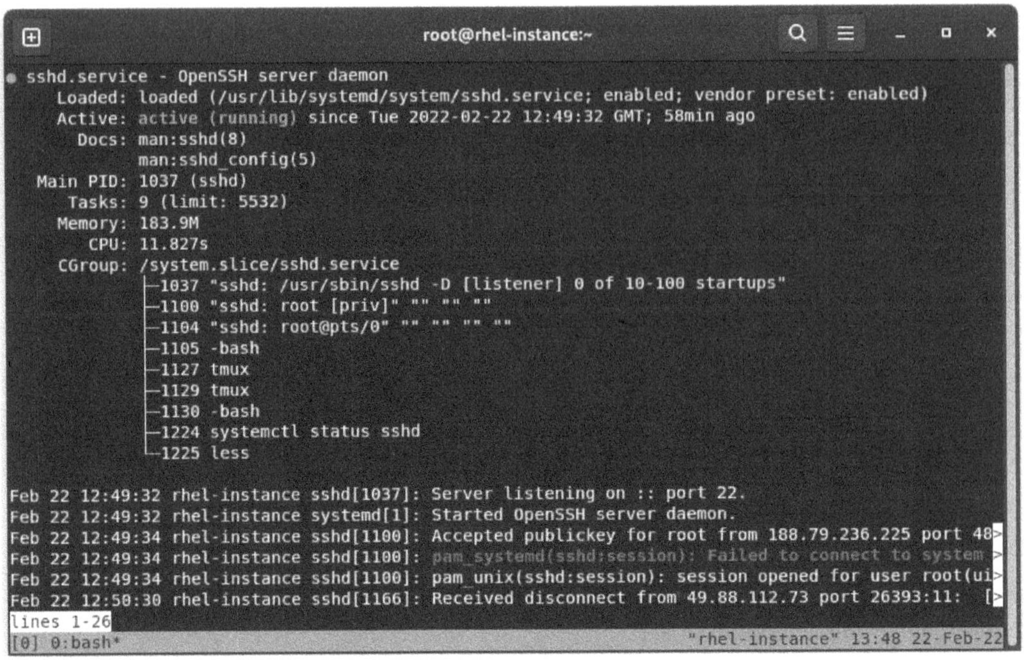

Figure 4.3 – Status of sshd daemon

This status information provides details about the unit file defining the service, its default status at boot, if it is running or not, its PID, some other details about its resource consumption, and some of the most recent log entries for the service, which are quite useful when you're debugging simple service start failures.

One important thing to check is the output of `systemctl list-unit-files` as it reports the defined unit files in the system, as well as the current status and the vendor preset for each one.

Now that we have covered how to start/stop and status check services, let's work on managing the actual system boot status itself.

Managing boot targets

The default status we have defined at boot is important when it comes to talking about **runlevels**.

A runlevel defines a predefined set of services based on usage; that is, they define which services will be started or stopped when we're using a specific functionality.

For example, there are runlevels that are used to define the following:

- **Halt mode**
- **Single-user mode**

- **Multi-user mode**
- **Networked multiuser**
- **Graphical UI**
- **Reboot**

Each of those runlevels allows a predefined set of services to be started/stopped when the runlevel is changed with `init $runlevel`. Of course, levels used to be based on each other and were very simple, as outlined here:

- Halt stopped all the services and then halted or powered off the system.
- Single-user mode started a shell for one user.
- Multi-user mode enabled regular login daemons on the virtual terminals.
- Networked was like multi-user but with the network started.
- Graphical was like networked but with graphical login via display manager (gdm or others).
- Reboot was like halt, but at the end of processing services, it issued a reboot instead of a halt.

These runlevels (and the default one when the system is booted) used to be defined in /etc/inittab, but the file placeholder reminds us of the following:

```
# inittab is no longer used.
#
# ADDING CONFIGURATION HERE WILL HAVE NO EFFECT ON YOUR SYSTEM.
#
# Ctrl-Alt-Delete is handled by /usr/lib/systemd/system/ctrl-
alt-del.target
#
# systemd uses 'targets' instead of runlevels. By default,
there are two main targets:
#
# multi-user.target: analogous to runlevel 3
# graphical.target: analogous to runlevel 5
#
# To view current default target, run:
# systemctl get-default
#
# To set a default target, run:
# systemctl set-default TARGET.target
```

So, by making this change to `systemd`, a new way to check the available boot targets and define them is in place.

We can find the available system targets by listing this folder:

```
#ls -l /usr/lib/systemd/system/*.target
```

Or, more correctly, we can use `systemctl`, like so:

```
#systemctl list-unit-files *.target
```

When you examine the output on your system, you will find some compatibility aliases for runlevels 0 to 6 that provide compatibility with the traditional ones.

For example, for regular server usage, the default target will be `multi-user.target` when you're running without graphical mode or `graphical.target` when you're using it.

We can define, as instructed in the placeholder at `/etc/inittab`, the new runlevel to use by executing the following command:

```
#systemctl set-default TARGET.target
```

We can verify the active one by using the following command:

```
#systemctl get-default
```

This brings us to the next question: *What does a target definition look like?* Let's examine the output in the following screenshot:

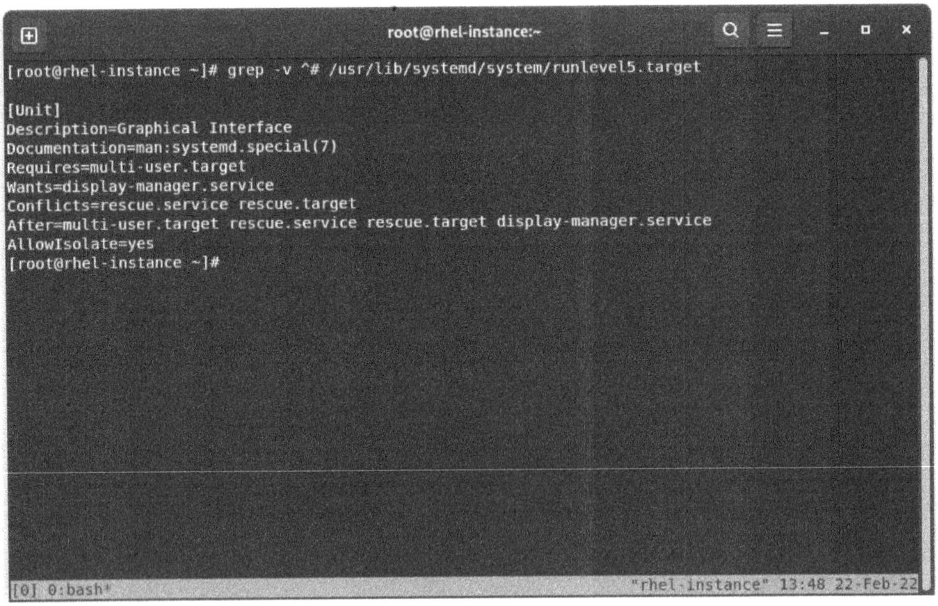

Figure 4.4 – Contents of runlevel 5 from its target unit definition

As you can see, it is set as a dependency of another target (`multi-user.target`) and has some requirements on other services, such as `display-manager.service`, and also other conflicts, and the target can only be reached when other targets have completed.

In this way, `systemd` can select the proper order of services to start and the dependencies to reach the configured boot target.

Just in case, we can switch to a new runlevel with `systemctl isolate targetname.target`.

With that, we have covered the service's status, as well as how to start, stop, and enable it on boot, but there are other tasks we should execute in our system, but in a periodic way. Let's get further into this topic.

Scheduling tasks with cron and systemd

The skills you will learn in this section will be concerned with scheduling periodic tasks in the system for business services and maintenance.

For regular system usage, there are tasks that need to be executed periodically, ranging from temporary folder cleanup, updating the cache's refresh rate, and performing check-in with inventory systems, among other things.

The traditional way to set them up is via `cron`, which is provided in RHEL 9 via the `cronie` package.

`cronie` implements a daemon that's compatible with the traditional *Vixie cron* and allows us to define both user and system crontabs.

A crontab defines several parameters for a task that must be executed. Let's see how it works.

System-wide crontab

A system-wide crontab can be defined in `/etc/crontab` or in individual files at `/etc/cron.d`. Other additional folders exist, such as `/etc/cron.hourly`, `/etc/cron.daily`, `/etc/cron.weekly`, and `/etc/cron.monthly`.

In the folders for *hourly*, *daily*, *weekly*, or *monthly*, you can find scripts or symbolic links to them. When the period since the preceding execution is met (1 hour, 1 day, 1 week, 1 month), the script will be executed.

In contrast, in `/etc/crontab` or `/etc/cron.d`, as well as in the user crontabs, the standard definition of jobs is used.

Jobs are defined by specifying the parameters that are relevant to the execution period, the user that will be executing the job (except for user crontabs), and the command to execute, as illustrated in the following code snippet:

```
# Run the hourly jobs
SHELL=/bin/bash
PATH=/sbin:/bin:/usr/sbin:/usr/bin
MAILTO=root
01 * * * * root run-parts /etc/cron.hourly
```

By looking at the standard /etc/crontab file, we can check the meaning of each field, as follows:

```
# Example of job definition:
# .-------------- minute (0 - 59)
# | .------------ hour (0 - 23)
# | | .---------- day of month (1 - 31)
# | | | .-------- month (1 - 12) OR jan,feb,mar,apr ...
# | | | | .----- day of week (0 - 6) (Sunday=0 or 7) OR
sun,mon,tue,wed,thu,fri,sat
# | | | | |
# * * * * * user-name  command to be executed
```

Based on this, if we check the initial example, 01 * * * * root run-parts /etc/cron.hourly, we can deduce the following:

- Run at minute 01
- Run every hour
- Run every day
- Run every month
- Run every day of the week
- Run as root
- Execute the run-parts /etc/cron.hourly command

This, in brief, means that the job will run on the first minute of every hour as the root user and will run any script located in that folder.

Sometimes, it is possible to see an indication, such as */number, which means that the job will be executed every multiple of that number. For example, */3 will run every 3 minutes if it is on the first column, every 3 hours if it's on the second, and so on.

Any command we might execute from the command line can be executed via cron, and the output will be—by default—mailed to the user running the job. It is a common practice to either define the user that will receive the email via the MAILTO variable in the crontab file or to redirect them to the appropriate log files for the standard output and standard error (stdout and stderr).

User crontab

As with the system-wide **crontab**, users can define their own crontabs so that tasks are executed by the user. This is, for example, useful for running periodic scripts both for a human user or a system account for a service.

The syntax for user crontabs is the same as it is system-wide. However, the column for the username is not there, since it is always executed as the user is defining the crontab itself.

A user can check their crontab via crontab -l, as follows:

```
[root@rhel-instance ~]# crontab -l
no crontab for root
```

A new one can be created by editing it via crontab -e, which will open a text editor so that a new entry can be created.

Let's work with an example by creating an entry, like this:

```
*/2 * * * * date >> datecron
```

When we exit the editor, it will reply as follows:

```
crontab: installing new crontab
```

This will create a file in the /var/spool/cron/ folder with the name of the user that created it. It is a text file, so you can check its contents directly.

After some time (at least 2 minutes), we'll have a file in our $HOME folder that contains the contents of each execution (because we're using the *append* redirect; that is, >>). You can see the output here:

```
[root@rhel-instance ~]# cat datecron
Mon Jan 11 21:02:01 GMT 2021
Mon Jan 11 21:04:01 GMT 2021
```

Now that we've covered the traditional crontab, let's learn about the systemd way of doing things; that is, using timers.

systemd timers

Apart from the regular `cron` **daemon**, a `cron`-style `systemd` feature is to use **timers**. A timer allows us to define, via a unit file, a job that will be executed.

We can check the ones that are already available in our system with the following code:

```
#systemctl list-unit-files *.timer
...
timers.target                          static
dnf-makecache.timer                    enabled
fstrim.timer                           disabled
systemd-tmpfiles-clean.timer           static
...
```

Let's see, for example, `fstrim.timer`, which is used on **solid-state drives** (**SSDs**) to perform a trim at `/usr/lib/systemd/system/fstrim.timer`:

```
[Unit]
Description=Discard unused blocks once a week
Documentation=man:fstrim

..
[Timer]
OnCalendar=weekly
AccuracySec=1h
Persistent=true

...
[Install]
WantedBy=timers.target
```

The preceding timer sets a weekly execution for `fstrim.service`, as follows:

```
[Unit]
Description=Discard unused blocks
[Service]
Type=oneshot
ExecStart=/usr/sbin/fstrim -av
```

As the `fstrim -av` command shows, we are only executing this once.

One of the advantages of having the service timers as unit files, similar to the service itself, is that they can be deployed and updated via the `/etc/cron.d/` files with the regular `cron` daemon, which is handled by `systemd`.

We now know a bit more about how to schedule tasks, but to get the whole picture, scheduling always requires proper timing, so we'll cover this next.

Learning about time synchronization with chrony and NTP

In this section, you will understand the importance of **time synchronization** and how to configure the service.

With connected systems, it is important to keep a **source of truth** (SOT) as regards timing (think about bank accounts, incoming transfer wires, outgoing payments, and more that must be correctly timestamped and sorted). Also, consider tracing logs between users connecting, issues happening, and so on; they all need to be in sync so that we can diagnose and debug between all the different systems involved.

You might think that the system clock, which is defined when the system is provisioned, should be OK, but setting the system clock is not enough as the clocks tend to drift; internal batteries can cause the clock to drift or to even reset, and even intense CPU activity can affect it. To keep clocks accurate, they need to be regularly synced against a reference clock that fixes the drift and tries to anticipate future drifts before the local clock is compared against the remote reference.

The system clock can be synced against a **Global Positioning System** (GPS) unit, for example, or more easily against other systems that have connections to more precise clocks (other GPS units, atomic clocks, and so on). NTP is an internet protocol that's used over the **User Datagram Protocol** (UDP) to maintain communication between the clients and the servers.

> **Tip**
>
> NTP organizes servers by stratum. A stratum 0 device is a GPS device or an atomic clock that directly sends a signal to a server, a stratum 1 server (primary server) is connected to a stratum 0 device, a stratum 2 server is connected to stratum 1 servers, and so on... This hierarchy allows us to reduce the usage of higher-stratum servers but also keep a reliable time source for our systems.

Clients connect to servers and compare the times that are received to reduce the effects of network latency.

Let's see how the NTP client works.

NTP client

In RHEL 9, `chrony` acts as both the server (when enabled) and the client (via the `chronyc` command), and it comes with some features that make it suitable for current hardware and user needs, such as fluctuating networks (laptop is suspended/resumed or flaky connections).

One interesting feature is that `chrony` does not **step** the clock after its initial sync, which means that the time doesn't *jump*. Instead, the system clock runs faster or slower so that, after a period of time, it will be in sync with the reference clock it's using. This makes the time to be a continuum from the operating system and application's point of view: the seconds are going faster or slower than what they should be if compared against a clock until they match the reference clock.

`chrony` is configured via `/etc/chrony.conf` and acts as a client, so it connects to servers to check if they're eligible to be the time source. The main difference between the traditional **server** directive and the **pool** is that the latter can receive several entries while the former only uses one. It is possible to have several servers and pools because, in effect, the servers will be added to the list of possible sources once duplicates have been removed.

For pool or server directives, there are several options available (described in `man chrony.conf`), such as `iburst`, which enables faster checks so that they can quickly transition to a synchronized status.

The actual sources for time can be checked with `chronyc sources`, as illustrated in the following screenshot:

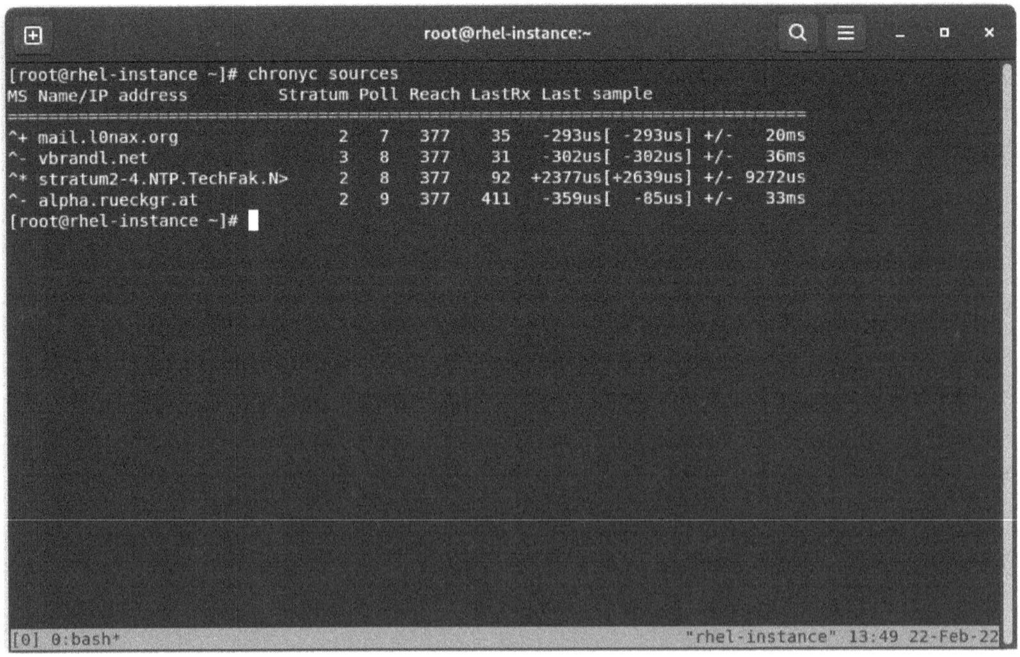

Figure 4.5 – chronyc sources output

As we can see, we know what the status is for each server based on the first column (`M`), as outlined here:

- `^`: This is a server
- `=`: This is a peer

In the second column (`S`), we can see the different statuses for each entry, as follows:

- `*`: This is our current synchronized server
- `+`: This is another acceptable time source
- `?`: This is used to indicate sources that have lost network connectivity
- `x`: This server is considered a false ticker (its time is considered inconsistent compared to other sources)
- `~`: A source that has a high variability (it also appears during daemon startup)

So, we can see that our system is connected to a server that is considering the reference at `ts1.sct.de`, which is a stratum 2 server.

More detailed information can be checked via the `chronyc tracking` command, as illustrated in the following screenshot:

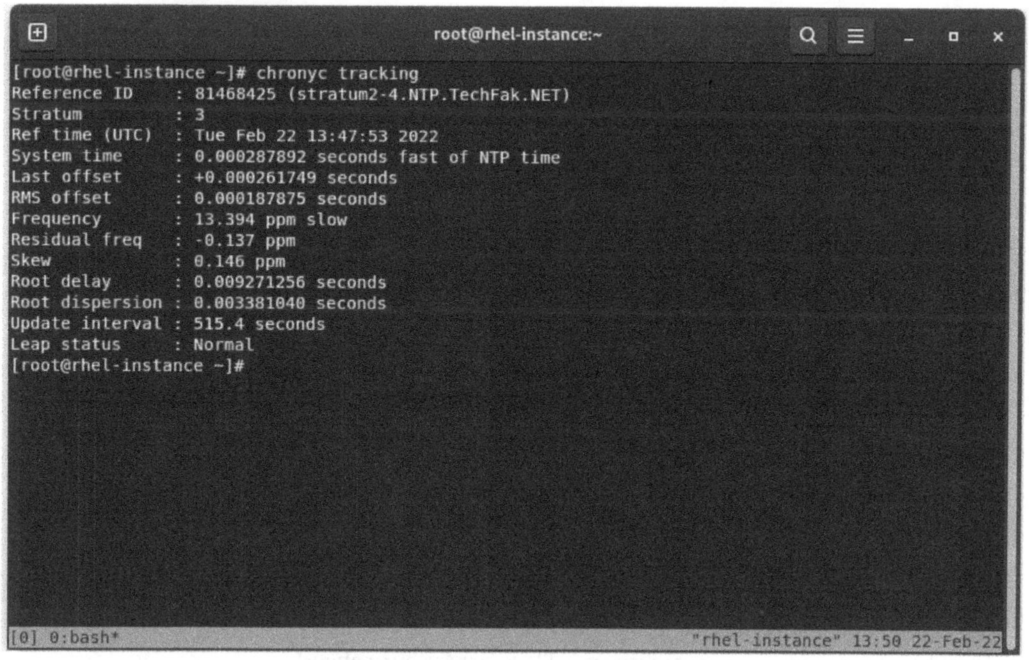
root@rhel-instance:~
[root@rhel-instance ~]# chronyc tracking
Reference ID   : 81468425 (stratum2-4.NTP.TechFak.NET)
Stratum   : 3
Ref time (UTC) : Tue Feb 22 13:47:53 2022
System time   : 0.000287892 seconds fast of NTP time
Last offset   : +0.000261749 seconds
RMS offset   : 0.000187875 seconds
Frequency   : 13.394 ppm slow
Residual freq   : -0.137 ppm
Skew   : 0.146 ppm
Root delay   : 0.009271256 seconds
Root dispersion : 0.003381040 seconds
Update interval : 515.4 seconds
Leap status   : Normal
[root@rhel-instance ~]#
[0] 0:bash*
"rhel-instance" 13:50 22-Feb-22

Figure 4.6 – chronyc tracking output

This provides more detailed information about our clock and our reference clock. Each field in the preceding screenshot has the following meaning:

- `Reference ID`: ID and name/**Internet Protocol (IP)** address of the server that the system has synchronized.

- `Stratum`: Our stratum level. In this example, our synchronized server is a stratum 3 clock.

- `Ref time (UTC)`: The last time the reference was processed.

- `System time`: When running in normal mode (without time skip), this references how far away or behind the system is from the reference clock.

- `Last offset`: Estimated offset on the last clock update. If it's positive, this indicates that our local time was ahead of our source.

- `RMS offset`: **Long-term average (LTA)** of the offset value.

- `Frequency`: This is the rate at which the system clock would be wrong if `chronyd` does not fix it, expressed in **parts per million (ppm)**.

- `Residual freq`: Reflects any difference between the measurements for the current reference clock.

- `Skew`: Estimated error on the frequency.

- `Root delay`: Total network delays to the stratum -1 synchronized server.

- `Root dispersion`: Total dispersion accumulated through all the computers connected to the stratum -1 server we're synchronized to.

- `Update interval`: Interval between the last two clock updates.

- `Leap status`: This can be `Normal`, `Insert`, `Delete`, or `Not synchronized`. It reports the leap status.

Tip

Don't underestimate the information sources you have at your fingertips. Remember that when you're preparing for **Red Hat Certified System Administrator (RHCSA)** exams, the information that's available in the system can be checked during the exam: man pages, documentation included with the program (`/usr/share/doc/program/`), and more. For example, more detailed information about each field listed here can be found via the `man chronyc` command.

To configure the client with additional options, other than the ones provided at install time or via the kickstart file, we can edit the `/etc/chrony.conf` file.

Let's learn how to convert our system into an NTP server for our network.

NTP server

As we introduced earlier, `chrony` can also be configured as a server for your network. In this mode, our system will be providing accurate clock information to other hosts without consuming external bandwidth or resources from higher-stratum servers.

This configuration is also performed via the `/etc/chrony.conf` file, which is where we will be adding a new directive; that is, `allow`. You can see an illustration of this here:

```
# Allow NTP client access from all hosts
allow all
```

This change enables `chrony` to listen on all host requests. Alternatively, we can define a subnet or host to listen to, such as `allow 1.1.1.1`. More than one directive can be used to define different subnets. Alternatively, you can use the `deny` directive to block specific hosts or subnets from reaching our NTP server.

The serving time starts from the base that our server is already synchronized with, as well as an external NTP server, but let's think about an environment without connectivity. In this case, our server will not be connected to an external source and it will not serve time.

`chrony` allows us to define a fake stratum for our server. This is done via the `local` directive in the configuration file. This allows the daemon to get a higher local stratum so that it can serve the time to other hosts. You can see an example of this here:

```
local stratum 3 orphan
```

With this directive, we're setting the local stratum to 3 and we're using the `orphan` option, which enables a special mode in which all servers with an equal local stratum are ignored unless no other source can be selected, and its reference ID is smaller than the local one. This means that we can set several NTP servers in our disconnected network, but only one of them will be the reference.

Now that we have covered time synchronization, we are going to dive into resource monitoring. Later, we'll look at logging. All of this is related to our time reference for the system.

Checking for free resources – memory and disk (free and df)

In this section, you will check the availability of system **resources** such as **memory** and **disk**.

Keeping a system running smoothly means using monitoring so that we can check that the services are running and that the system provides the resources for them to do their tasks.

There are simple commands we can use to monitor the most basic use cases, such as the following ones:

- Disk
- CPU
- Memory
- Network

This includes several ways of monitoring, such as one-shot monitoring, continuously, or even for a period of time to diagnose performance better.

Memory

Memory can be monitored via the `free` command. It provides details on how much **random-access memory (RAM)** and swap memory are available and in use, which also indicates how much memory is used by shares, buffers, or caches.

Linux tends to use all available memory; any unused RAM is directed toward caches or buffers and memory pages that are not being used. These are swapped out to disk if available. You can see an example of this here:

```
# free
              total       used        free        shared
  buff/cache   available
  Mem:         823112      484884      44012       2976
  294216       318856
  Swap:        8388604     185856      8202748
```

In the preceding output, we can see that the system has a total of 823 **megabytes (MB)** of RAM and that it's using some swap and some memory for buffers. This system is not swapping heavily as it's almost idle (we'll check the load average later in this chapter), so we should not be concerned about it.

When RAM usage gets high and there's no more swap available, the kernel includes a protection mechanism called **Out-of-Memory Killer (OOM Killer)**. It determines— based on time in execution, resource usage, and more—which processes in the system should be terminated to recover the system so that it's functional. This, however, comes at a cost, as the kernel knows about the processes that may have gone out of control. However, the killer may kill databases and web servers and leave the system in an unstable state. For production servers, it is sometimes typical—instead of letting the OOM-Killer start killing processes in an uncontrolled way—to either tune the values for some critical process so that those are not killed or to cause a system crash.

A system crash is used to collect debug information that can later be analyzed via a dump containing information about what caused the crash, as well as a memory dump that can be diagnosed.

We will come back to this topic in *Chapter 16, Kernel Tuning and Managing Performance Profiles with tuned*. Let's move on and check the disk space that's in use.

Disk space

Disk space can be checked via the df tool. df provides data as output for each filesystem. This indicates the filesystem and its size, available space, percent of utilization, and mount point.

Let's check this in our example system, as follows:

```
# df
Filesystem              1K-blocks      Used Available Use% Mounted
on
devtmpfs                   442628         0    442628   0% /dev
tmpfs                      489100         0    489100   0% /dev/
shm
tmpfs                      195640      5380    190260   3% /run
/dev/mapper/rhel-root    40935908  18334816  22601092  45% /
/dev/sda2                 1038336    565624    472712  55% /boot
/dev/sda1                  102182      7704     94478   8% /boot/
efi
```

By using this, it's easy to focus on filesystems with higher utilization and less free space to prevent issues.

> **Important Note**
> If a file is being written, such as by a process logging its output, removing the file will just unlink the file from the filesystem, but since the process still has the file handle open, the space is not reclaimed until the process is stopped. In case of critical situations where disk space must be made available as soon as possible, it's better to empty the file via a redirect, such as echo " " > filename. This will recover the disk space immediately while the process is still running. Doing this with the rm command will require the process to be finalized.

We'll check out CPU usage next.

CPU

When it comes to monitoring the CPU, we can make use of several tools, such as ps. You can see this in use here:

```
root         1004  0.0  0.3    7844   3192 ?        S<     12:49   0:00 /usr/sbin/sedispatch
root         1027  0.0  0.1   79120   1588 ?        Ssl    12:49   0:00 /usr/sbin/irqbalance --foregrou
root         1035  0.0  0.5  160680   5444 ?        Ssl    12:49   0:00 /usr/sbin/rsyslogd -n
chrony       1036  0.0  0.2   83972   2548 ?        S      12:49   0:00 /usr/sbin/chronyd -F 2
root         1037  0.0  0.9   16128   9624 ?        Ss     12:49   0:00 sshd: /usr/sbin/sshd -D [listen
root         1039  0.0  0.0    7316    168 ?        Ss     12:49   0:00 /usr/bin/rhsmcertd
root         1040  0.0  0.3   54932   3760 ?        Ssl    12:49   0:00 /usr/sbin/gssproxy -D
root         1055  0.0  0.1    4648   1532 ?        Ss     12:49   0:00 /usr/sbin/atd -f
root         1056  0.0  0.3    8524   3560 ?        Ss     12:49   0:00 /usr/sbin/crond -n
root         1070  0.0  0.1    3048   1096 tty1     Ss+    12:49   0:00 /sbin/agetty -o -p -- \u --nocl
root         1071  0.0  0.1    5616   1088 ttyS0    Ss+    12:49   0:00 /sbin/agetty -o -p -- \u --keep
root         1100  0.0  1.2   19376  11940 ?        Ss     12:49   0:00 sshd: root [priv]
root         1104  0.0  0.7   19720   7476 ?        S      12:49   0:00 sshd: root@pts/0
root         1105  0.0  0.5    9072   5712 pts/0    Ss     12:49   0:00 -bash
root         1127  0.0  0.3    5464   3620 pts/0    S+     12:49   0:00 tmux
root         1129  0.0  0.3    6140   3276 ?        Rs     12:49   0:00 tmux
root         1130  0.0  0.5    8836   5668 pts/1    Ss     12:49   0:00 -bash
root         1188  0.0  0.0       0      0 ?        I      13:04   0:00 [kworker/1:0-mm_percpu_wq]
root         1194  0.0  0.0       0      0 ?        I      13:26   0:00 [kworker/u4:0-events_unbound]
root         1197  0.0  0.0       0      0 ?        I      13:29   0:00 [kworker/0:1-events]
root         1203  0.0  0.0       0      0 ?        I      13:39   0:00 [kworker/u4:2-events_unbound]
root         1204  0.0  0.0       0      0 ?        I      13:41   0:00 [kworker/0:0-events_power_effic
root         1226  0.0  0.0       0      0 ?        I      13:48   0:00 [kworker/u4:1-events_unbound]
root         1251  0.0  0.0       0      0 ?        I      13:49   0:00 [kworker/0:2-events]
root         1282  0.0  0.0       0      0 ?        I      13:50   0:00 [kworker/u4:3]
root         1308  0.0  0.3   10176   3392 pts/1    R+     13:50   0:00 ps aux
[root@rhel-instance ~]#
[0] 0:bash*                                                   "rhel-instance" 13:50 22-Feb-22
```

Figure 4.7 – Output of the ps aux command (every process in the system)

The ps command is the de facto standard for checking which process is running, as well as resource consumption usage.

As for any other command, we could write a lot about all the different command arguments we could use (so, again, check the man page for details), but as a rule, try to learn about their basic usage or the ones that are more useful for you. For anything else, check the manual. For example, ps aux provides enough information for normal usage (every process in the system).

The top tool, as shown in the following screenshot, refreshes the screen regularly and can sort the output of running processes, such as CPU usage, memory usage, and more. In addition, top also shows a five-line summary of memory usage, load average, running processes, and so on:

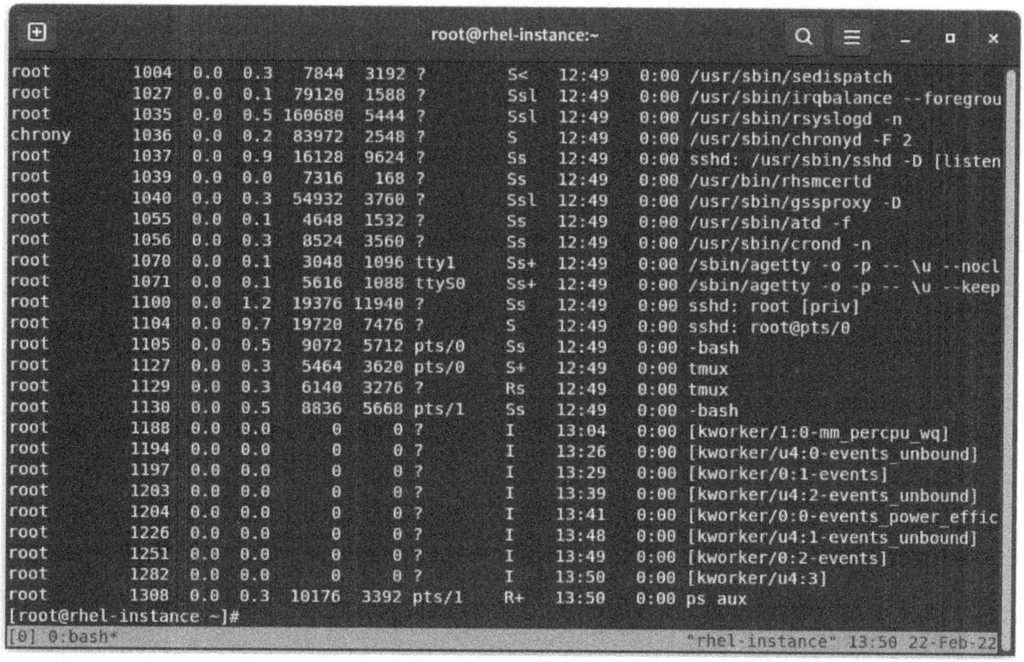

Figure 4.8 – top execution on our test system

CPU usage is not the only thing that may keep our system sluggish. Now, let's learn a bit about load average indicators.

Load average

Load average is usually provided as a group of three numbers, such as load average: 0.81, 1.00, 1.17, which is the average that's calculated for 1, 5, and 15 minutes, respectively. This indicates how busy a system is; the higher it is, the worse it will respond. The values that are compared for each time frame give us an idea of whether the system load is increasing (higher values in 1 or 5 and lower on 15) or if it is going down (higher at 15 minutes, lower at 5 and 1), so it becomes a quick way to find out if something happened or if it is ongoing. If a system usually has a high load average (over 1.00 per CPU), it would be a good idea to dig a bit into the possible causes (too much demand for its power, not many resources available, and so on).

Now that we have covered the basics, let's move on and look at some extra checks we can perform on our system resources' usage.

Other monitoring tools

For monitoring network resources, we can check the packages that are sent/received for each card via `ifconfig` (installed via the `net-tools` package), for example, and match the values that are received for transmitted packages, received packages, errors, and so on.

When the goal is to perform more complete monitoring, we should ensure that the `sysstat` package is installed. It includes some interactive tools such as `iostat`, which can be used to check disk performance, but the most important thing is that it also sets up a job that will collect system performance data on a periodical basis (the default is every 10 minutes). This will be stored in `/var/log/sa/`.

Historical data that's recorded and stored per day (##) at `/var/log/sa/sa##` and `/var/log/sa/sar##` can be queried so that we can compare it against other days. By running the data collector (which is executed by a `systemd` timer) with a higher frequency, we can increase the granularity for specific periods while an issue is being investigated.

However, it appears the `sar` file is showing lots of data, as we can see here:

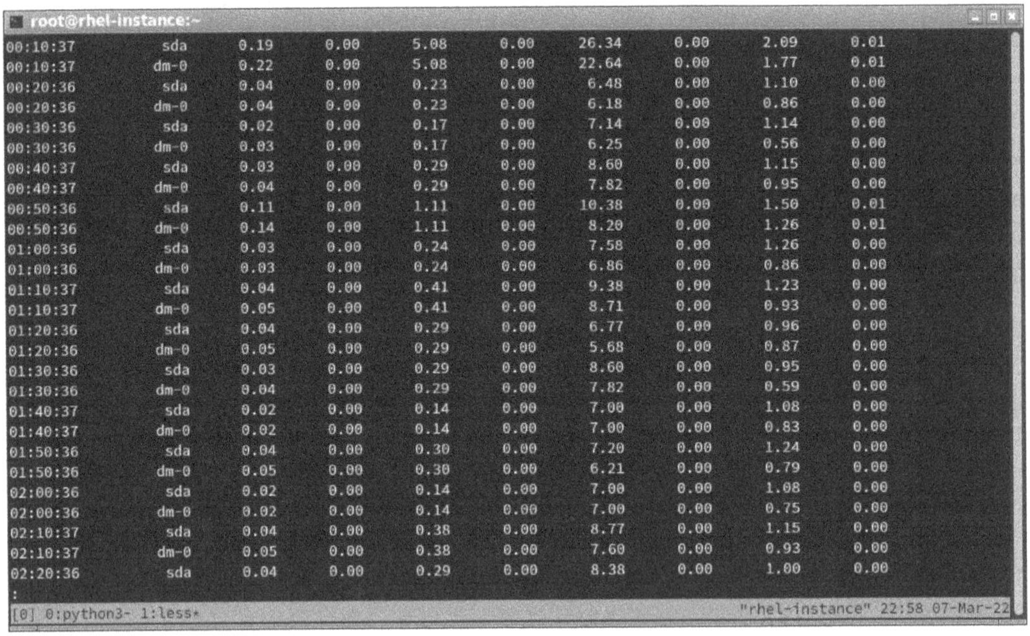

```
root@rhel-instance:~
00:10:37        sda    0.19    0.00    5.08    0.00   26.34    0.00    2.09    0.01
00:10:37       dm-0    0.22    0.00    5.08    0.00   22.64    0.00    1.77    0.01
00:20:36        sda    0.04    0.00    0.23    0.00    6.48    0.00    1.10    0.00
00:20:36       dm-0    0.04    0.00    0.23    0.00    6.18    0.00    0.86    0.00
00:30:36        sda    0.02    0.00    0.17    0.00    7.14    0.00    1.14    0.00
00:30:36       dm-0    0.03    0.00    0.17    0.00    6.25    0.00    0.56    0.00
00:40:37        sda    0.03    0.00    0.29    0.00    8.60    0.00    1.15    0.00
00:40:37       dm-0    0.04    0.00    0.29    0.00    7.82    0.00    0.95    0.00
00:50:36        sda    0.11    0.00    1.11    0.00   10.38    0.00    1.50    0.01
00:50:36       dm-0    0.14    0.00    1.11    0.00    8.20    0.00    1.26    0.01
01:00:36        sda    0.03    0.00    0.24    0.00    7.58    0.00    1.26    0.00
01:00:36       dm-0    0.03    0.00    0.24    0.00    6.86    0.00    0.86    0.00
01:10:37        sda    0.04    0.00    0.41    0.00    9.38    0.00    1.23    0.00
01:10:37       dm-0    0.05    0.00    0.41    0.00    8.71    0.00    0.93    0.00
01:20:36        sda    0.04    0.00    0.29    0.00    6.77    0.00    0.96    0.00
01:20:36       dm-0    0.05    0.00    0.29    0.00    5.68    0.00    0.87    0.00
01:30:36        sda    0.03    0.00    0.29    0.00    8.60    0.00    0.95    0.00
01:30:36       dm-0    0.04    0.00    0.29    0.00    7.82    0.00    0.59    0.00
01:40:37        sda    0.02    0.00    0.14    0.00    7.00    0.00    1.08    0.00
01:40:37       dm-0    0.02    0.00    0.14    0.00    7.00    0.00    0.83    0.00
01:50:36        sda    0.04    0.00    0.30    0.00    7.20    0.00    1.24    0.00
01:50:36       dm-0    0.05    0.00    0.30    0.00    6.21    0.00    0.79    0.00
02:00:36        sda    0.02    0.00    0.14    0.00    7.00    0.00    1.08    0.00
02:00:36       dm-0    0.02    0.00    0.14    0.00    7.00    0.00    0.75    0.00
02:10:37        sda    0.04    0.00    0.38    0.00    8.77    0.00    1.15    0.00
02:10:37       dm-0    0.05    0.00    0.38    0.00    7.60    0.00    0.93    0.00
02:20:36        sda    0.04    0.00    0.29    0.00    8.38    0.00    1.00    0.00
:
[0] 0:python3- 1:less*                                    "rhel-instance" 22:58 07-Mar-22
```

Figure 4.9 – Contents of /var/log/sar02 on the example system

Here, instead of seeing sda, we might see references to devices such as 8-0. In this case, the device's name is using the values for the major/minor, which we can check in the /dev/ folder. We can see this by running ls -l /dev/* | grep 8, as shown in the following screenshot:

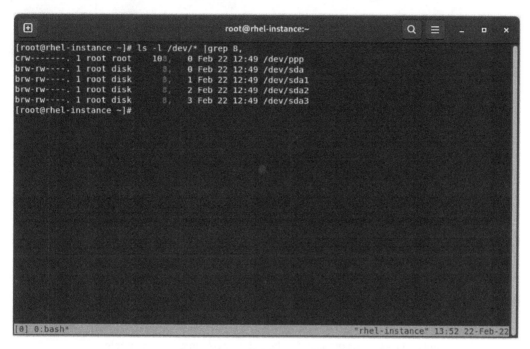

```
[root@rhel-instance ~]# ls -l /dev/* |grep 8,
crw-------. 1 root root    108,  0 Feb 22 12:49 /dev/ppp
brw-rw----. 1 root disk      8,  0 Feb 22 12:49 /dev/sda
brw-rw----. 1 root disk      8,  1 Feb 22 12:49 /dev/sda1
brw-rw----. 1 root disk      8,  2 Feb 22 12:49 /dev/sda2
brw-rw----. 1 root disk      8,  3 Feb 22 12:49 /dev/sda3
[root@rhel-instance ~]#
```

Figure 4.10 – Directory listing for /dev/ for locating the device corresponding to major 8 and minor 0

Here, we can see that this corresponds to the full hard-drive statistics at /dev/sda.

> **Tip**
>
> Processing the data via sar is a good way to get insights into what's going on with our system, but since the sysstat package has been around for a long time in Linux, there are tools such as sarstats (https://github.com/mbaldessari/sarstats) that help us process the data that's recorded and present it graphically as a **Portable Document File** (**PDF**) file.

In the following screenshot, we can see the system service times for the different drives, along with a label at the time the system crashes. This helps us identify the system's activity at that point:

I/O

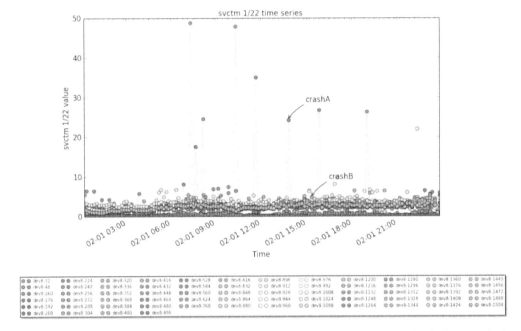

svctm - The average service time (in milliseconds) for I/O requests that were issued to the device. Warning! Do not trust this field any more. This field will be removed in a future sysstat version.

Figure 4.11 – sarstats graphics for the disk service's time in the example
PDF at https://acksyn.org/software/sarstats/sar01.pdf

Modern tooling for monitoring the system's resources has evolved, and **Performance Co-Pilot** (pcp and, optionally, the pcp-gui package) can be set up for more powerful options. Just bear in mind that pcp requires us to also start the data collector on the system.

RHEL 9 also includes **Cockpit**, which is installed by default when we do a server installation. This package provides a set of tools that enable web management for the system, and it can also be made part of other products via plugins that extend its functionality.

The web service provided by Cockpit can be reached at your host IP at port 9090, so you should access https://localhost:9090 to get a login screen so that we can use our system credentials to log in.

> **Important Tip**
>
> If Cockpit is not installed or available, make sure that you execute `dnf install cockpit` to install the package and use `systemctl enable --now cockpit.cocket` to start the service. If you are accessing the server remotely, instead of using `localhost`, use the server hostname or IP address after allowing the firewall to connect via `firewall-cmd --add-service=cockpit`, if you haven't done so previously.

After logging in, we will see a dashboard showing the relevant system information and links to other sections, as shown in the following screenshot:

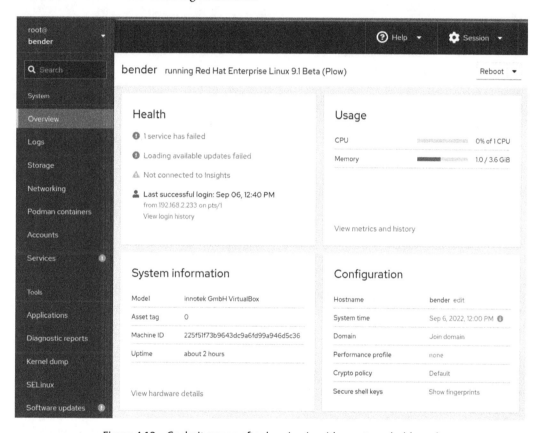

Figure 4.12 – Cockpit screen after logging in with a system dashboard

As you can see, Cockpit includes several tabs that can be used to view the status of the system and even perform some administration tasks, such as **Security-Enhanced Linux (SELinux)** tasks, software updates, subscriptions, and more.

For example, we can check the system graphs on performance, as shown in the following screenshot:

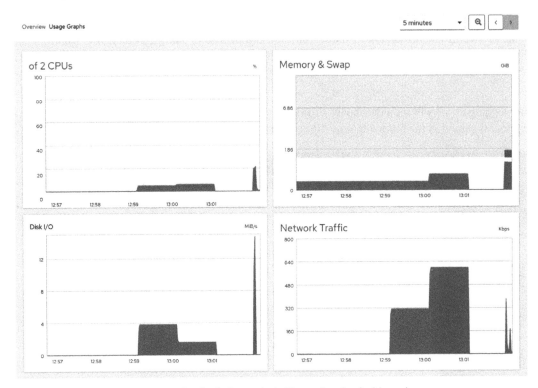

Figure 4.13 – Cockpit graphs in Usage Graphs dashboard

Cockpit allows us to check a service's status and package upgrade status, plus other configuration settings from a graphical interface that can also connect remotely to other systems. These can be selected from the lateral menu on the left.

There are better tools suited for large deployment monitoring and management, such as *Ansible* and *Satellite*, so it is important to get used to the tools we have for troubleshooting and simple scripts we can build. This allows us to combine what we've learned so far to quickly generate hints about things that require our attention.

With that, we have covered some of the basics of checking resource usage. Now, let's check out how to find information about the running services and errors we can review.

Finding logs, using journald, and reading log files, including log preservation and rotation

In this section, you will learn how to review a system's status via logs.

Previously in this chapter, we learned how to manage system services via systemd, check their status, and check their logs. Traditionally, the different daemons and system components used to create files under the /var/log/ folder are based on the name of the daemon or service. If the service is used to create several logs, it would do so inside a folder for the service (for example, httpd or samba).

The rsyslogd system log daemon has a new systemd partner—named systemd-journald. service—that also stores logs, but instead of using the traditional plain text format, it uses the binary format, which can be queried via the journalctl command.

It's really important to get used to reading the log files as it's the basis for troubleshooting, so let's learn about general logging and how to use it.

Logs contain status information for the services that generate them. They might have some common formatting and can often be configured, but they tend to use several common elements, such as the following:

- Timestamp
- Module generating the entry
- Message

Here is an example:

```
Jan 03 22:36:47 rhel-instance sshd[50197]: Invalid user admin
from 49.232.135.77 port 47694
```

In this case, we can see that someone attempted to log in to our system as the admin user from IP address 49.232.135.77.

We can correlate that event with additional logs, such as the ones for the login subsystem via journalctl -u systemd-logind. In this example, we cannot find any login for the admin user (this is expected, as the admin user was not defined in this system).

Additionally, we can see the name of the host (rhel-instance), the service generating it (sshd), a PID of 50197, and the message that's been logged by that service.

In addition to `journalctl`, there are additional logs that we can look at when we wish to check the system's health. Let's look at an example with `/var/log/messages`:

```
⊞                        root@rhel-instance:~                    Q  ≡  _  □  ×
virt:rhel:8050020211110001900:b4937e53.x86_64
Feb 22 13:31:48 rhel-instance dnf[1198]: Metadata cache created.
Feb 22 13:31:48 rhel-instance systemd[1]: dnf-makecache.service: Deactivated successfully.
Feb 22 13:31:48 rhel-instance systemd[1]: Finished dnf makecache.
Feb 22 13:31:48 rhel-instance systemd[1]: dnf-makecache.service: Consumed 1.505s CPU time.
Feb 22 13:50:53 rhel-instance systemd[1]: Reloading.
Feb 22 13:50:53 rhel-instance systemd-rc-local-generator[1295]: /etc/rc.d/rc.local is not marked e
xecutable, skipping.
Feb 22 13:50:53 rhel-instance systemd[1]: Queuing reload/restart jobs for marked units…
Feb 22 13:52:01 rhel-instance systemd[1]: Started /usr/bin/systemctl start man-db-cache-update.
Feb 22 13:52:01 rhel-instance systemd[1]: Starting man-db-cache-update.service...
Feb 22 13:52:01 rhel-instance systemd[1]: Reloading.
Feb 22 13:52:01 rhel-instance systemd-rc-local-generator[1353]: /etc/rc.d/rc.local is not marked e
xecutable, skipping.
Feb 22 13:52:02 rhel-instance systemd[1]: Queuing reload/restart jobs for marked units…
Feb 22 13:52:04 rhel-instance systemd[1]: man-db-cache-update.service: Deactivated successfully.
Feb 22 13:52:04 rhel-instance systemd[1]: Finished man-db-cache-update.service.
Feb 22 13:52:04 rhel-instance systemd[1]: run-r61c4153a4aba46679600e9bcc03447a7.service: Deactivat
ed successfully.
Feb 22 13:54:27 rhel-instance subscription-manager[1662]: Added subscription for 'Content Access'
contract 'None'
Feb 22 13:54:27 rhel-instance subscription-manager[1662]: Added subscription for product ' Content
 Access'
Feb 22 13:54:27 rhel-instance subscription-manager[1662]: Removed subscription for 'Content Access
' contract 'None'
Feb 22 13:54:27 rhel-instance subscription-manager[1662]: Removed subscription for product ' Conte
:
[0] 0:bash*                                            "rhel-instance" 13:54 22-Feb-22
```

Figure 4.14 – Excerpt of /var/log/messages

In this example, we can see how the system ran some commands while following a similar output to the initial lines. For example, in the preceding example, we can see how `sysstat` has been executed every 10 minutes, as well as how the `dnf` cache has been updated.

Let's look at a list of important logs that are available in a standard system installation (note that the filenames are relative to the `/var/log` folder):

- `boot.log`: Stores the messages that are emitted by the system during boot. It might contain escape codes that are used to provide colorized output.

- `audit/audit.log`: Contains the stored messages that have been generated by the kernel audit subsystem.

- `secure`: Contains security-related messages, such as failed `sshd` login attempts.

- `dnf.log`: Logs generated by the **Dandified YUM (DNF)** package manager, such as cache refreshes.

- `firewalld`: Output generated by the `firewalld` daemon.

- `lastlog`: This is a binary file that contains information about the last few users logging in to the system (to be queried via the `last` command).

- `messages`: The default logging facility. This means that anything that is not a specific log will go here. Usually, this is the best place to start checking what happened with a system.

- `maillog`: The log for the mail subsystem. When enabled, it attempts to deliver messages. Any messages that are received will be stored here. It's common practice to configure outgoing mail from servers so that system alerts or script outputs can be delivered.

- `btmp`: Binary log for failed access to the system.

- `wtmp`: Binary log for access to the system.

- `sa/sar*`: Text logs for the `sysstat` utility (the binary ones, named `sa`, plus the day number, are converted via a `cron` job at night).

Additional log files might exist, depending on the services that have been installed, the installation method that was used, and so on. It is very important to get used to the available logs and—of course—review their contents to see how the messages are formatted, how many logs are created every day, and what kind of information they produce.

Using the information that's been logged, we will get hints on how to configure each individual daemon. This allows us to adjust the log level between showing just errors or being more verbose about debugging issues. This means we can configure the required log rotation to avoid risking system stability because all the space has been consumed by logs.

Log rotation

During regular system operation, lots of daemons are in use, and the system itself generates logs that are used for troubleshooting and system checks.

Some services might allow us to define a log file to write to based on the date, but usually, the standard is to log to a file named like the daemon in the `/var/log` directory; for example, `/var/log/cron`. Writing to the same file will cause the file to grow until the drive holding the logs is filled, which might not make sense as after a while (sometimes, under company-defined policies), logs are no longer useful.

The `logrotate` package provides a script with a `cron` entry that simplifies the log rotation process. It is configured via `/etc/logrotate.conf` and is executed on a daily basis, as shown here:

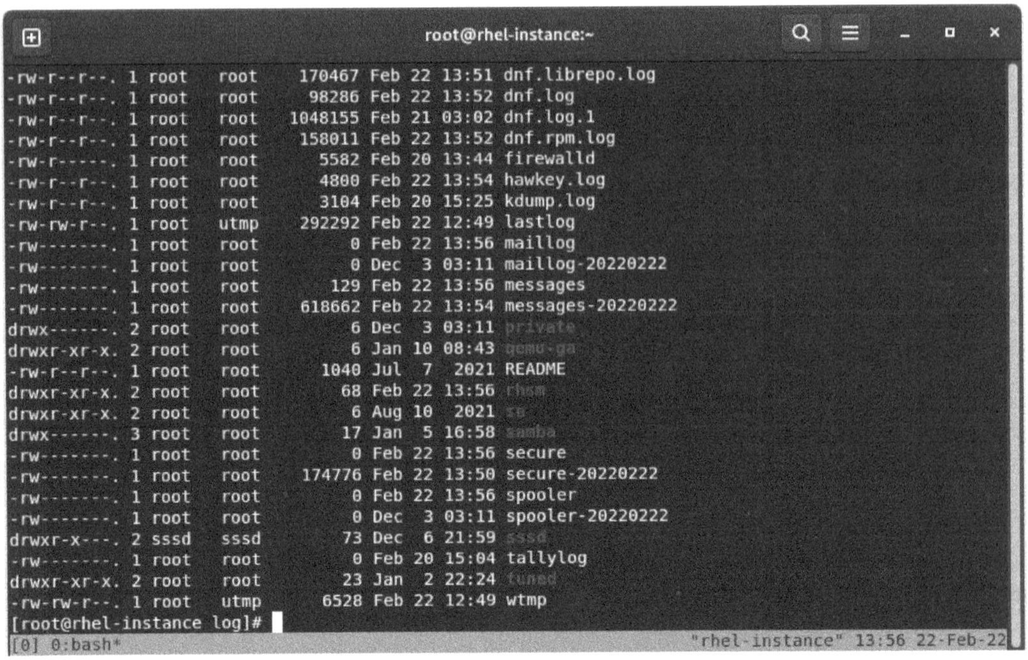

Figure 4.15 – Example listing of logs and rotated logs (using date extension)

If we check the contents of the configuration file, we will see that it includes some file definitions either directly there or via drop-in files in the `/etc/logrotate.d/` folder, which allows each program to drop its own requirements without it affecting others when packages are installed, removed, or updated.

Why is this important? Because—if you remember from some of the tips earlier in this chapter (while speaking about disk space)—if `logrotate` just deleted the files and created a new one, the actual disk space would not be freed, and the daemon writing to the log would continue to write to the file it was writing to (via the file handle). To overcome this, each definition file can define a post-rotation command. This signals the process of log rotation so that it can close and then reopen the files it uses for logging. Some programs might require a signal such as `kill -SIGHUP PID` or a special parameter on execution, such as `chronyc cyclelogs`.

With these definitions, `logrotate` will be able to apply the configuration for each service and, at the same time, keep the service working in a sane state.

The configuration can also include special directives, such as the following:

- `missingok`
- `nocreate`
- `copytruncate`
- `notifempty`

You can find out more about them (and others) on the **man page** for `logrotate.conf` (yes—some packages also include a man page for the configuration files, so try checking `man logrotate.conf` to get the full details!).

The remaining general configuration in the main file allows us to define some common directives, such as how many days of logs to keep, if we want to use the date in the file extension for the rotated log files, if we want to use compression on the rotated logs, how frequently we want to have the rotation executed, and so on.

Let's look at some examples.

The following example will rotate on a daily basis, keep 30 rotated logs, compress them, and use an extension with `date` as part of its trailing filename:

```
rotate 30
daily
compress
dateext
```

In this example, it will keep 4 logs rotated on a weekly basis (so 4 weeks) and will compress the logs, but use a sequence number for each rotated log (this means that each time a rotation happens, the sequence number is increased for the previously rotated logs too):

```
rotate 4
weekly
compress
```

One of the advantages of this approach (not using `dateext`) is that the log naming convention is predictable as we have `daemon.log` as the current one, `daemon.1.log` as the prior one, and so on. This makes it easier to script log parsing and processing.

Summary

In this chapter, we learned about `systemd` and how it takes care of booting the required system services in an optimized way. We also learned how to check a service's status, how to enable, disable, start, and stop it, and how to make the system boot into the different targets that we boot our system into.

Time synchronization was introduced as a must-have feature, and it ensures our service functions properly. It also allows us to determine the status of our system clock and how to act as a clock server for our network.

We also used system tools to monitor resource usage, learned how to check the logs that are created by our system to find out about the functional status of the different tools, and how we can ensure that logs are maintained properly so that older entries are discarded when they are no longer relevant.

In the next chapter, we will dive into securing the system with different users, groups, and permissions.

5

Securing Systems with Users, Groups, and Permissions

Security is a key part of managing a system and understanding the security concepts in order to provide the right access to the right resource to the right user, or group of users, is required for any system administrator.

In this chapter, we will review the basics of security in **Red Hat Enterprise Linux** (**RHEL**). We will add new users to a system and change their attributes. We will also add a user to a group and reviewing groups before making changes will be seen in this chapter. We will review how to handle user passwords and change the age requirements for them, locking or restricting user access. We will use sudo as a way to assign admin privileges to different users in the system (and even disable the root account). We will also take a deeper look into file permissions and how to change them, using an extended capability to enable commands to be run by a different user or group or to simplify group collaboration in directories.

We will cover the following topics:

- Creating, modifying, and deleting local user accounts and groups
- Managing groups and reviewing assignments
- Adjusting password policies
- Configuring sudo access for administrative tasks
- Checking, reviewing, and modifying file permissions
- Using special permissions

Let's get started in the world of permissions and security with user accounts and groups.

Creating, modifying, and deleting local user accounts and groups

One of the first tasks that a system administrator has to do when preparing a system to be accessed by users is to create new user accounts for the people accessing the system. In this section, we will review how local accounts are created and deleted, as well as how they are assigned to groups.

The first step is to create a new user account in the system. That is done by using the `useradd` command. Let's add `user01` to the system by running the following:

```
[root@rhel-instance ~]# useradd user01
[root@rhel-instance ~]# grep user01 /etc/passwd
user01:x:1001:1001::/home/user01:/bin/bash
[root@rhel-instance ~]# id user01
uid=1001(user01) gid=1001(user01) groups=1001(user01)
```

With that, the user is created.

> **Important Note**
> To be able to add users, we need administrative privileges. In the current configuration, we do that by running the commands as `root`.

The account is created using the default options configured in the system, such as the following:

- **No password assigned**: The new user will not be able to log in using a password. However, we can switch to that account by using `su` as `root`. We will see how to add a password to the user next.

- **User ID (UID)**: The first number over 999 available. In the command we ran before, for `user01`, the UID is `1001`.

- **Group ID (GID)**: The same number as the UID. In this case, the GID is `1001`.

- **Description**: No description is added when creating the user. This field is left empty.

- **Home**: A home directory is created in `/home/$USER` – in this case, `/home/user01`. This will be the default and main directory for the user and is where their personal preferences and files will be stored. Initial contents are copied from `/etc/skel`.

- **Shell**: The default shell is `bash`.

> **Tip**
> The default options applied when a new user is created are defined in the `/etc/default/useradd` file.

Once the user is created, we can add (or change) the password by running, as `root`, the command `passwd` followed by the username to change it:

```
[root@rhel-instance ~]# passwd user01
Changing password for user user01.
New password: redhat
BAD PASSWORD: The password is shorter than 8 characters
Retype new password: redhat
passwd: all authentication tokens updated successfully
```

Now, the user has the new password assigned. Note two things:

- The `root` user can change the password to any user without knowing the previous one (a full password reset). This is useful when a user comes back from their holidays and doesn't remember their password.

- In the example, we show the password assigned, `redhat`, but that is not shown on the screen. The password is too simple and does not meet the default complexity criteria – however, as `root` we can still assign it.

Let's check the new user with the `id` command we learned before:

```
[root@rhel-instance ~]# id user01
uid=1001(user01) gid=1001(user01) groups=1001(user01)
```

After the steps taken in this section, we now have the user in the system and ready to be used. The main options we could have used to customize the user creation with `useradd` are the following:

- `-u` or `--uid`: Assign a specific UID to the user.
- `-g` or `--gid`: Assign a main group to the user. It can be specified by number (GID) or by name. The group needs to be created first.
- `-G` or `--groups`: Make the user part of other groups by providing a comma-separated list of them.
- `-c` or `--comment`: Provide a description for the user, specified between quotes if you want to use spaces.
- `-d` or `--home-dir`: Define the home directory for the user.
- `-s` or `--shell`: Assign a custom shell to the user.
- `-p` or `--password`: A way to provide a password to the user. The password should be already encrypted to use this method. It is *not* recommended to use this option, as there are ways to capture the encrypted password. Please use `passwd` instead.
- `-r` or `--system`: To create a system account instead of a user account.

What if we need to change any of the user's properties, such as, for example, the description? The tool for that is `usermod`. Let's modify the description to `user01`:

```
[root@rhel-instance ~]# usermod -c "User 01" user01
[root@rhel-instance ~]# grep user01 /etc/passwd
user01:x:1001:1001:User 01:/home/user01:/bin/bash
```

The `usermod` command uses the same options as `useradd`. It will be easy to customize your current users now.

Let's create `user02` as an example of how to use the options:

```
[root@rhel-instance ~]# useradd --uid 1002 --groups wheel \
--comment "User 02" --home-dir /home/user02 \
--shell /bin/bash user02
[root@rhel-instance ~]# grep user02 /etc/passwd
user02:x:1002:1002:User 02:/home/user02:/bin/bash
[root@rhel-instance ~]# id user02
uid=1002(user02) gid=1002(user02) groups=1002(user02),10(wheel)
```

> **Tip**
> When the command line is too long, the \ character can be added, then press *Enter* and continue the command on a new line.

Now, we know how to create a user, but we may need to create a group too and add our users to it. Let's create the `finance` group with the `groupadd` command:

```
[root@rhel-instance ~]# groupadd finance
[root@rhel-instance ~]# grep finance /etc/group
finance:x:1003:
```

We can add the `user01` and `user02` users to the `finance` group:

```
[root@rhel-instance ~]# usermod -aG finance user01
[root@rhel-instance ~]# usermod -aG finance user02
[root@rhel-instance ~]# grep finance /etc/group
finance:x:1003:user01,user02
```

> **Important Note**
> We are using the -aG option to add the user to the group, instead of modifying the groups
> the user belongs to.

Once we know how to create users and groups, let's check how to delete them with the `userdel`
command:

```
[root@rhel-instance ~]# userdel user01
[root@rhel-instance ~]# grep user01 /etc/passwd
[root@rhel-instance ~]# id user01
id: 'user01': no such user
[root@rhel-instance ~]# grep user02 /etc/passwd
user02:x:1002:1002:User 02:/home/user02:/bin/bash
[root@rhel-instance ~]# id user02
uid=1002(user02) gid=1002(user02)
groups=1002(user02),10(wheel),1003(finance)
[root@rhel-instance ~]# ls /home/
user   user01   user02
[root@rhel-instance ~]# rm -rf /home/user01/
```

As you can see, we needed to manually delete the home directory. This way of removing a user is
good if we want to keep its data for future use.

To fully remove a user, we shall apply the -r option. Let's try it with `user02`:

```
[root@rhel-instance ~]# userdel -r user02
[root@rhel-instance ~]# ls /home/
user
[root@rhel-instance ~]# grep user02 /etc/passwd
[root@rhel-instance ~]# id user02
id: 'user02': no such user
```

Now, let's remove the `finance` group with the `groupdel` command:

```
[root@rhel-instance ~]# groupdel finance
[root@rhel-instance ~]# grep finance /etc/group
```

As we have seen, it's simple and easy to create users and groups in RHEL and make simple assignments.
In the next section, let's check in more depth how to manage groups and assignments to them.

Managing groups and reviewing assignments

We have seen how to create a group with `groupadd` and delete it with `groupdel`. Let's see how to modify a created group with `groupmod`.

Let's create a group to work with. We will create the misspelled `acounting` group by running the following:

```
[root@rhel-instance ~]# groupadd -g 1099 acounting
[root@rhel-instance ~]# tail -n1 /etc/group
acounting:x:1099:
```

You see we made a mistake in the name by not spelling it `accounting`. We may even have added some user accounts to it and we need to modify it. We can do so using `groupmod` and running the following:

```
[root@rhel-instance ~]# groupmod -n accounting acounting
[root@rhel-instance ~]# tail -n1 /etc/group
accounting:x:1099:
```

Now, we've seen how we modify the group name. We can modify not just the name but the GID by using the `-g` option:

```
[root@rhel-instance ~]# groupmod -g 1111 accounting
[root@rhel-instance ~]# tail -n1 /etc/group
accounting:x:1111:
```

We can see which groups are assigned to a user by running the `groups` command:

```
[root@rhel-instance ~]# groups user
user : user wheel
```

With this, we are ready to manage groups and users in a Linux system. Let's move on to password policies.

Adjusting password policies

As was mentioned in *Chapter 3, Basic Commands and Simple Shell Scripts*, users are stored in the /etc/passwd file while the encrypted passwords, or **password hashes**, are stored in the /etc/shadow file.

> **Tip**
>
> A hashing algorithm is made so that it generates a precise string of characters, or a hash, from a provided piece of data (that is, a file or a word). It does it in a way that it will always generate the same hash from the same original data, but the original data is almost impossible to recreate from the hash. That's why they are used to store passwords or verify the integrity of a downloaded file.

Let's take a look at one example by running the `grep user` as `root` against `/etc/shadow`:

```
user:$6$tOT/cvZ4PWRc18XX$0v3.ADE/ibzlUGbDLer0ZYaMPNRJ5gK17LeKno
MfKK9 .nFz8grN3IafmHvoHPuh3XrU81nJu0.is5znztB64Y/:18650:0:99999
:7:3:19113:
```

As with the password file, the data stored in `/etc/shadow` has an entry per line and the fields are separated by colons (`:`):

- `user`: The account name. It should be the same one as in `/etc/passwd`.

- `$6$tOT/cvZ4PWRc18XX$0v3.ADE/ibzlUGbDLer0ZYaMPNRJ5gK17LeKnoMfKK 9.nFz8grN3IafmHvoHPuh3XrU81nJu0.is5znztB64Y/:` The password hash. It contains three parts separated by `$`:

    - `$6`: The algorithm used to encrypt the file. In this case, the value 6 indicates that it is SHA-512. The number 1 is for the old, now insecure, MD5 algorithm.

    - `$tOT/cvZ4PWRc18XX`: The `salt` password. This token is used to improve password encryption.

    - `$0v3.ADE/ibzlUGbDLer0ZYaMPNRJ5gK17LeKnoMfKK9. nFz8grN3IafmHvoHPuh3XrU81nJu0.is5znztB64Y/`: An encrypted password hash. Using `salt` and the SHA-512 algorithm, this token is created. When the user validates, the process is run again and if the same hash is generated, the password is validated and access is granted.

- `18650`: The time and date when the password was last changed. The format is the number of days since `1970-01-01 00:00` UTC (this date is also known as **the epoch**).

- `0`: The minimum number of days until the user can change the password again.

- `99999`: The maximum number of days until the user has to change the password again. If empty, it won't expire.

- `7`: The number of days the user will be warned that the password is about to expire.

- `3`: The number of days the user can still log in even when the password has expired.

- 19113: The date on which the password should expire. If empty, it won't expire on a specific date.

- <empty>: The last colon is left to allow us to add new fields easily.

> **Tip**
> To convert the date field to a human-readable date, you can run the following command:
> date -d '1970-01-01 UTC + 18650 days'.

How do we change the expiration dates for passwords? The tool to do so is chage, for **change age**. Let's first review the options that can be used in the same order as they are stored in /etc/shadow:

- -d or --lastday: The time and date when the password was last changed. The format for it is YYYY-MM-DD.

- -m or --mindays: The minimum number of days until the user can change the password again.

- -W or --warndays: The number of days the user will be warned that the password is about to expire.

- -I or --inactive: The number of days, once the password has expired, that will have to pass before the account is locked.

- -E or --expiredate: The date after which the user's account will be locked. The date should be expressed in the YYYY-MM-DD format.

Let's try it. First, we create the usertest account:

```
[root@rhel-instance ~]# adduser usertest
[root@rhel-instance ~]# grep usertest /etc/shadow
usertest:!!:19062:0:99999:7:::
```

> **Important Note**
> adduser and useradd have similar functionality in RHEL 9. Feel free to type it the way you feel most comfortable.

You'll notice in the previous example from the two exclamation marks (!!) in bold that the password is not set and we are using the defaults. Let's change the password and check the difference. Use any password you like:

```
[root@rhel-instance ~]# passwd usertest
Changing password for user usertest.
New password:
```

```
Retype new password:
passwd: all authentication tokens updated successfully.
[root@rhel-instance ~]# grep usertest /etc/shadow
usertest:$6$4PEVPj7M4GD8CH.4$VqiYY.
IXetwZA/g54bFP1ZJwQ/yc6bnaFauHGA1 1eFzsGh/
uFbJwxZCQTFHIASuamBz.27gb4ZpywwOA840eI.:19062:0:99999:7:::
```

The password hash is created and the date for the last change is just kept the same as the current date. Let's establish some options:

```
[root@rhel-instance ~]# chage --mindays 0 --warndays 7
--inactive 3 --expiredate 2030-01-01 usertest
[root@rhel-instance ~]# grep usertest /etc/shadow
usertest:$6$4PEVPj7M4GD8CH.4$VqiYY.IXetwZA/g54bFP1ZJwQ/
yc6bnaFauHGA1 1eFzsGh/uFbJwxZCQTFHIASuamBz.27gb4ZpywwOA
840eI.:19062:0:99999:7:3:21915:
[root@rhel-instance ~]# date -d '1970-01-01 UTC + 21915 days'
Tue Jan  1 01:00:00 AM CET 2030
```

Please notice the changes in the /etc/shadow file corresponding to the values specified for chage. We can check the changes with the -l option for chage:

```
[root@rhel-instance ~]# chage -l usertest
Last password change                    : mar 11, 2022
Password expires                        : never
Password inactive                       : never
Account expires                         : Jan 01, 2030
Minimum number of days between password change   : 0
Maximum number of days between password change   : 99999
Number of days of warning before password expires: 7
```

> **Important Note**
>
> In RHEL 9 some of the password controls are also configured using the /etc/security/ pwquality.conf file, for example, the minimal acceptable size for the new password (minlen parameter) or the maximum credit for having uppercase or lowercase characters. You may want to check this file in case you need to change these parameters.

To change the default values, we will edit `/etc/login.defs`. Let's check the section for the most common changes:

```
# Password aging controls:
#
#       PASS_MAX_DAYS       Maximum number of days a password may be
used.
#       PASS_MIN_DAYS       Minimum number of days allowed between
password changes.
#       PASS_MIN_LEN        Minimum acceptable password length.
#       PASS_WARN_AGE       Number of days warning given before a
password expires.
#
PASS_MAX_DAYS       99999
PASS_MIN_DAYS       0
PASS_WARN_AGE       7
```

Please take some minutes to review the options in `/etc/login.defs`.

Now, we could have a situation in which a user has left the company. How can we lock the account so the user cannot access the system? The `usermod` command has the `-L` option, for **lock**, to do so. Let's try it. First, let's log into the system:

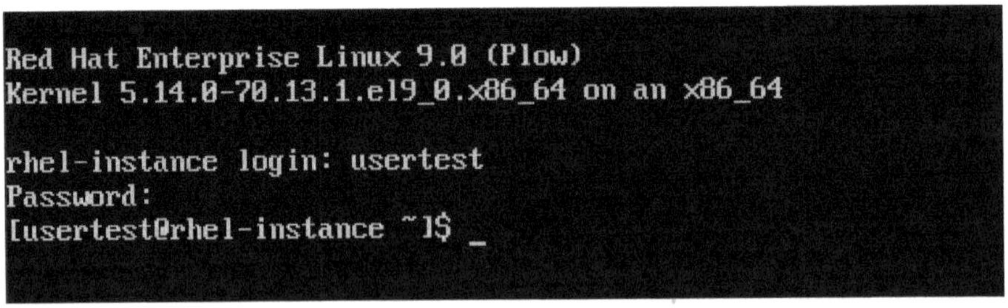

```
Red Hat Enterprise Linux 9.0 (Plow)
Kernel 5.14.0-70.13.1.el9_0.x86_64 on an x86_64

rhel-instance login: usertest
Password:
[usertest@rhel-instance ~]$ _
```

Figure 5.1 – The usertest user account logging into the system

Now, let's lock the account:

```
[root@rhel-instance ~]# usermod -L usertest
[root@rhel-instance ~]# grep usertest /etc/shadow
usertest:!$6$4PEVPj7M4GD8CH.4$VqiYY.
IXetwZA/g54bFP1ZJwQ/yc6bnaFauHGA 11eFzsGh/
uFbJwxZCQTFHIASuamBz.27gb4ZpywwOA840eI.:18651:0:99999:7:3
:21915:
```

Notice that there is a ! character added before the password hash. This is the mechanism used to lock it. Let's try to log in again:

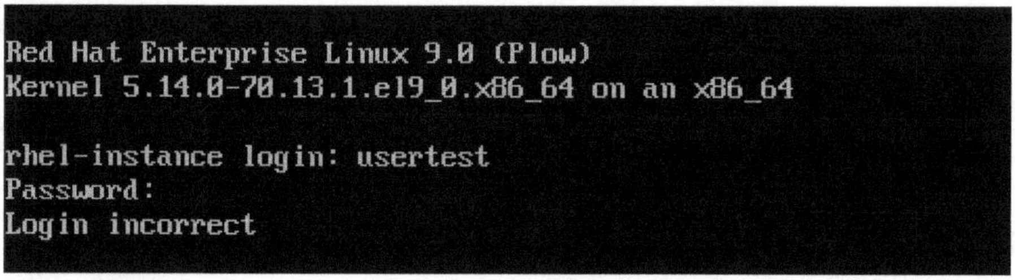

```
Red Hat Enterprise Linux 9.0 (Plow)
Kernel 5.14.0-70.13.1.el9_0.x86_64 on an x86_64

rhel-instance login: usertest
Password:
Login incorrect
```

Figure 5.2 – The usertest user account not being able to log into the system

The account can be unlocked by using the -U option:

```
[root@rhel-instance ~]# usermod -U usertest
[root@rhel-instance ~]# grep usertest /etc/shadow
usertest:$6$4PEVPj7M4GD8CH.4$VqiYY.
IXetwZA/g54bFP1ZJwQ/yc6bnaFauHGA1 1eFzsGh/
uFbJwxZCQTFHIASuamBz.27gb4ZpywwOA840eI.:19062:0:99999:7:3:21
915:
```

Now, you can see that the ! character is removed. Feel free to try logging in again.

> **Important Note**
> To fully lock the account from access, not just from logging in with a password (there are other mechanisms), we should set the expiry date to 1.

Another common use case is when you want users to access the system, such as having a network shared directory (that is, via NFS or CIFS, as explained in *Chapter 12, Managing Local Storage and Filesystems*), but you do not want them to be able to run commands in the system. For that, we can use a very special shell – the nologin shell. Let's assign that shell to the usertest user account using usermod:

```
[root@rhel-instance ~]# usermod -s /sbin/nologin usertest
[root@rhel-instance ~]# grep usertest /etc/passwd
usertest:x:1001:1001::/home/usertest:/sbin/nologin
[root@rhel-instance ~]# su - usertest
This account is currently not available.
[root@rhel-instance ~]# usermod -s /bin/bash usertest
```

```
[root@rhel-instance ~]# su - usertest
[usertest@rhel-instance ~]$
```

Note that we are reviewing the changes in /etc/passwd this time, as it is where the modification is applied.

As you can see, it's easy to set the values for password aging for any user, lock them, or restrict access to the system. Let's move on to more administrative tasks and how to delegate admin access.

Configuring sudo access for administrative tasks

There is a way to delegate administrative access to users in RHEL, and it is done with a tool called **sudo**, which stands for **Super User Do**.

It not only allows you to grant full administrative privileges to users or groups but also to be very granular about the privileged commands that some users may be able to execute.

Let's start by understanding the default configuration and how to change it.

Understanding sudo configuration

The tool has its main configuration file in /etc/sudoers and includes this part in the default configuration:

```
root ALL=(ALL)   ALL
%wheel     ALL=(ALL)   ALL
## Read drop-in files from /etc/sudoers.d (the # here does not
mean a comment)
#includedir /etc/sudoers.d
```

Let's analyze these four specific lines one by one to understand what they do:

1. The first line enables the root user to use sudo for any command that they want to run:

    ```
    root ALL=(ALL)   ALL
    ```

2. The second line enables the user in the wheel group to use sudo for any command that they want to run. We will explain the details of the syntax later:

    ```
    %wheel     ALL=(ALL)   ALL
    ```

> **Important Note**
>
> Please do not disable the `wheel` group directive unless there is an important reason to do so. This behavior is expected by other programs to be available and disabling it may cause some problems.

3. The third line, and all the lines starting with #, are considered comments, and they are only intended to add descriptive content with no effect on the final configuration. It is possible to include other `sudoers` files from within the `sudoers` file currently being parsed using the `@include` and `@includedir` directives:

   ```
   ## Read drop-in files from /etc/sudoers.d (the # here
   does not mean a comment)
   ```

4. The fourth line is the only exception to the previous rule. This line enables the `/etc/sudoers.d` directory as a source for configuration files. We can drop a file in that folder and it will be used by `sudo`. For compatibility with `sudo` versions prior to 1.9.1, `#include` and `#includedir` are also accepted:

   ```
   #includedir /etc/sudoers.d
   ```

   The exception to this last rule is files that end with ~ or contain a . (dot) character.

As you have seen, the default configuration enables `root` and the members of the `wheel` group to run any command as an administrator using `sudo`.

The easiest way to use it is to add a user to the `wheel` group to grant that user full admin privileges. An example of how to modify the `usertest` account to make it an admin account is as follows:

```
[root@rhel-instance ~]# usermod -aG wheel usertest
[root@rhel-instance ~]# groups usertest
usertest : usertest wheel
```

> **Important Note**
>
> For cloud instances, the account root does not have a valid password assigned. To be able to manage the mentioned cloud instance, in some clouds such as **Amazon Web Services (AWS)**, a user is created by default and added to the `wheel` group. In the case of AWS, the default user account is `ec2-user`. In other clouds, a custom user is also created and also added to the `wheel` group.

To edit the `/etc/sudoers` file, as happens with other sensitive files, there is a tool that helps not only ensure that two admins are not editing it simultaneously but also that the syntax is correct. In this case, the tool to edit it is `visudo`.

Using sudo to run admin commands

We will use the `user` account in these examples. As you may remember, in *Chapter 1, Getting RHEL Up and Running*, we enabled the checkbox in which we requested the account to be the administrator. Under the hood, the account was added to the `wheel` group, so we can start using `sudo` to run admin commands.

Let's log in with the `user` account and try to run an administrative command such as `adduser`:

```
[root@rhel-instance ~]# su - user
[user@rhel-instance ~]$ adduser john
adduser: Permission denied.
adduser: cannot lock /etc/passwd; try again later.
```

As you can see, we receive a `Permission denied` error message. To be able to run it with `sudo`, we only need to add it to the beginning of the command line:

```
[user@rhel-instance ~]$ sudo adduser john
We trust you have received the usual lecture from the local
System
Administrator. It usually boils down to these three things:
    #1) Respect the privacy of others.
    #2) Think before you type.
    #3) With great power comes great responsibility.
[sudo] password for user:
[user@rhel-instance ~]$ id john
uid=1002(john) gid=1002(john) groups=1002(john)
```

In this case, we see that we have been shown a warning message that is shown the first time we run `sudo` successfully. Then, we are asked for *our own password* – not the admin password, as there may not even be one, but the one we have for the user running `sudo`. Once the password is typed correctly, the command is run and registered in the system journal:

```
mar 11 19:44:26 rhel-instance.example.com useradd[1710]: user :
TTY=pts/0 ; PWD=/home/user ; USER=root ; COMMAND=/sbin/adduser
john
```

> **Important Note**
>
> Once you have run `sudo` successfully, it will remember that validation for 15 minutes (as the default behavior). This is done so you don't have to type your password again and again if you need to run more than one administrative command in a session. To increase it to 30 minutes, we can add the following line using `visudo`:
>
> `Defaults:USER timestamp_timeout=30.`

Sometimes, you want to have an interactive session so that there is no need to type `sudo` again and again. For that, the `-i` option is really useful. Let's try it:

```
[user@rhel-instance ~]$ sudo -i
[sudo] password for user:
[root@rhel-instance ~]#
```

Let's now move on to customizing the configuration of `sudo` in the `sudoers` file.

Configuring sudoers

We have seen the details of the default `/etc/sudoers` file in the previous section. Let's see a couple of examples of how to make a more granular configuration.

Let's start by making `sudo` run admin commands without requesting a password for the users in the `wheel` group. We can run `visudo` and make the line that starts with `%wheel` look as follows:

```
%wheel          ALL=(ALL)          NOPASSWD: ALL
```

Save it. Note that there is a commented line in the configuration file with that configuration. Now, let's try it:

```
[user@rhel-instance ~]$ sudo adduser ellen
[user@rhel-instance ~]$ id ellen
uid=1003(ellen) gid=1003(ellen) groups=1003(ellen)
```

We can now create a file with your favorite editor to make the new user account, `ellen`, able to run admin commands. Let's create the `/etc/sudoers.d/ellen` file with this content:

```
ellen ALL=(ALL)    ALL
```

With this, we are using the `/etc/sudoers.d` directory to extend the `sudo` configuration.

We will review the detailed configuration of `sudoers` here despite it not being part of the RHCSA exam. As you can see, there are three fields separated by spaces or tabs to define policies in the configuration files. Let's review them:

- The first field is to specify who is affected by the policy:

    - We can add users by simply putting the username in the first field.

    - We can add groups by using the `%` character before the name of the group in the first field.

- The second field is for where the policy applies:

  - We have so far used `ALL=(ALL)` to specify everything.

  - In the first part of this field, we can define a group of computers to be run, such as `SERVERS=10.0.0.0/255.255.255.0`.

  - In the second part, we can specify commands such as `NETWORK=/usr/sbin/ip`.

  - Between parentheses is the user account that can be used to run the command.

- The third field is to specify which commands will use the password and which won't.

  The syntax goes as follows:

  ```
  user  hosts = (run-as) commands
  ```

  Let's see an example:

  ```
  Runas_AliasDB = oracle
  Host_Alias SERVERS=10.0.0.0/255.255.255.0
  Cmnd_Alias NETWORK=/usr/sbin/ip
  pete  SERVERS=NETWORK
  julia SERVERS=(DB)ALL
  ```

We have already seen how to provide administrative access to users in RHEL and even how to do it in a very granular manner. Let's move on now to the section on working with file permissions.

Checking, reviewing, and modifying file permissions

So far, we have learned how to create users and groups and even provide administrative capabilities to them. It's now time to see how permissions work at the file and directory level.

As you'll remember, in *Chapter 3, Basic Commands and Simple Shell Scripts*, we already saw how to see the permissions that are applied to a file. Let's review them now and dive deeper.

Let's get the permissions info for some example files by listing it with -1, for the long option. Remember to run this as the `root` user (or using `sudo`):

```
[root@rhel-instance ~]# ls -l /usr/bin/bash
-rwxr-xr-x. 1 root root 1390064 Aug 9  2021 /usr/bin/bash
[root@rhel-instance ~]# ls -l /etc/passwd
-rw-r--r--. 1 root root 1740 Mar 11 21:35 /etc/passwd
[root@rhel-instance ~]# ls -l /etc/shadow
----------. 1 root root 1170 Mar 11 21:35 /etc/shadow
[root@rhel-instance ~]# ls -ld /tmp
drwxrwxrwt. 5 root root 4096 Mar 11 17:35 /tmp
```

Remember that, in Linux, *everything is a file*.

Now, let's review the five different blocks of information that the permissions include by using the ones for /usr/bin/bash:

```
-rwxr-xr-x.
```

The blocks are as follows:

Block 1	Block 2	Block 3	Block 4	Block 5
-	rwx	r-x	r-x	.

Table 5.1 – File permissions by block

Let's review them again, as they are very important:

- **Block 1** is for the special permissions that the file may have. If it is a regular file and has no special permissions (as in this case), it will appear as -:

  - Directories will appear with d.

  - Links, usually symbolic links, will appear with an l.

  - Special permissions to run a file as a different user or group, called **setuid** or **setgid**, will appear as s.

  - Special permissions for directories so that the owner can only remove or rename the file, called a **sticky bit**, will appear as t.

- **Block 2** is the permissions for the *user* owning the file, and consists of three characters:

  - The first one, r, is the read permission assigned.

  - The second one, w, is the write permission assigned.

  - The third one, x, is the executable permission (note that the executable permission for directories means being able to enter them.)

- **Block 3** is permissions for the *group*. It consists of the same three characters for read, write, and execute (rwx). In this case, write is missing.

- **Block 4** is the permissions for *others*. It also consists of the same three characters for read, write, and execute (rwx) as before. As in the previous block, write is missing.

- **Block 5** indicates that there is an **SELinux** context applied to the file. More on this topic in *Chapter 10, Keeping Your System Hardened with SELinux.*

To change permissions for a file, we will use the chmod command.

First, let's create a file:

```
[root@rhel-instance ~]# touch file.txt
[root@rhel-instance ~]# ls -l file.txt
-rw-r--r--. 1 root root 0 Mar 11 22:30 file.txt
```

As you can see, the file is created with your username as the owner, your main group as the group, and a default set of permissions. The default set of permissions is defined by umask, and in RHEL, the defaults for newly created file permissions are as follows:

- **User**: Read and write
- **Group**: Read
- **Others**: Read

To change permissions using chmod, we specify the changes with three characters:

- The first one, which determines whom the change affects:
  - u: User
  - g: Group
  - o: Others
- The second one to add or remove permissions:
  - +: Add
  - -: Remove
- The third one, which determines the permission to be changed:
  - r: Read
  - w: Write
  - x: Execute

So, to add write permissions to the group, we can run the following:

```
[root@rhel-instance ~]# chmod g+w file.txt
[root@rhel-instance ~]# ls -l file.txt
-rw-rw-r--. 1 root root 0 Mar 11 22:30 file.txt
```

And to remove read permissions from others, we run the following:

```
[root@rhel-instance ~]# chmod o-r file.txt
[root@rhel-instance ~]# ls -l file.txt
-rw-rw----. 1 root root 0 Mar 11 22:30 file.txt
```

The permissions are stored in four octal digits. This means that special permissions are stored in a number from 0 to 7, the same way user, group, and other permissions are stored, each one of them with a number from 0 to 7.

Some examples are shown as follows:

Alphabetic	Octal	Description
-rwxr-xr--	0754	User: read, write, execute. Group: read, execute. Others: read.
-rw-r-----	0640	User: read, write. Group: read. Others: nothing.
-r--------	0400	User: read Group: nothing Others: nothing.

Table 5.2 – Example file permissions

How does it work? We assign a number (power of 2) for each permission:

- **Read:** $2^2 = 4$
- **Write:** $2^1 = 2$
- **Execute:** $2^0 = 1$
- **Nothing:** 0

We add them:

```
rwx = 4 + 2 + 1 = 7
rw- = 4 + 2 = 6
r-x = 4 + 1 = 5
r-- = 4
--- = 0
```

This is how we can assign permissions using numbers. Now, let's try it:

```
[root@rhel-instance ~]# chmod 0755 file.txt
[root@rhel-instance ~]# ls -l file.txt
-rwxr-xr-x. 1 root root 0 Mar 11 22:30 file.txt
[root@rhel-instance ~]# chmod 0640 file.txt
```

```
[root@rhel-instance ~]# ls -l file.txt
-rw-r-----. 1 root root 0 Mar 11 22:30 file.txt
[root@rhel-instance ~]# chmod 0600 file.txt
[root@rhel-instance ~]# ls -l file.txt
-rw-------. 1 root root 0 Mar 11 22:30 file.txt
```

As we said before, the default configuration of permissions is set by umask. We can very easily see the value:

```
[root@rhel-instance ~]# umask
0022
[root@rhel-instance ~]# umask -S
u=rwx,g=rx,o=rx
```

All the newly created files have execute permissions removed (1).

With this umask, 0022, the one provided by default in RHEL, we will have write permissions for group and others also removed (2).

Even when it is not recommended to change umask, we could give it a try to learn how it works. Let's start by using the most permissive umask, 0000, to see how all read and write permissions are assigned to newly created files:

```
[root@rhel-instance ~]# umask 0000
[root@rhel-instance ~]# touch file2.txt
[root@rhel-instance ~]# ls -l file2.txt
-rw-rw-rw-. 1 root root 0 Mar 11 22:33 file2.txt
```

Now, let's use the more restrictive umask for group and others permissions:

```
[root@rhel-instance ~]# umask 0066
[root@rhel-instance ~]# touch file3.txt
[root@rhel-instance ~]# ls -l file3.txt
-rw-------. 1 root root 0 Mar 11 22:33 file3.txt
```

If we try a higher number, it won't work and will return an error:

```
[root@rhel-instance ~]# umask 0088
-bash: umask: 0088: octal number out of range
```

You can see that the effect of 0066 and 0077 is the same:

```
[root@rhel-instance ~]# umask 0077
[root@rhel-instance ~]# touch file4.txt
[root@rhel-instance ~]# ls -l file4.txt
-rw-------. 1 root root 0 Mar 11 22:35 file4.txt
```

Let's re-establish umask in our session to the defaults to continue practicing:

```
[root@rhel-instance ~]# umask 0022
```

Now, we may find ourselves with the need to create a directory for a specific user or group or to change the owner of a file. To be able to change the ownership of a file or directory, the chown or chgrp tools are used. Let's see how it works. Let's move to /var/tmp and create the folders for finance and accounting:

```
[root@rhel-instance ~]# cd /var/tmp/
[root@rhel-instance tmp]# mkdir finance
[root@rhel-instance tmp]# mkdir accounting
[root@rhel-instance tmp]# ls -l
total 0
drwxr-xr-x. 2 root root 6 Mar 11 22:35 accounting
drwxr-xr-x. 2 root root 6 Mar 11 22:35 finance
```

Now, let's create the groups for finance and accounting:

```
[root@rhel-instance tmp]# groupadd finance
[root@rhel-instance tmp]# groupadd accounting
groupadd: group 'accounting' already exists
```

In this example, the accounting group was already created. Let's change the group for each directory with chgrp:

```
[root@rhel-instance tmp]# chgrp accounting accounting/
[root@rhel-instance tmp]# chgrp finance finance/
[root@rhel-instance tmp]# ls -l
total 0
drwxr-xr-x. 2 root accounting 6 Mar 11 22:35 accounting
drwxr-xr-x. 2 root finance    6 Mar 11 22:35 finance
```

Now, we create users for `sonia` and `matilde`, and assign them to `finance` and `accounting` respectively:

```
[root@rhel-instance tmp]# adduser sonia
[root@rhel-instance tmp]# adduser matilde
[root@rhel-instance tmp]# usermod -aG finance sonia
[root@rhel-instance tmp]# usermod -aG accounting matilde
[root@rhel-instance tmp]# groups sonia
sonia : sonia finance
[root@rhel-instance tmp]# groups matilde
matilde : matilde accounting
```

Now, we can create a personal folder for each under their group folder:

```
[root@rhel-instance tmp]# cd finance/
[root@rhel-instance finance]# mkdir personal_sonia
[root@rhel-instance finance]# chown sonia personal_sonia
[root@rhel-instance finance]# ls -l
total 0
drwxr-xr-x. 2 sonia root 6 Mar 11 22:44 personal_sonia
[root@rhel-instance finance]# chgrp sonia personal_sonia/
[root@rhel-instance finance]# ls -l
total 0
drwxr-xr-x. 2 sonia sonia 6 Mar 11 22:44 personal_sonia
```

There is a way to specify a user and group to `chown` – using the : separator. Let's use it with `matilde`:

```
[root@rhel-instance tmp]# cd ../accounting
[root@rhel-instance accounting]# mkdir personal_matilde
[root@rhel-instance accounting]# chown matilde:matilde \
personal_matilde
[root@rhel-instance accounting]# ls -l
total 0
drwxr-xr-x. 2 matilde matilde 6 Mar 11 22:46 personal_matilde
```

If we want to change the permissions for a full branch, we can use `chown` with `-R`, the recursive option. Let's copy a branch and change its permissions:

```
[root@rhel-instance accounting]# cp -rv /usr/share/doc/audit
personal_matilde/
```

```
'/usr/share/doc/audit' -> 'personal_matilde/audit'
'/usr/share/doc/audit/ChangeLog' -> 'personal_matilde/audit/
ChangeLog'
'/usr/share/doc/audit/README' -> 'personal_matilde/audit/
README'
'/usr/share/doc/audit/auditd.cron' -> 'personal_matilde/audit/
auditd.cron'
[root@rhel-instance accounting]# chown -R matilde:matilde \
personal_matilde/audit
[root@rhel-instance accounting]# ls -l personal_matilde/audit/
total 20
-rw-r--r--. 1 matilde matilde  271 Mar 11 22:56 auditd.cron
-rw-r--r--. 1 matilde matilde 8006 Mar 11 22:56 ChangeLog
-rw-r--r--. 1 matilde matilde 4953 Mar 11 22:56 README
```

With this, we have a good understanding of permissions in RHEL, their default behaviors, and how to work with them.

Let's move on to some more advanced topics about permissions.

Using special permissions

As we've seen in the previous section, there are special permissions that could be applied to files and directories. Let's start by reviewing **Set-UID** (or **SUID**) and **Set-GUID** (or **SGID**).

Understanding and applying SUID

Let's review how SUID applies to files and directories:

- **SUID permission applied to a file**: When applied to an executable file, this file will run as if the owner of the file was running it, applying the permissions.

- **SUID permission applied to a directory**: No effect.

Let's check a file with SUID:

```
[root@rhel-instance ~]# ls -l /usr/bin/passwd
-rwsr-xr-x. 1 root root 32648 Aug 10  2021 /usr/bin/passwd
```

> Tip
> In this example, s in the executable bit of the user block is lowercase because the executable bit is set – if the executable bit isn't set, it will be in uppercase (S).

The `passwd` command requires `root` permissions to change hashes in the `/etc/shadow` file.

To apply this permission, we can use the `chmod` command, applying `u+s` permissions:

```
[root@rhel-instance ~]# touch testsuid
[root@rhel-instance ~]# ls -l testsuid
-rw-r--r--. 1 root root 0 Mar 11 23:16 testsuid
[root@rhel-instance ~]# chmod u+s testsuid
[root@rhel-instance ~]# ls -l testsuid
-rwsr--r--. 1 root root 0 Mar 11 23:16 testsuid
```

> **Tip**
> Be very careful when assigning SUID to files as `root`. If you leave write permissions on the file, any user will be able to change the content and execute anything as `root`.

Understanding and applying SGID

Let's review how SGID applies to files and directories:

- **SGID permission applied to a file**: When applied to an executable file, this file will run with the group permissions of the file.

- **SGID permission applied to a directory**: New files created in that directory will have the group of the directory applied to them.

Let's check a file with SGID:

```
[root@rhel-instance ~]# ls -l /usr/bin/write
-rwxr-sr-x. 1 root tty 24456 Aug 20  2021 /usr/bin/write
```

We can try applying the permission to a file with `chmod` using `g+s`:

```
[root@rhel-instance ~]# touch testgid
[root@rhel-instance ~]# chmod g+s testgid
[root@rhel-instance ~]# ls -l testgid
-rw-r-sr--. 1 root root 0 Mar 11 23:23 testgid
```

Now, let's try it with a directory. Let's go to our previous example:

```
[root@rhel-instance ~]# cd /var/tmp/
[root@rhel-instance tmp]# ls
accounting   finance
```

```
[root@rhel-instance tmp]# chmod g+s accounting finance
[root@rhel-instance tmp]# ls -l
total 0
drwxr-sr-x. 3 root accounting 30 Mar 11 23:46 accounting
drwxr-sr-x. 3 root finance    28 Mar 11 23:44 finance
[root@rhel-instance tmp]# touch finance/testfinance
[root@rhel-instance tmp]# ls -l finance/testfinance
-rw-r--r--. 1 root finance 0 Mar 11 23:47 finance/testfinance
[root@rhel-instance tmp]# touch accounting/testaccounting
[root@rhel-instance tmp]# ls -l accounting/testaccounting
-rw-r--r--. 1 root accounting 0 Mar 11 23:47 accounting/
testaccounting
```

You can see how, after applying SGID to the folders, they show the s permission for the group (in bold). Also, when creating new files in those directories, the group assigned to them is the same as the group that the parent directory has (also in bold). This way, we ensure group permissions are properly assigned.

Using the sticky bit

The last of the permissions to be used is the **sticky bit**. It only affects directories and what it does is simple: when a user creates a file in a directory with the sticky bit, only that user can edit or delete that file.

Let's check an example:

```
[root@rhel-instance ~]# ls -ld /tmp
drwxrwxrwt. 8 root root 4096 Mar 11 23:31 /tmp
```

We could apply those to the previous example, also with chmod using o+t:

```
[root@rhel-instance ~]# cd /var/tmp/
[root@rhel-instance tmp]# ls -l
total 0
drwxr-sr-x. 3 root accounting 52 Mar 11 23:47 accounting
drwxr-sr-x. 3 root finance    47 Mar 11 23:47 finance
[root@rhel-instance tmp]# chmod o+t accounting finance
[root@rhel-instance tmp]# ls -l
total 0
drwxr-sr-t. 3 root accounting 52 Mar 11 23:47 accounting
drwxr-sr-t. 3 root finance    47 Mar 11 23:47 finance
```

Let's give it a try. We will add the `sonia` user to the `accounting` group. We will grant a write permission to the group for the `/var/tmp/accounting` directory. Then, we will create a file with the `matilde` user and try to delete it with the `sonia` user. Let's go:

```
[root@rhel-instance ~] # usermod -aG accounting sonia
[root@rhel-instance ~]# cd /var/tmp/
[root@rhel-instance tmp]# chmod g+w accounting
[root@rhel-instance tmp]# ls -l
total 0
drwxrwsr-t. 3 root accounting 52 Mar 11 23:47 accounting
drwxr-sr-t. 3 root finance    47 Mar 11 23:47 finance
[root@rhel-instance tmp]# su - matilde
[matilde@rhel-instance ~]$ cd /var/tmp/accounting/
[matilde@rhel-instance accounting]$ touch teststickybit
[matilde@rhel-instance accounting]$ exit
logout
[root@rhel-instance tmp]# su - sonia
[sonia@rhel-instance ~]$ cd /var/tmp/accounting/
[sonia@rhel-instance accounting]$ ls -l teststickybit
-rw-rw-r--. 1 matilde accounting 0 Mar 11 23:50 teststickybit
[sonia@rhel-instance accounting]$ rm -f teststickybit
rm: cannot remove 'teststickybit': Operation not permitted
```

> **Tip**
> The numeric values for special permissions are SUID = 4, SGID = 2, and sticky bit = 1.

With this, we have completed how to manage permissions in RHEL.

Summary

In this chapter, we have reviewed the permission management system in RHEL, implemented using traditional permissions. We have learned how to create user accounts and groups and how to ensure that passwords are managed correctly. We have also learned how passwords are stored in the system and even how to block shell access to a user. We have created files and folders, assigning permissions to them and ensuring that users can collaborate with an enforced set of rules.

These are the basics of managing access in RHEL and will be very useful to avoid security issues when managing systems. As this is such an important topic, we recommend reviewing this chapter carefully, reading the man pages for the commands shown, and making an effort to get a really good understanding of the topic, as it will avoid any uncomfortable situations in the future.

Now, you are ready to start providing services to users and managing their access, which is what we will cover in the next chapter. Remember to thoroughly practice and test the lessons learned here.

6

Enabling Network Connectivity

When we installed our system in the first chapter, we enabled the network interface. However, network configuration is (or can be) even more than that.

A server connected to a network might require additional interfaces for configuring other networks reaching backup servers, performing internal services from other servers, or even accessing storage that is not presented directly via a **Storage Array Network (SAN)** as local drives but as, for example, **Internet Small Computer System Interface (iSCSI)** drives.

Additionally, a server might use redundant network capabilities to ensure that, in the event of a failure in one of the cards or switches, the server can still be reached and perform properly.

In this chapter, we will learn about how to define network configuration for our RHEL 9 machine using different methods and perform some basic network troubleshooting.

This knowledge will be key since servers are commonly used to provide services to other systems and we need networking for that purpose.

In this chapter, we will cover the following topics:

- Exploring network configuration in RHEL
- Getting to know the configuration files and NetworkManager
- Configuring network interfaces with IPv4 and IPv6
- Configuring hostname and hostname resolutions (DNS)
- An overview of firewall configuration
- Testing network connectivity

Let's get hands-on with networking!

Technical requirements

You can continue using the virtual machine we created at the beginning of this book in *Chapter 1, Getting RHEL Up and Running*. Additionally, to test network communication, it might be useful to create a second virtual machine or reuse the one we created in the previous chapters for testing the **Network Time Protocol** (**NTP**) configuration, as we will use this to check for connectivity. Any additional packages that are required will be indicated in the text. Any additional files that are required for this chapter can be downloaded from `https://github.com/PacktPublishing/Red-Hat-Enterprise-Linux-RHEL-9-Administration`.

Exploring network configuration in RHEL

A network is made of different devices that have been interconnected so that information and resources can be shared among them – for example, internet access, printers, and files.

Networks have been present since the beginning of computing. Initially, the most common were non-IP-based ones, which were generally used for sharing data across computers in the local network, but with the expansion of internet services and the requirements for applications or remote services, IP networks were expanded and the concept of the intranet was introduced, where **Transmission Control Protocol** or **Internet Protocol** (**TCP** or **IP**) was used as transport, and the applications started to be more like internet services (or were even based on them).

> **Important Note: Some Changes in RHEL 9 Compared to RHEL 8**
>
> With each new release, the software life cycle may drive some changes, adding new software, and removing or changing the **Technology Preview** classification for them. In RHEL 9, there are some relevant changes in regards to networking: **Wireguard** is added as an unsupported Technology Preview; `teamd` and `libteam` are now deprecated and a `bond` should be used instead; `iptables-nft` and `ipset` are also deprecated (which provided the `iptables`, `ip6tables`, `ebtables` and `arptables` commands) and `nftables` should be used instead. The new **MultiPath TCP daemon** (**mptcpd**) can be used to configure MPTCP endpoints without using `iproute2`. NetworkManager is now used for configuring networking (`network-scripts` have been deprecated) and defaults to using key files to store the connections, but the `ifcfg` files can still be used.

The migration to IP-based networks allowed for other means of transport of topologies such as **Network Basic Input/Output System** (**NetBIOS**) so that they could run on top of it (it was working on top of **NetBIOS Extended User Interface** (**NetBEUI**) before, and even if other networks such as **InfiniBand** or **Remote Direct Memory Access** (**RDMA**) are still in use, they are not as common as TCP or IP).

TCP/ IP, of course, are built on top of other protocols. You can check the OSI layer definition at `https://www.redhat.com/sysadmin/osi-model-bean-dip`. However, some concepts are still involved. We will cover these when we become familiar with TCP/IP and the networks.

Before we get into the actual details, we need to clarify a few common TCP/IP and networking keywords we'll be using from now on:

- **IP address**: This is the address that's used for interacting with other devices on the network.

- **Netmask**: This is used to determine which devices are in the neighborhood. It can be expressed via a mask or a network size, such as 255.255.255.0 or /24.

- **Gateway**: This is the IP address of the device that will get all our traffic when the target device is outside our netmask so that we cannot reach it directly.

- **DNS**: This is the IP address of a server or servers that translates **domain names** into IP addresses so that the hosts can connect to them.

- **MAC address**: This is the physical interface address. It is unique for each card and helps identify the card in the network so that the proper traffic is sent to it.

- **Network Interface Card** (**NIC**): This card allows our device to connect to the network. It can be wireless or wired.

- **Extended Service Set Identification** (**ESSID**): This is how a wireless network is named.

- **Virtual Private Network** (**VPN**): This is a virtual network that is created between the client and the server. Once established, it allows you to direct connection to the services as if they were local, even if the client and the server are in different places. For example, a VPN is used to allow remote workers to connect to their corporate network using their private internet connection.

- **Virtual Local Area Network** (**VLAN**): This allows us to define virtual networks on top of the actual wiring. We can then use a specific header field so that the network equipment can correctly understand and process them.

- **IPv6**: This is the replacement protocol for **IPv4**, which is still the predominant protocol in networks today. IPv4 uses 32-bit numbers for addresses while IPv6 has 128 bits (2^{32} vs 2^{128}), making IPv6 the answer to a growing number of devices – whether mobile, cars, or the **Internet of Things** (**IoT**).

In the following sections, we will use some of these terms when we explain how a network is set up and defined within **Red Hat Enterprise Linux** (**RHEL**) systems.

In general, when systems are connected, some relationships between the devices on the network are established. Sometimes, some hosts are providers of services, often called servers, and the consumers are known as clients. When the systems in the network perform roles, these networks are known as **Peer-to-Peer** (**p2p**) networks.

In the following section, we'll become familiar with the configuration files and the different approaches for configuring networking in our system.

Getting to know the configuration files and NetworkManager

Now that we have learned about some of the keywords and concepts of networking, it's time to look at where we can use them to network our system.

> **Important Note**
> Although the `network-scripts` package used to be available, in RHEL 9, it has been removed – use NetworkManager instead.

NetworkManager is a utility that was created in 2004 to make network configuration and its usage easier for desktop users. At that point, all configuration was done via text files and it was more or less static. Once a system was connected to a network, the information barely changed at all. With the adoption of wireless networks, more flexibility was required to automate and ease the connection to different networks via different profiles or VPNs.

NetworkManager was created to cover these gaps and aimed to be a component that could be used in many distributions but from a new standpoint – for example, it queries the **Hardware Abstraction Layer** (**HAL**) at startup to learn about the available network devices and any changes to them.

Imagine a laptop system. It can be connected to a wired cable, disconnected when you're moving it to another location or cubicle, and can connect to a wireless network. All those events are relayed to NetworkManager, which takes care of reconfiguring network interfaces and routes, authenticating with the wireless network, and making the user's life a lot easier than it traditionally used to be.

> **Tip**
> The hardware that is connected to the system can be queried with several commands, depending on how the hardware is connected – for example, via utilities such as `lsusb`, `lspci`, or `lshw` (provided by installing the `usbutils`, `pciutils`, and `lshw` packages, respectively).

In the following screenshot, we can see the available packages related to NetworkManager, as obtained via the `dnf search network manager` command:

Figure 6.1 – The NetworkManager-related packages available for installation in a RHEL 9 system

NetworkManager is configured with the files located in the /etc/NetworkManager folder, especially NetworkManager.conf and the files available in that folder:

- conf.d
- dispatcher.d
- dnsmasq-shared.d
- dnsmasq.d
- system-connections

Can't remember what a dispatcher is? Remember to use man networkmanager to get the details on this!

The man page of NetworkManager explains that those scripts are executed in alphabetical order based on network events and will receive two arguments – the name of the device for the event and the action.

There are several actions you can perform, as follows:

- pre-up: The interface gets connected to a network but is not activated yet. The script must be executed before the connection can be notified as activated.

- up: The interface has been activated.

- pre-down: The interface is being deactivated but hasn't been disconnected from the network yet. In the case of forced disconnections (as in, a lost wireless connection or lost carrier), this will not be executed.

- down: The interface has been deactivated.

- vpn-up, vpn-down, vpn-pre-up, or vpn-pre-down: Similar to the preceding interfaces but for VPN connections.

- hostname: The hostname has been changed.

- dhcp4-change or dhcp6-change: The DHCP lease has changed (has been renewed or rebounded, for example).

- connectivity-change: Connectivity transitions such as no connection or the system going online, for example.

Now that we have learned a bit about NetworkManager and how it works and was designed, let's learn how to configure network interfaces.

Configuring network interfaces with IPv4 and IPv6

There are several approaches to configuring network interfaces and several network configurations. These will help us determine what we need to do and the required parameters and settings.

Let's look at some examples:

- A server might have two or more **Network Interface Cards** (**NICs**) for redundancy, but only one of them is active at a time.

- A server might use a trunk network and require that we define VLANs on top for accessing or providing the different services in the networks.

- Two or more NICS can be combined to provide increased output and redundancy via teaming.

Configuration can be performed in several ways, too:

- `nmtui`: A text-based interface for configuring a network
- `nmcli`: The command-line interface for NetworkManager
- `nm-connection-editor`: The graphical tool available for graphical environments
- Using text configuration files

> **Important Note**
>
> Before editing your network configuration, ensure that you can reach the system that is being configured in another way. In the case of a server, this can be done via a remote management card or physical console access. A mistake in the configuration might leave the system unreachable.

Before we move on, let's learn a bit more about IPv4 and IPv6.

IPv4 and IPv6 – what do they mean?

IPv4 was created in 1983 and uses a 32-bit address space, which provides 2^{32} unique addresses (4,294,967,296), but from those possible ones, large blocks are reserved for special usage. IPv6, ratified as Internet Standard in 2017, is the latest version at the time of writing and uses a 128-bit address space instead – that is, 2^{128} (3.4 x 10^{38} addresses).

Long story short, the number of IPv4 addresses seemed huge at the time, but today, where phones, tablets, computers, laptops, servers, lightbulbs, smart plugs, and all other IoT devices require an IP address, the number of public IP addresses, available has been depleted meaning that it's not possible to assign more. This has caused some **Internet Service Providers** (**ISPs**) to use techniques such as **Carrier-Grade Network Address Translation** (**CGNAT**), which is similar to what private networks do. It causes all the traffic from several devices to appear to come from only one IP and has the device interacting on both networks (a router) to do the proper routing from outgoing and incoming packages to the original requestors.

Why no IPv6, then? The main problem is that IPv4 and IPv6 are not interoperable, and even though IPv6 was a draft in 1998, not all network equipment is compatible with it and might not have been tested yet. Check out `https://www.ripe.net/support/training/videos/ipv6/transition-mechanisms` for more details.

In the following section, we will learn about how to configure network interfaces using a text-based user interface to NetworkManager named `nmtui`.

Configuring interfaces with nmtui

`nmtui` provides a text-based interface for configuration. This is the initial screen you'll see when it is executed by running `nmtui` on a Terminal:

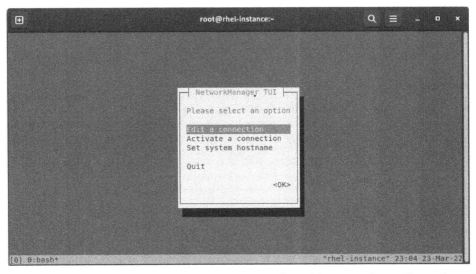

Figure 6.2 – The nmtui welcome screen showing a menu of possible actions that can be performed

Let's explore the available options for our interface. In this case, let's select **Edit a connection**. On the screen that appears, move down and edit the **Wired Connection** option that we have in our system to get to the following screen:

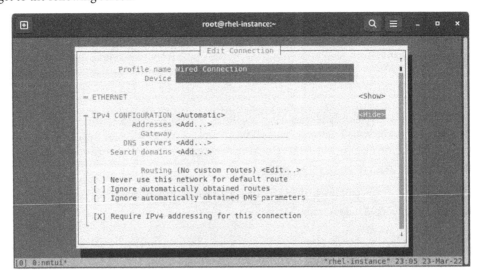

Figure 6.3 – The Edit Connection page with the IPv4 options expanded

It will be hard to show screenshots for each step, as one of the advantages of the text interface is that we can condense a lot of options into a simple screen. However, the preceding screenshot makes it easy to understand each of the required parameters:

- IP address

- Netmask

- Gateway

- Search domain

- Routing

As you can see, there are also checkboxes for ignoring routes or DNS parameters that are obtained when the connection is set to `Automatic`. Additionally, there are other options for interfaces: `Disabled`, `Link-Local`, `Manual`, and `Shared`.

Let's discuss the `Automatic` option, which means that the interface will be set to autoconfiguration. This is one of the most common settings for configuration. It does not mean that everything is done magically, though. Let's dig into this a bit more.

In a network (corporate, private, or otherwise), it is typical to have a special service or server for **Dynamic Host Routing Protocol (DHCP)**. DHCP is a protocol that runs on top of TCP/IP that allows you to configure hosts dynamically, using the configuration that was made previously either by the network administrator or some appliance and its default settings.

DHCP allows you to autoconfigure (from the client side) many aspects of network configuration, such as the IP, netmask, gateway, DNS, search domain, and time server. The configuration that's received is given a lease that is valid for some time. After that, the system attempts to renew it or if the system is powered off or disconnected, the lease is released.

Usually, the DHCP configuration is considered to be tied to dynamic IPs, but keep in mind that a DHCP server can use two different approaches: a pool of IPs that can be reused by different connected systems or fixed mappings of MAC addresses to static IPs.

Let's, for example, think about a **Small Office/Home Office (SOHO)** network with a private IP range in the `192.168.1.0/24` subnet.

We can define our ISP router to be at an IP of `192.168.1.1` because of the subnet (`/24`), which means that the last part of the IPv4 address can range from 0 to 255.

Using that IP range, we can set up hosts to get dynamic configuration and a dynamic IP from a pool in the last 100 IPs and leave the ones at the beginning for fixed equipment (even if they get the configuration dynamically), such as printers or storage devices.

As we mentioned previously, we can create reservations for servers, but in general, for devices that are always going to have the same address, it is also common practice to configure static addressing. In this way, if the DHCP server becomes unavailable, the servers will be still reachable from other services with either a valid lease or other servers and devices with the static addresses configured.

> **Tip**
>
> Just to get familiar with this concept, IP addresses are represented in IPv4 with a dotted notation separating four groups of numbers, such as `192.168.2.12`, while in IPv6, numbers are separated with `:` – for example, `2001:db8:0:1::c000:207`.

Configuring interfaces with nm-connection-editor

If our system has the graphical environment installed, which is not the case for our test system, we can use the graphical configuration tool instead. If it is not installed, proceed to execute `dnf install nm-connection-editor` in a shell console inside your graphical session.

> **Tip**
>
> To install the graphical interface, you can run the `dnf groupinstall "Server with GUI" -y` command or select it during installation.

In the following screenshot, we can see the window that was opened by executing `nm-connection-editor`. It's similar to the text interface shown by `nmtui` earlier in this chapter:

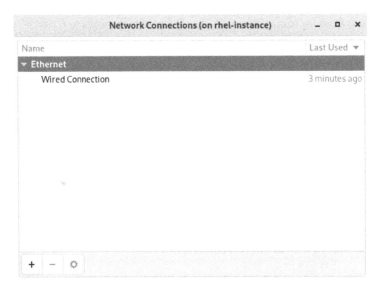

Figure 6.4 – The initial screen for nm-connection-editor

Here, we can see the +, -, and *gear* buttons, which are used to add, remove, or configure the connection that's highlighted, respectively.

Let's click on our **Wired Connection** option and then click on the gear icon to open the details:

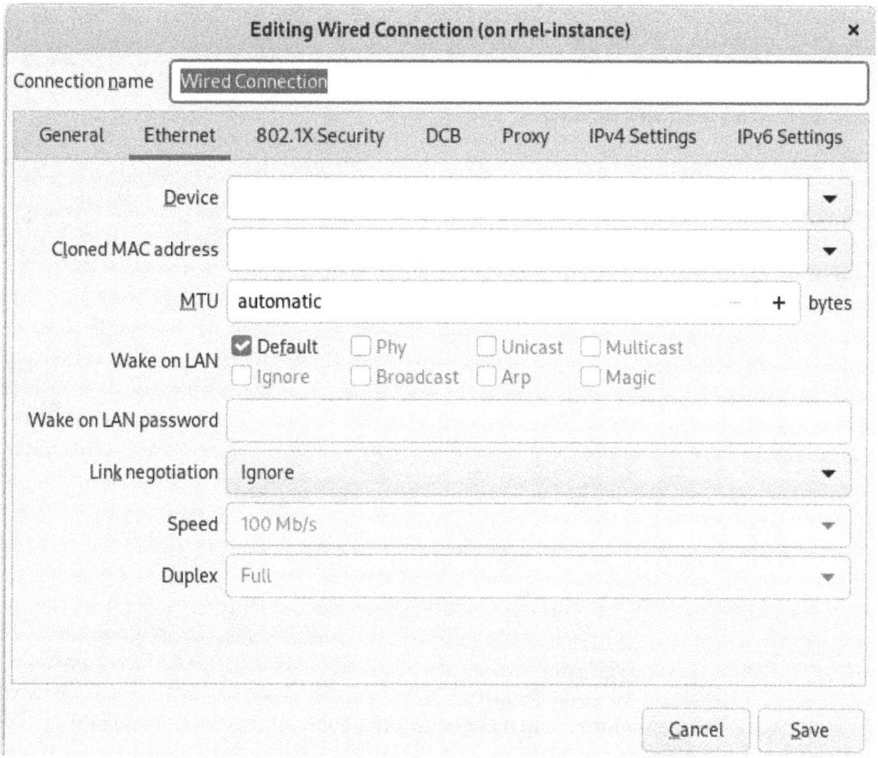

Figure 6.5 – The dialog for editing a network connection

In the dialog, we can see the fields we had in the simpler command-line configuration tool, plus extra fields and different tabs for each group of options.

The important fields to remember are those that are used to **Connect automatically with priority** in the **General** tab. This enables our system to automatically enable that NIC when a connection is available:

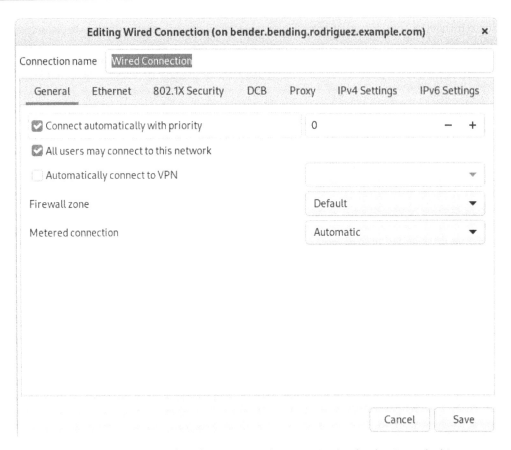

Figure 6.6 – The dialog for editing a network connection (under the General tab)

As you can find by examining the different tabs, there are lots of choices, such as marking a connection to be metered. This means that, for example, if a connection is made via a mobile phone, extra charges may be specified if network usage is not controlled.

When we're creating extra networks, we can define physical or virtual devices based on the packages we have installed in our system (if you recall the list of packages we saw when searching for NetworkManager, we had packages for different VPNs, Wi-Fi, and others), as we can see in the following screenshot:

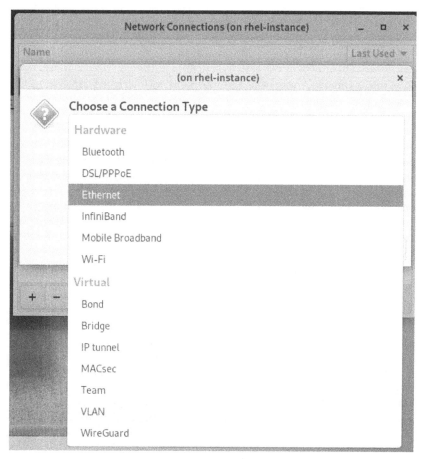

Figure 6.7 – nm-connection-editor with the plugins for Wi-
Fi, OpenVPN, PPTP, Bluetooth, and more installed

For server environments, the most common network types are **Bond**, **Bridge**, and **Team** (a part of **Ethernet**), while for desktops, the most common network types are **Ethernet**, **Wi-Fi**, and **Mobile Broadband**.

Each type of connection has some requirements. For example, for bonds, bridges, and teams, we need more than one network interface that can be combined.

Now, let's move on to review nmcli usage in the following section.

Configuring interfaces with nmcli

nmcli is the command-line interface for NetworkManager. It allows us not only to check but also to configure the network interfaces in our system, and even if using it requires more memory skills than nmtui requires, it empowers users and administrators with scripting capabilities to automate the network setup of their systems.

> **Tip**
>
> Most commands allow us to use autocompletion – that is, pressing the *Tab* key will use the autocompletion lists on the command line to suggest the syntax. For example, typing nmcli dev on the command line and pressing *Tab* will autocomplete the command to nmcli device. In this case, it might not be as critical, as nmcli takes both arguments as valid, but in other cases, it's mandatory to spell it properly for the code to work.

Let's start by checking the available connections in our system with nmcli dev and then using nmcli con show to check out its details:

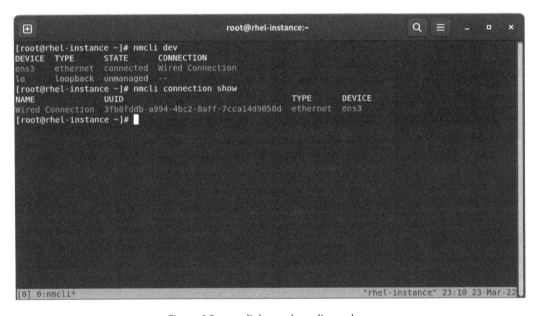

Figure 6.8 – nmcli dev and nmcli con show

When controlling a network connection, for example, when using `nmcli con up "Wired Connection"` or when disabling it with `nmcli con down ens3`, we should bear in mind what we explained about NetworkManager – if the connection is available in the system, NetworkManager might reactivate it after being disconnected because the connection and the devices required are available in our system.

Now, let's create a new interface to illustrate the process of adding a new connection via IPv4:

```
nmcli con add con-name eth0 type ethernet \
ifname eth0 ipv4.address 192.168.1.2/24 \
ipv4.gateway 192.168.1.254
```

We can do the same with IPv6:

```
nmcli con add con-name eth0 type ethernet \
ifname eth0 ipv6.address 2001:db8:0:1::c000:207/64 \
ipv6.gateway 2001:db8:0:1::1 ipv4.address \
192.0.1.3/24 ipv4.gateway 192.0.1.1
```

Note that a new connection will be created with the preceding commands, as we're not specifying the ID for modifying it but running two different connection `add` commands.

Once the preceding commands have been executed, we can check the network connections that have been defined with `nmcli connection show eth0` and validate that the proper settings were applied (or, of course, via `nmtui`, `nm-connection-editor`, or the text files that were created on disk, as the information is shared and stored in the system).

When we reviewed the output of `nmcli connection show interface`, the output contained some keys separated by dots, such as the following:

- `ipv4.address`
- `ipv4.gateway`
- `ipv6.address`
- `ipv6.gateway`
- `connection.id`

We can use these keys to define new values via `nmcli con mod $key $value`, as shown in the following example:

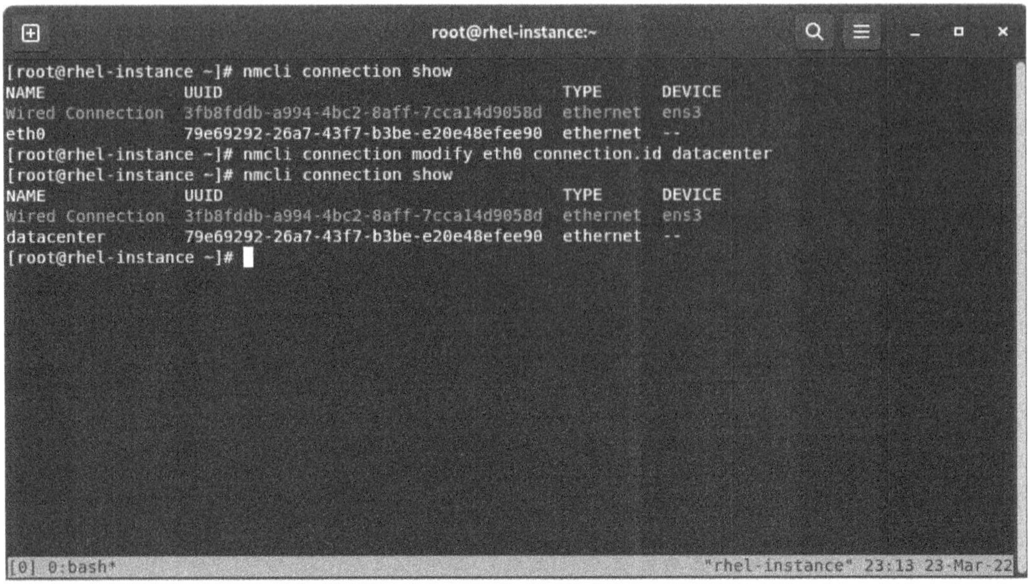

Figure 6.9 – An example of modifying a network connection to
change the name of the connection ID and IP address

Of course, after doing the preceding tests, we can also remove the connection to avoid problems in our system with `nmcli con del datacenter`.

The following commands can be used to modify connections with the `nmcli` tool:

- `nmcli con show`: Shows the status of the connections.

- `nmcli con show NAME`: Shows the details of a connection named NAME.

- `nmcli dev status`: Shows the statuses of the devices in the system. Note that this means **devices**, not connections that might be using those devices.

- `nmcli con add con-NAME`: Adds a new connection.

- `nmcli con mod NAME`: Modifies a connection.

- `nmcli con up NAME`: Brings up a connection.

- `nmcli con down NAME`: Brings down a connection (which can still be reenabled by NetworkManager).

- `nmcli con del NAME`: Removes a connection definition from the system.

> **Tip**
>
> Check `man nmcli-examples` to find more examples that are included in the documentation for the system.

Where does NetworkManager store configuration?

In previous versions, the usage of `ifcfg` files allowed us to configure and tune network settings, but as discussed at the beginning of this chapter, these have now been deprecated, leaving us with NetworkManager to handle all the configuration.

By default, NetworkManager stores the configuration in the `/etc/NetworkManager/system-connections/` folder as key files (pairs of key-value grouped under `[groups]`). Feel free to examine the ones on our system after we've used `nmcli` to check how it affected the files for the interfaces.

Now that we have reviewed the different ways to configure networking in our system, let's learn about naming resolutions.

Configuring hostname and hostname resolutions (DNS)

Remembering IP addresses, whether they are IPv4 or IPv6 addresses, can become a nightmare. To make things easier, a more human approach was used for the hostnames and the DNS, in that we can translate these easier-to-remember names into the IP addresses that our systems use for connections.

Hostnames are the names we assign to a host to identify them, but when they're used in addition to a DNS server, we must have other hosts that can *resolve* them into IP addresses they can connect to.

We can use the `hostname` command to see or temporarily modify the current hostname, as shown in the following screenshot:

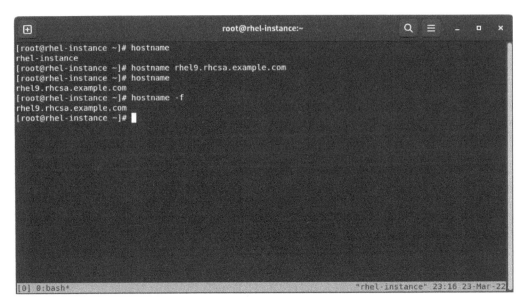

Figure 6.10 – Querying and changing the hostname for our host

Bear in mind that this change is only temporary – whenever we restart the server, it will use the configured one.

To define a newly configured hostname, we will use the `hostnamectl set-hostname` command, as shown in the following screenshot:

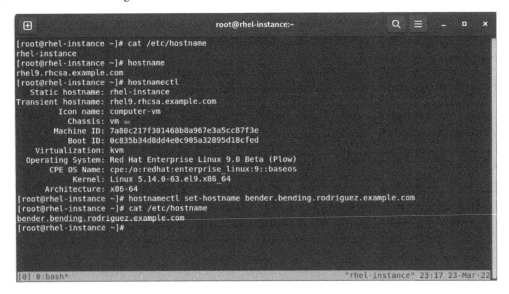

Figure 6.11 – Checking the previously configured hostname and
the definition of a new one via hostnamectl

Note in the preceding example how we have `Transient hostname` versus `Static hostname`, which refers to the temporary status of the name that was defined with `hostname` instead of `hostnamectl`.

When it comes to name resolution, there are several approaches we can take. One, of course, is to use DNS servers, which we will explain later in this section, but there are other ways.

In general, systems have several resolvers, and those are defined in the `/etc/nsswitch.conf` configuration file. These resolvers are not only for network naming but also, for example, for resolving users, where, for example, a corporate **Lightweight Directory Access Protocol (LDAP)** server might be used to define users and passwords. By default, `nsswitch.conf` instructs our system to use for host resolution this entry: `hosts: files dns myhostname`.

This means that we are using the files in our `/etc/` directory as our first source. In the case of hostnames, this refers to the `/etc/hosts` file. If an entry is defined in that file, the value that was specified will be used; if not, the `/etc/resolv.conf` file will determine how to proceed with its resolution. These files, and especially `resolv.conf`, are configured when the system is deployed and when a connection is activated. NetworkManager takes care of updating the values that were obtained via DHCP if autoconfiguration was used or the specified DNS servers if a manual configuration was used.

In the following screenshot, we can see the entries that have been defined in our `/etc/hosts` file, how pinging a host fails because the name does not exist, and how, after manually adding an entry to the `/etc/hosts` file, our system can reach it:

Figure 6.12 – Adding a static host entry to our local system

As we mentioned previously, DNS resolution is done via the configuration at `/etc/resolv.conf`, which, by default, contains a `search` parameter and a `nameserver` parameter. If we check the man page of `resolv.conf`, we can obtain descriptions for the common parameters:

- `nameserver`: Contains the IP of the nameserver to use. Currently, only a maximum of three entries (each on its own line) will be used by the `resolv` library in the system. The resolution is performed in order each time, so if one server fails, it will time out, try the next one, and so on.

- `domain`: The local domain name. It allows us to use short names for hosts that are relative to the local domain in our host. If it's not listed, it's calculated based on the hostname of our system (everything after the first ".").

- `search`: By default, this contains the local domain name, and it's the list of domains we can attempt to use to resolve the short name that's provided. It's limited to 6 domains and 256 characters. `domain` and `search` are mutually exclusive since it's just the last one in the file that's used.

> **Tip**
> DNS resolution works by asking special servers (as in, DNS servers) for the relevant data for a domain. This happens hierarchically, with the topmost general servers being called **root servers**. DNS servers not only contain registers or entries for converting hostnames into IPs but also information about the mail server to use when sending an email, verification details for security, or reverse entries. DNS servers can also be used to block access to services by returning invalid IPs for some domains or to speed up internet navigation by using faster DNS servers than the ones provided by the ISP. When a domain name is registered, a new entry is created in the root tables for the domain pointing to the DNS server. This will take care of that domain resolution, and later, those entries will be populated and cached across the internet for faster resolution.

If we want to modify the DNS servers defined for a connection, remember to use `nmcli con mod NAME ipv4.dns IP` (or IPv6 equivalent) and use a + symbol beforehand, as in, `+ipv4.dns`, to add a new entry to the list of DNS servers. Any manual changes to `resolv.conf` might be overwritten.

Now that we have learned about how DNS works and how our system uses it, let's look at how to secure system network access.

An overview of firewall configuration

When a system is connected to a network, many of the running services can be reached from other systems. That is the goal behind having systems connected. However, we also want to keep systems secure and away from unauthorized usage.

A **firewall** is a software layer that sits between the network cards and the services and allows us to fine-tune what is allowed or not.

We cannot completely block all the incoming connections to our system as the incoming connection is often a response to a request that our system made.

The connections are blocked via a kernel framework named **netfilter**, which is used by the firewall software to modify how the packets are processed. **nftables** is a new filter and packet classifier subsystem that enhances parts of the netfilter code, but retains the architecture and provides faster processing between other features using only one interface (**Netfilter Tables – nft**), thus deprecating old frameworks such as iptables, ip6tables, ebtables, and arptables, which have now been fully deprecated in RHEL 9.

> **Important Note**
>
> As we explained earlier regarding network configuration, a bad configuration in a firewall can lock you out of the system, so be extremely careful when you're setting restrictive rules so that you can log into the system again if you are remotely accessing it.

firewalld is a frontend to the **nftables** framework that on the vast majority of occasions, should be able to deal with the requirements for filtering, so it is the recommended frontend for editing rules. It can be installed on your system by installing the firewalld package, which should be included in a base installation. It will provide the firewall-cmd command once installed for interacting with the service.

firewalld uses the concept of zones, which allows us to predefine a set of rules for each of those zones. These can also be assigned to network connections. This is more relevant, for example, for laptops that might be roaming across connections and have some default settings for when you're using home or corporate connections. However, they will default to a more secure one when you're using Wi-Fi from a cafeteria, for example.

firewalld also uses predefined services so that the firewall knows what ports and protocols should be enabled based on the services and zones they have been enabled on.

Let's check out the available zones and some more details about the home zone:

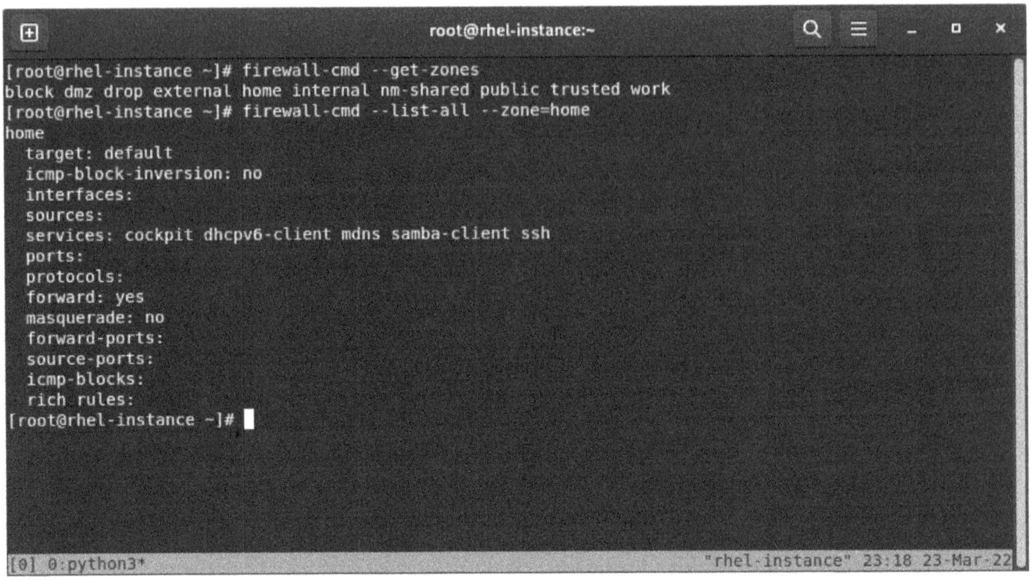

```
[root@rhel-instance ~]# firewall-cmd --get-zones
block dmz drop external home internal nm-shared public trusted work
[root@rhel-instance ~]# firewall-cmd --list-all --zone=home
home
  target: default
  icmp-block-inversion: no
  interfaces:
  sources:
  services: cockpit dhcpv6-client mdns samba-client ssh
  ports:
  protocols:
  forward: yes
  masquerade: no
  forward-ports:
  source-ports:
  icmp-blocks:
  rich rules:
[root@rhel-instance ~]# 
[0] 0:python3*                                        "rhel-instance" 23:18 23-Mar-22
```

Figure 6.13 – Available zones and configuration for the home zone

As we can see, several zones have been defined:

- public: This is the default zone for newly added interfaces. It allows us to cockpit SSH and DHCP clients and rejects all incoming traffic not related to the outgoing traffic.

- block: Rejects all incoming traffic unless it's related to outgoing traffic.

- dmz: Rejects all incoming traffic unless it's related to outgoing or SSH connections.

- drop: Drops all incoming packets that are not related to outgoing ones (not even ping).

- external: Blocks all incoming traffic except that related to outgoing traffic. It also allows SSH and it masquerades traffic as originating from this interface.

- home: In addition to public, it allows smb and mdns.

- internal: Based on the home zone.

- trusted: Allows all incoming traffic.

- work: Blocks all incoming traffic except that related to outgoing or SSH, cockpit, or DHCP traffic.

Next, we'll learn how to use those zones when we're configuring the firewall.

Configuring the firewall

As shown in the introduction to this section, a firewall can be configured via the `firewall-cmd` command (as well as the cockpit web interface, which was described earlier in this book in *Chapter 4, Tools for Regular Operations*). The most common command options that are used are as follows:

- `firewall-cmd --get-zones`: Lists the available zones
- `firewall-cmd --get-active-zones`: Lists the active zones and interfaces that have been assigned
- `firewall-cmd --list-all`: Dumps the current configuration
- `firewall-cmd --add-service`: Adds a service to the current zone
- `firewall-cmd --add-port`: Adds a port or protocol to the current zone
- `firewall-cmd --remove-service`: Removes the service from the current zone
- `firewall-cmd --remove-port`: Removes the port or protocol from the current zone

> **Important Note**
> Note that you need to mention the port number and service name after the preceding commands to add or remove a service or port.

- `firewall-cmd --reload`: Reloads the configuration from the saved data, thus discarding the runtime configuration
- `firewall-cmd –get-default-zone`: Gets the default zone
- `firewall-cmd --set-default-zone`: Defines the default zone to use

For example, when we install an HTTP server in our system (for serving web pages), port 80 on TCP must be enabled.

Let's try this in our sample system by installing, running, and opening the HTTP port:

```
dnf -y install httpd
systemctl enable httpd
systemctl start httpd
firewall-cmd --add-service=http
curl localhost
```

The last command will make a petition to the local `http` server to grab the results. If you have access to an additional system, you can try to connect to the IP of the server that we have been using to watch the default web page be served by the system.

In the following screenshot, we can see the output of the `curl localhost` command:

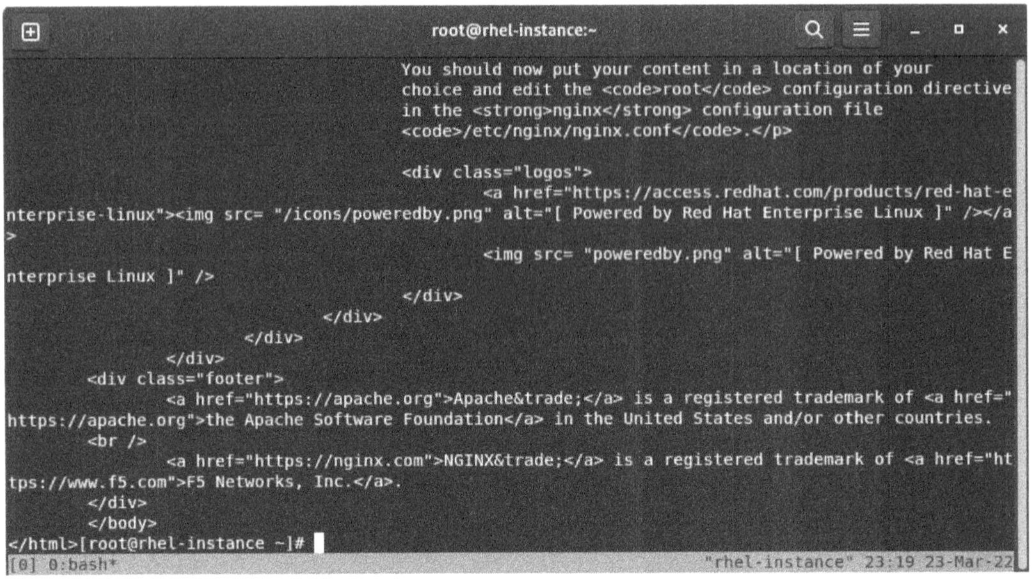

Figure 6.14 – The output of curl when requesting the web page hosted by our system

At this point, we have reviewed how to configure some basic firewall rules, so we are ready to check the network's connectivity.

Testing network connectivity

In the previous sections, we were interacting with network interfaces, addresses, and firewall rules that define, limit, or allow connections to our system. In this section, we will review some of the basic tools that can be used to validate that network connectivity exists.

Note that the following commands assume that the firewall is not set to strict mode and that we can use the **Internet Control Message Protocol** (**ICMP**) to reach the servers hosting the service. In secured networks, the service might be working but not answering `ping` – it may only be answering the service queries themselves.

There are several commands that we can use here, so consider these suggestions for diagnosing issues:

- Check the local interface's IP address, netmask, and gateway.
- Use the `ping` command with the IP address of the gateway to validate the proper network configuration.
- Use the `ping` command to ping the DNS servers in `/etc/resolv.conf` to see if those are reachable. Alternatively, use the `host` or `dig` command to query the DNS servers.

- If there's supposedly external network connectivity, try to reach external DNS servers such as 8.8.8.8 or 1.1.1.1 or use curl or wget to request some of the web pages of known services – for example, curl nasa.gov.

This should give you a rough idea of where a problem might be based on how far you reach into the tests. Remember that there are other tools, such as tracepath, that will show the hops a TCP packet does before reaching the destination. The man pages for each command will give you hints about and examples of their usage.

In the following screenshot, you can see the output of tracepath against one web server:

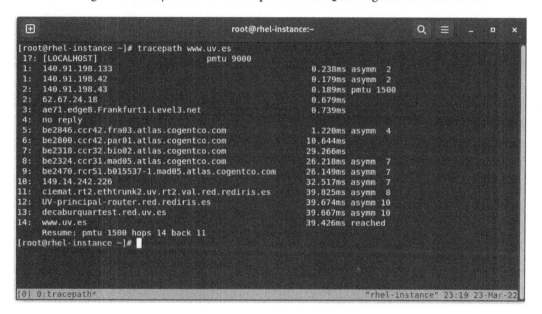

Figure 6.15 – The output of the tracepath command against the
web server for the University of Valencia, Spain

As we can see, 14 steps were performed across different servers until our data package reached the destination host. This allowed us to learn about how a package traverses the internet to reach target systems.

Summary

In this chapter, we learned about configuring network interfaces using different approaches, either via manual interaction or via methods that allow us to script or automate the configuration.

Some troubleshooting for network issues was also introduced to help us find some of the basic errors that might occur.

As we mentioned in this chapter's introduction, networking is the basis for our system to reach other services and provide services to other systems. We also introduced the idea of more complex network setups that fall outside the scope of this RHCSA level, but it's interesting to at least be familiar with the keywords that we'll be using in our careers.

In the next chapter, we will cover some important topics related to security, such as adding to, patching, and managing the software in our systems.

7

Adding, Patching, and Managing Software

Maintaining a system's software to solve security issues, apply fixes, and keep the system up to date is an essential task in systems management. In this chapter, we will review how the **Red Hat Subscription Management system** works, how to ensure that packages are verified, and other software management tasks to keep the system fresh.

Going a bit more into the details, in this chapter, we will go over how the subscription system works and how to use your developer subscription for self-training or installing a personal server. We will also check how to manage the software origins, also known as repositories, that your system will use. That includes learning the role of signatures in package management, to ensure the software installed is the software Red Hat provides. We will also learn about key tasks such as adding and removing packages and package groups, using different software versions with modularity, and reviewing and rolling back changes.

To simplify expanding your knowledge, enabling you to prepare your own labs, we will see how to have a full local copy of all the **Red Hat Enterprise Linux** (**RHEL**) repositories on your system.

And last but not least, we need to understand **Red Hat Package Manager** (**RPM**), now changed to RPM Package Manager, by learning the basics of how the internals of package management work.

In summary, we will cover the following topics in this chapter:

- RHEL subscription registration and management
- Managing repositories and signatures with Yum/DNF
- Doing software installations, updates, and rollbacks with Yum/DNF
- Creating and syncing repositories with createrepo and reposync
- Understanding RPM internals

Now, let's start managing the software in our systems.

RHEL subscription registration and management

RHEL is a fully **open source operating system**, which means that all the source code used to build it is available to access, modify, redistribute, and learn from. Pre-built binaries are, on the other hand, delivered as a service, and accessible via a subscription. As seen in *Chapter 1*, *Getting RHEL Up and Running*, we can have, for our own personal use, a developer subscription. That subscription provides access to ISO images, but also to the updated, signed packages that are part of RHEL 9. These are the exact same bits that are used in production by so many companies worldwide.

Let's see how to use that subscription with our own system.

First, let's take a look at the **Red Hat Customer Portal** at `https://access.redhat.com` and click **LOG IN**:

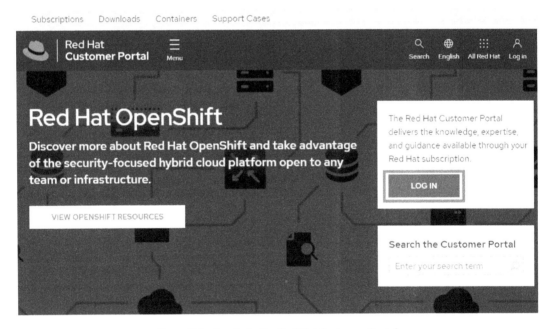

Figure 7.1 – Log in to the Red Hat Customer Portal

Once we click on **LOG IN**, we will be redirected to the **Single Sign-On** (**SSO**) page for all the Red Hat services. There, we will need to use the username we created in *Chapter 1*, *Getting RHEL Up and Running*. In the following screenshot, we are using `student` as an example:

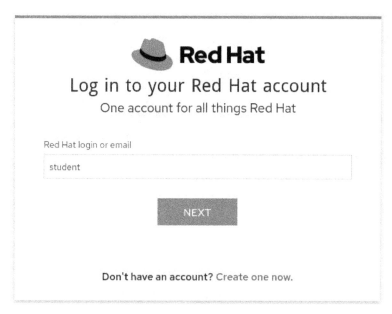

Figure 7.2 – Entering our username in Red Hat single sign-on

Now it's time to type our password to log in:

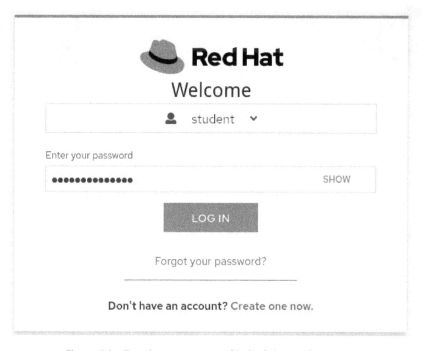

Figure 7.3 – Entering our password in Red Hat single sign-on

Once logged in, we shall go to the **Red Hat subscriptions page** by clicking on the **My Subscriptions** link in the top bar:

Figure 7.4 – Accessing the subscriptions page in the Red Hat customer portal

The subscriptions page will look like this for a user with one physical machine subscribed:

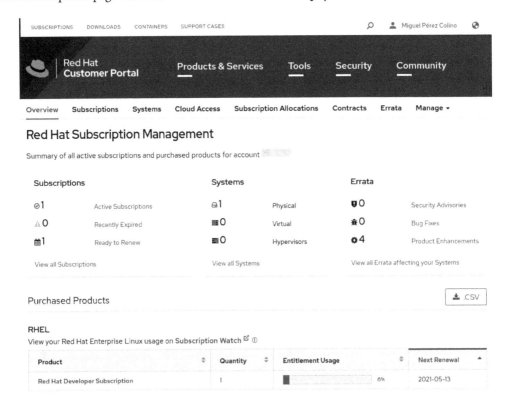

Figure 7.5 – Subscription page example in the Red Hat customer portal

> **Tip**
> The **Red Hat Developer Subscription** was updated in January 2021 to support up to 16 systems. You can use your account for more than one single system to simulate larger production-like deployments.

Now let's register our new system:

```
[root@rhel-instance ~]# subscription-manager register
Registering to: subscription.rhsm.redhat.com:443/subscription
Username: student
Password:
The system has been registered with ID: 2f230fd7-d529-4591-
af3e-d6a25cf9537e
The registered system name is: rhel-instance.example.com
```

With this, our system will be registered in the Red Hat **Content Delivery Network** (**CDN**) but will still not have a subscription assigned.

Let's go to the subscriptions page and refresh to see the new system there. We will click on **View all Systems** to continue:

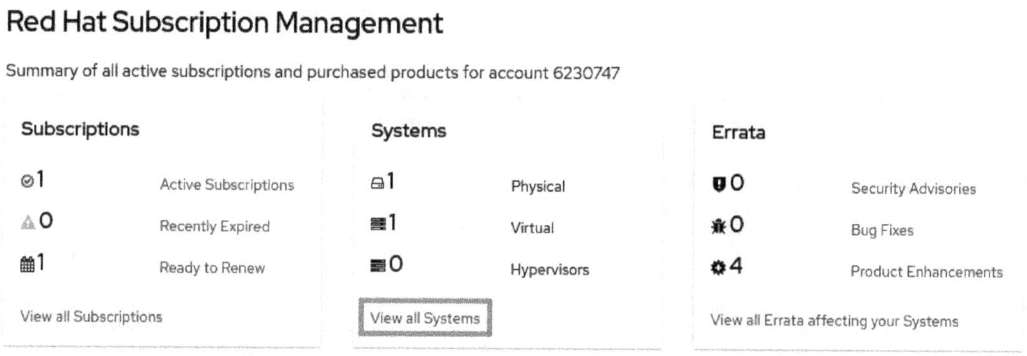

Red Hat Subscription Management

Summary of all active subscriptions and purchased products for account 6230747

Subscriptions		Systems		Errata	
⊘1	Active Subscriptions	⊟1	Physical	⬦0	Security Advisories
⚠0	Recently Expired	▤1	Virtual	✳0	Bug Fixes
🏛1	Ready to Renew	▤0	Hypervisors	⚙4	Product Enhancements
View all Subscriptions		View all Systems		View all Errata affecting your Systems	

Figure 7.6 – Subscriptions page with the new subscribed system

We can see our new system on the page, `rhel9.example.com`, with the question mark next to it to show it does not have a subscription attached to it. Let's click on the name of the system to see the details:

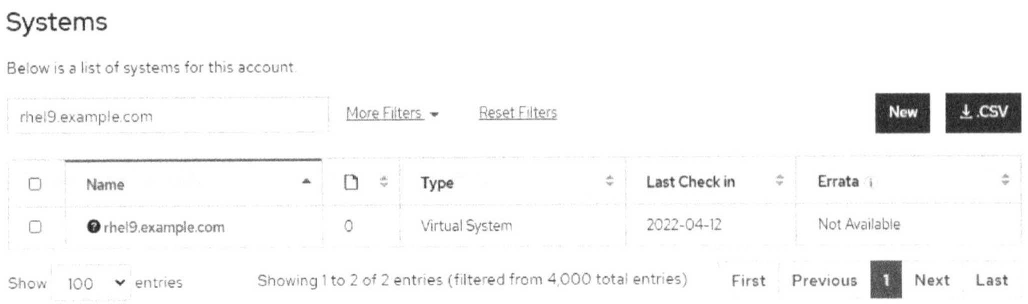

Systems

Below is a list of systems for this account.

	Name	▲	🗋	⬍	Type	⬍	Last Check in	⬍	Errata ⓘ		⬍
☐	❷ rhel9.example.com		0		Virtual System		2022-04-12		Not Available		

rhel9.example.com   More Filters ▾   Reset Filters   New   ⬇ .CSV

Show 100 ⌄ entries   Showing 1 to 2 of 2 entries (filtered from 4,000 total entries)   First   Previous   **1**   Next   Last

Figure 7.7 – Subscription page with the new subscribed system

Once on the specific system page, we see all the details for the system. We click on **Subscriptions** to see the attached ones:

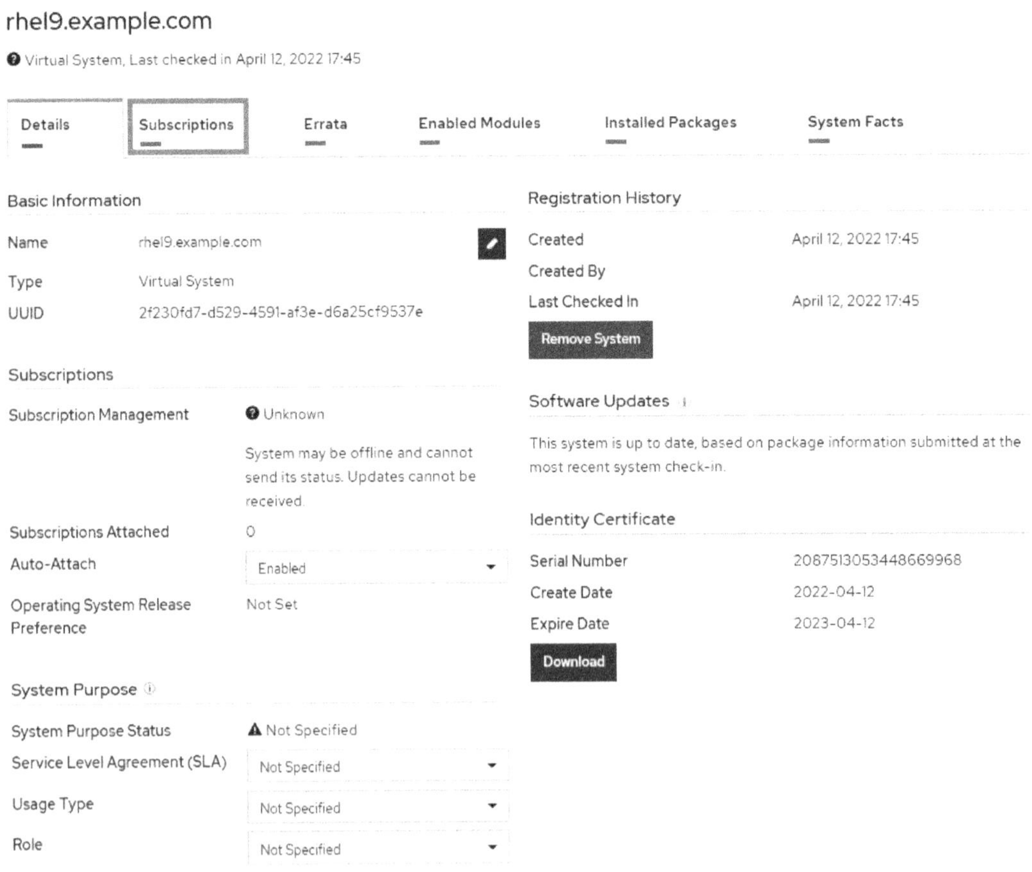

rhel9.example.com

❷ Virtual System, Last checked in April 12, 2022 17:45

Details	Subscriptions	Errata	Enabled Modules	Installed Packages	System Facts

Basic Information

Name	rhel9.example.com	✏
Type	Virtual System	
UUID	2f230fd7-d529-4591-af3e-d6a25cf9537e	

Registration History

Created	April 12, 2022 17:45
Created By	
Last Checked In	April 12, 2022 17:45

Remove System

Subscriptions

Subscription Management	❷ Unknown
	System may be offline and cannot send its status. Updates cannot be received.
Subscriptions Attached	0
Auto-Attach	Enabled ▾
Operating System Release Preference	Not Set

Software Updates ⓘ

This system is up to date, based on package information submitted at the most recent system check-in.

Identity Certificate

Serial Number	2087513053448669968
Create Date	2022-04-12
Expire Date	2023-04-12

Download

System Purpose ⓘ

System Purpose Status	⚠ Not Specified
Service Level Agreement (SLA)	Not Specified ▾
Usage Type	Not Specified ▾
Role	Not Specified ▾

Figure 7.8 – Subscriptions page with the new subscribed system's details

We can see on the page that there are no attached subscriptions for this system:

rhel9.example.com

❓ Virtual System, Last checked in April 12, 2022 17:45

Details	Subscriptions	Errata	Enabled Modules	Installed Packages	System Facts

Subscriptions attached to this system

There are no subscriptions to display

Attach subscriptions for this system to receive updates for installed products
Learn more about attaching subscriptions to systems

Attach Subscriptions   Run Auto-Attach

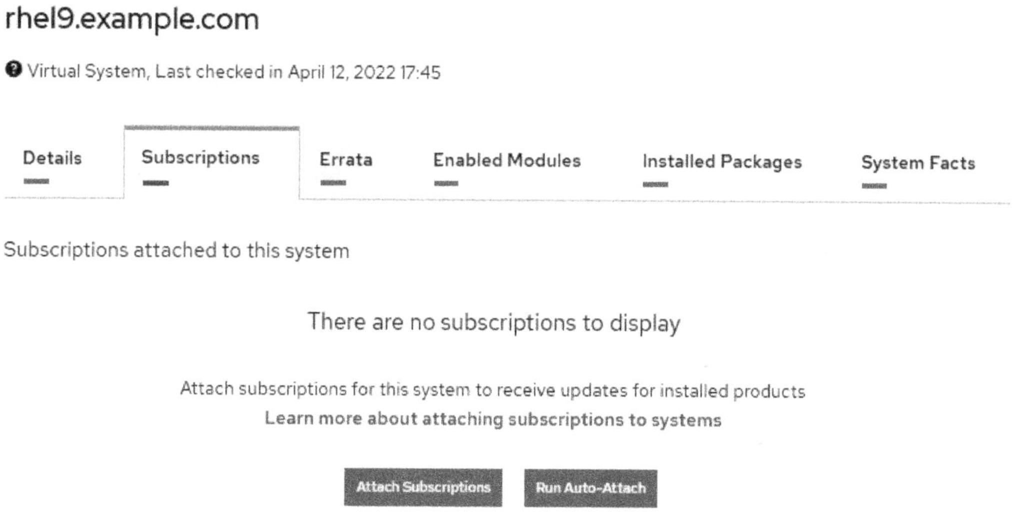

Figure 7.9 – Subscriptions page with the new subscribed system, with no subscription attached

Let's attach a subscription to our system using `subscription-manager attach`:

```
[root@rhel-instance ~]# subscription-manager attach --auto
Installed Product Current Status:
Product Name: Red Hat Enterprise Linux for x86_64
Status:         Subscribed
```

The result of the command shows that the system is now registered and has a subscription attached to it for Red Hat Enterprise Linux for x86_64. Let's refresh the **Subscriptions** page to ensure the subscription attachment runs properly:

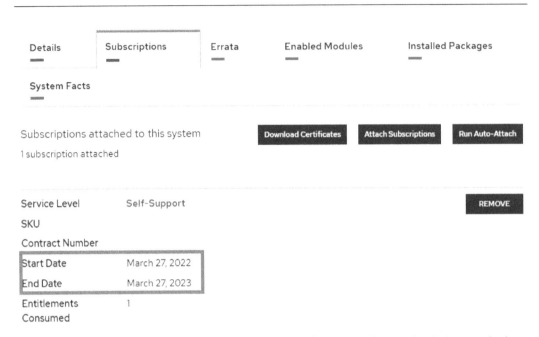

Figure 7.10 – Subscriptions page with the new subscribed system, with one subscription attached

With this, we know for sure that the system is properly registered and subscribed to the Red Hat CDN and it's ready to access all the software, patches, and updates available from it.

Also, in the system, we can see that a new file with the information on the software **repositories**, or **repos** for short, has been created:

```
[root@rhel-instance ~]# ls -l /etc/yum.repos.d/redhat.repo
-rw-r--r--. 1 root root 61372 Apr  13 00:56 /etc/yum.repos.d/
redhat.repo
```

Now we know how to manage available subscriptions and assign them to a running system so it has access to the software binaries built by Red Hat. Let's learn more about how to use the provided repositories in the next section.

Managing repositories and signatures with yum/dnf

RHEL, like many other Linux distributions, has a mechanism to provide software based on repos. These contain a list of software packages (which could be end user applications such as Firefox, or components for them such as GTK3), a list of dependencies between the packages, and other useful metadata.

Once we finish subscribing the system, we can take a look at the repositories available in the system using yum or dnf:

```
[root@rhel-instance ~]# yum repolist
Updating Subscription Management repositories.
repo id                                repo name
rhel-8-for-x86_64-appstream-rpms       Red Hat Enterprise Linux 9
for x86_64 - AppStream (RPMs)
rhel-8-for-x86_64-baseos-rpms          Red Hat Enterprise Linux 9
for x86_64 - BaseOS (RPMs)
[root@rhel-instance ~]# dnf repolist
Updating Subscription Management repositories.
repo id                                repo name
rhel-9-for-x86_64-appstream-rpms       Red Hat Enterprise Linux 9
for x86_64 - AppStream (RPMs)
rhel-9-for-x86_64-baseos-rpms          Red Hat Enterprise Linux 9
for x86_64 - BaseOS (RPMs)
```

As you can see, the output of yum and dnf is exactly the same. As a matter of fact, dnf is the evolution of yum, and in RHEL 9 the yum command is just a symbolic link to dnf:

```
[root@rhel-instance ~]# which yum
/usr/bin/yum
[root@rhel-instance ~]# ll /usr/bin/yum
lrwxrwxrwx. 1 root root 5 Ago 9  2021 /usr/bin/yum -> dnf-3
[root@rhel-instance ~]# which dnf
/usr/bin/dnf
[root@rhel-instance ~]# ll /usr/bin/dnf
lrwxrwxrwx. 1 root root 5 Ago 9  2021 /usr/bin/dnf -> dnf-3
```

They both can be used indistinctively in RHEL 9. From now on, we will use dnf only, but remember, if you prefer yum, feel free to use it.

> Tip
>
> **YUM** used to be an acronym for **Yellowdog Updater Modified**, a project that started as a Linux distribution for Macs called Yellowdog. **DNF** stands for **Dandified YUM**.

Now let's take a look at the repository definition created during the subscription attachment, `/etc/yum.repos.d/redhat.repo`. We can edit the file and go to the entry of the `BaseOS` repository shown above `rhel-9-for-x86_64-baseos-rpms`:

```
[rhel-9-for-x86_64-baseos-rpms]
name = Red Hat Enterprise Linux 9 for x86_64 - BaseOS (RPMs)
baseurl = https://cdn.redhat.com/content/dist/
rhel9/$releasever/x86_64/baseos/os
enabled = 1
gpgcheck = 1
gpgkey = file:///etc/pki/rpm-gpg/RPM-GPG-KEY-redhat-release
sslverify = 1
sslcacert = /etc/rhsm/ca/redhat-uep.pem
sslclientkey = /etc/pki/entitlement/2021426601576593026-key.pem
sslclientcert = /etc/pki/entitlement/2021426601576593026.pem
metadata_expire = 86400
enabled_metadata = 1
```

As you can see, every section in the file starts with the name of the section between brackets – in the previous case, `[rhel-9-for-x86_64-baseos-rpms]`. Now let's check all the entries under this section:

- **name**: A long descriptive name for the repository. It's the one displayed when we listed the repos in the previous example.

- **baseurl**: The main resource the packages will be obtained from. In this case, it is a single HTTPS source. It contains the `$releasever` variable that will be substituted before being accessed. Other methods are NFS, HTTP, and FTP.

- **enabled**: A variable to provide a simple way to have repositories enabled or disabled on the system. When set to `1`, it will be enabled, and when set to `0` it will be disabled.

- **gpgcheck**: A mechanism to verify packages. When set to `1`, it will be enabled and all the packages installed in the system with `dnf` / `yum` will be verified using their `gpg` signatures against a provided key.

- **gpgkey**: A key to verify, using `gpg`, the downloaded packages.

- **sslverify**: A mechanism to verify the machine subscription to the CDN. Enabled when set to `1`, and disabled when set to `0`.

- **sslcacert**: A certificate used as a certificate authority to validate client certificates.

- **sslclient key**: A client key used to habilitate the client certificate.

- **sslclientcert**: The client certificate that the machine will use to identify itself against the CDN.

- **metadata_expire**: The time in seconds after which to consider the retrieved metadata as expired. The default, as shown here, is 24 hours.

- **enabled_metadata**: An option to enable other tools (not `dnf`) to use the metadata as downloaded in this repository.

The minimal required options to have a running repository are: `name`, `baseurl`, and `gpgcheck`, having this last one set to `0`.

> **Important Note**
>
> Although it is possible to change the configuration of the repositories by editing the files, the best way to modify the Red Hat provided repos is by using commands that will be shown in this chapter. That's because the `redhat.repo` file will be overwritten by the subscription manager when refreshing data.

By running `dnf repolist`, we obtained the list of repositories enabled in the system. What if we want to see all the repositories, the ones enabled and the ones disabled? That can be done by running `dnf repolist --all`:

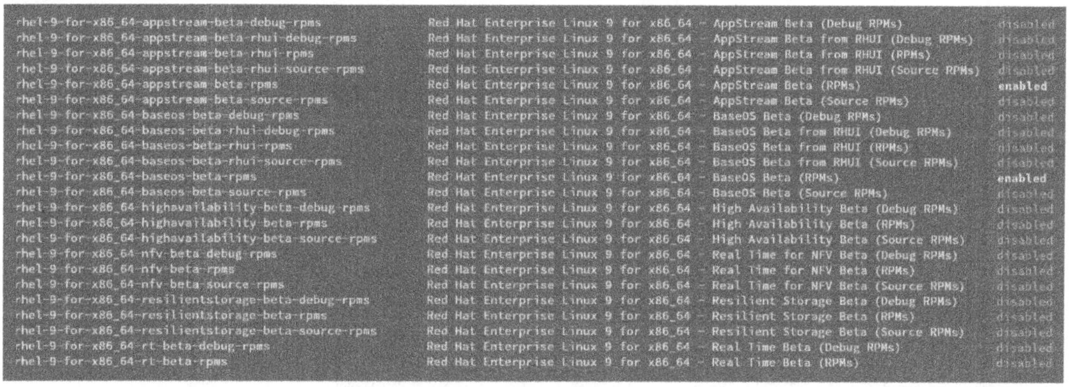

Figure 7.11 – Partial output of dnf repolist –all

The list is very extensive. It includes the repositories with the binaries as used in many production cases, from SAP to managing systems with Satellite. We can filter it with `grep` to search for `supplementary`:

```
[root@rhel-instance ~]# dnf repolist --all | grep supplementary
rhel-9-for-x86_64-supplementary-debug-
rpms                    disabled
rhel-9-for-x86_64-supplementary-eus-debug-
rpms              disabled
```

```
rhel-9-for-x86_64-supplementary-eus-
rpms                        disabled
rhel-9-for-x86_64-supplementary-eus-source-
rpms           disabled
rhel-9-for-x86_64-supplementary-
rpms                        disabled
rhel-9-for-x86_64-supplementary-source-
rpms                disabled
```

There are four different types of channels here:

- **Regular channels**: For example, `rhel-9-for-x86_64-supplementary-rpms`, which contains the packages, also known as `rpms`, ready to be installed on the system. These apply during the standard maintenance period.

- **Extended update support**: For example, `rhel-9-for-x86_64-supplementary-eus-rpms`, which contains `eus` in the name. These provide packages with backports to be able to keep the same minor version for a longer time. Don't use them unless it is required by a third-party vendor.

- **Source channels**: For example, `rhel-9-for-x86_64-supplementary-source-rpms`, which contains `source` in the name. They provide the source used to build the packages delivered in *regular* and *extended updated support* channels.

- **Debug channels**: For example, `rhel-9-for-x86_64-supplementary-debug-rpms`, containing `debug` in the name. These include the debugging information generated when building a package and are useful for in-deep troubleshooting of issues.

We can enable `rhel-9-for-x86_64-supplementary-rpms` by using the `config-manager` option of `dnf`, running the following:

```
[root@rhel-instance ~]# dnf config-manager --enable rhel-9-for-
x86_64-supplementary-rpms
Updating Subscription Management repositories.
[root@rhel-instance ~]# dnf repolist
Updating Subscription Management repositories.
repo id                                                      repo name
rhel-9-for-x86_64-appstream-rpms                             Red Hat
Enterprise Linux 9 for x86_64 - AppStream (RPMs)
rhel-9-for-x86_64-baseos-rpms                                Red Hat
Enterprise Linux 9 for x86_64 - BaseOS (RPMs)
rhel-9-for-x86_64-supplementary-rpms                         Red Hat
Enterprise Linux 9 for x86_64 - Supplementary (RPMs)
```

The repository is now enabled. You may want to try to enable and disable other repositories to practice.

Let's now try to add a repository for which we know only its URL, such as the **EPEL** repo. This repo contains **Extra Packages for Enterprise Linux**, and is specifically built for Linux, but not supported by Red Hat. As it is a well-known repo, it is replicated around the world and there is a local mirror at http://mirror.uv.es/mirror/fedora-epel/9/Everything/x86_64/ (you can find your local one in the mirror list here: https://admin.fedoraproject.org/mirrormanager/mirrors/EPEL). Now we can add this repo using dnf config-manager:

```
[root@rhel-instance ~]# dnf config-manager --add-repo="http://
mirror.uv.es/mirror/fedora-epel/9/Everything/x86_64/"
Updating Subscription Management repositories.
Adding repo from: http://mirror.uv.es/mirror/fedora-epel/9/
Everything/x86_64/
[root@rhel-instance ~]# dnf repolist
Updating Subscription Management repositories.
repo id                                                 repo name
mirror.uv.es_mirror_fedora-epel_9_Everything_x86_64_  created
by dnf config-manager from http://mirror.uv.es/mirror/fedora-
epel/9/Everything/x86_64/
rhel-9-for-x86_64-appstream-rpms                        Red Hat
Enterprise Linux 9 for x86_64 - AppStream (RPMs)
rhel-9-for-x86_64-baseos-rpms                           Red Hat
Enterprise Linux 9 for x86_64 - BaseOS (RPMs)
rhel-9-for-x86_64-supplementary-rpms                    Red Hat
Enterprise Linux 9 for x86_64 - Supplementary (RPMs)
```

We can inspect the newly created file – /etc/yum.repos.d/mirror.uv.es_mirror_fedora-epel_9_Everything_x86_64_.repo:

```
[mirror.uv.es_mirror_fedora-epel_9_Everything_x86_64_]
name=created by dnf config-manager from http://mirror.uv.es/
mirror/fedora-epel/9/Everything/x86_64/
baseurl=http://mirror.uv.es/mirror/fedora-epel/9/Everything/
x86_64/
enabled=1
```

You may have realized that there is an option missing in this repo, however, let's move ahead. We can search for a package available in EPEL, such as, for example, screen:

```
[root@rhel-instance ~]# dnf info screen
Updating Subscription Management repositories.
```

```
created by dnf config-manager from http://mirror.uv.es/mirror/
fedor  18 MB/s | 8.9 MB      00:00
Last metadata expiration check: 0:00:02 ago on Wed Apr 13
01:18:28 2022.
Available Packages
Name         : screen
Version      : 4.8.0
Release      : 6.el9
Architecture : x86_64
Size         : 649 k
Source       : screen-4.8.0-6.el9.src.rpm
Repository   : mirror.uv.es_mirror_fedora-epel_9_Everything_
x86_64_
Summary      : A screen manager that supports multiple logins
on one terminal
URL          : http://www.gnu.org/software/screen
License      : GPLv3+
Description  : The screen utility allows you to have multiple
logins on just one
             : terminal. Screen is useful for users who telnet
into a machine or are
             : connected via a dumb terminal, but want to use
more than just one
             : login.
             :
             : Install the screen package if you need a screen
manager that can
             : support multiple logins on one terminal.
```

The package is found, now let's try to install it:

```
[root@rhel-instance ~]# dnf install screen
Updating Subscription Management repositories.
You have enabled checking of packages via GPG keys. This is a
good thing.
However, you do not have any GPG public keys installed. You
need to download the keys for packages you wish to install and
install them.
You can do that by running the command:
```

```
    rpm --import public.gpg.key
```

Alternatively you can specify the url to the key you would like to use for a repository in the 'gpgkey' option in a repository section and DNF will install it for you.

For more information contact your distribution or package provider.

`[omitted]`

Important Note

If the error you get differs from the output shown above, it might be because of the cache and it can be cleared using the `dnf clean packages` command.

As we can see there is an error trying to install from this source, as it required the gpgcheck and gpgkey entries to be configured to have a properly secured repo (as gpg ensures that the content delivered is the same as the content created).

We can obtain gpgkey from the same mirror, at the URL `http://mirror.uv.es/mirror/fedora-epel/RPM-GPG-KEY-EPEL-9`, and drop it where dnf will search for it, `/etc/pki/rpm-gpg/`:

```
[root@rhel-instance ~]# curl -s http://mirror.uv.es/mirror/
fedora-epel/RPM-GPG-KEY-EPEL-9 > /etc/pki/rpm-gpg/RPM-GPG-KEY-
EPEL-9
[root@rhel-instance ~]# head -n 1 /etc/pki/rpm-gpg/RPM-GPG-KEY-
EPEL-9
-----BEGIN PGP PUBLIC KEY BLOCK-----
```

Now let's modify the file `/etc/yum.repos.d/mirror.uv.es_mirror_fedora-epel_9_Everything_x86_64_.repo` to look like this:

```
[mirror.uv.es_mirror_fedora-epel_9_Everything_x86_64_]
name=created by dnf config-manager from http://mirror.uv.es/
mirror/fedora-epel/9/Everything/x86_64/
baseurl=http://mirror.uv.es/mirror/fedora-epel/9/Everything/
x86_64/
enabled=1
gpgcheck=1
gpgkey=file:///etc/pki/rpm-gpg/RPM-GPG-KEY-EPEL-9
```

You can see we added both the gpgcheck and gpgkey entries in the file. Let's try installing the screen package again:

```
[root@rhel-instance ~]# dnf install screen
Updating Subscription Management repositories.
created by dnf config-manager from http://mirror.uv.es/mirror/
fedora-epel/9/Everything/x86_64/ 78 kB/s | 4.7 kB      00:00
Dependencies resolved.
================================================================Pac
kage Architecture Version Repository Size
================================================================
Installing:
screen x86_64 4.8.0-6.el9 mirror.uv.es_mirror_fedora-epel_9_
Everything_x86_64_ 649 k

[omitted]
Running transaction
  Preparing        :                                   1/1
  Running scriptlet: screen-4.8.0-6.el9.x86_64   1/1
  Installing       : screen-4.8.06.el9.x86_64     1/1
  Running scriptlet: screen-4.8.0-6.el9.x86_64   1/1
  Verifying        : screen-4.8.0-6.el9.x86_64    1/1
Installed products updated.
Installed:
  screen-4.8.0-6.el9.x86_64
Complete!
```

You'll notice that there is a step in which it asks you to confirm that the gpg key fingerprint is correct: FF8A D134 4597 106E CE81 3B91 8A38 72BF 3228 467C. To do so, you can go to the Fedora security page, as the Fedora project manages EPEL, and check. The URL for that page is https://getfedora.org/security/:

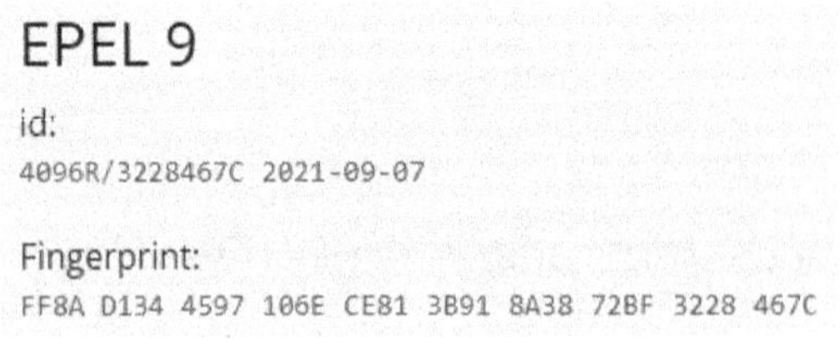

EPEL 9

id:
4096R/3228467C 2021-09-07

Fingerprint:
FF8A D134 4597 106E CE81 3B91 8A38 72BF 3228 467C

Figure 7.12 – Partial capture of the Fedora security page with an EPEL 9 gpg fingerprint

As you can see, it is correct. We have just verified that the signature we are using has the same fingerprint as announced by the project managing it, and now all the packages downloaded from this repo will be verified with it to avoid package tampering (which is someone changing the content before you receive it).

Let's review the commands that we used, which dnf provides to manage repos:

Command	Usage
dnf repolist	Shows repos configured and enabled
dnf repolist --all	Shows repos configured, enabled, or disabled
dnf config-manager --enable <reponame>	Enables an existing repo
dnf config-manager --add-repo=<repo_url>	Adds a new repo with the provided URL

Table 7.1 – dnf commands and parameters to manage repos

Now that we know how to securely manage repositories in RHEL, let's start adding more packages to our system, updating them, and undoing installations if we need to.

Doing software installations, updates, and rollbacks with YUM/DNF

In the previous section, we saw how to install a package. During the process, we saw a confirmation request to ensure that we were sure that we wanted to include new software in the system. Let's now install software with dnf install, but using the -y option to answer yes to all questions that the command will issue:

```
[root@rhel-instance ~]# dnf install zip -y
[omitted]
Installed:
zip-3.0-33.el9.x86_64
Complete!
```

As you can see, the zip package was installed, without asking questions. We also notice that dnf finds the dependent packages, resolves the **dependencies**, and installs all that is needed for a package to run. This way, the system is kept in a coherent status, making it more reliable and predictable.

We can see which packages are ready to be updated using the dnf check-update command:

```
[root@rhel-instance ~]# dnf check-update
Updating Subscription Management repositories.
Last metadata expiration check: 0:06:39 ago on Wed Apr 13
01:23:21 2022.
NetworkManager.x86_64 1:1.36.0-3.el9_0 rhel-9-for-x86_64-
baseos-rpms
NetworkManager-configserver.noarch 1:1.36.0-3.el9_0 rhel-9-for-
x86_64-baseos-rpms
NetworkManager-libnm.x86_64 1:1.36.0-3.el9_0 rhel-9-for-x86_64-
baseos-rpms
NetworkManager-team.x86_64 1:1.36.0-3.el9_0 rhel-9-for-x86_64-
baseos-rpms
[root@rhel-instance ~]# dnf update NetworkManager -y
[omitted]
Upgraded:
NetworkManager-1:1.36.0-3.el9_0.x86_64
NetworkManager-libnm-1:1.36.0-3.el9_0.x86_64
NetworkManager-team-1:1.36.0-3.el9_0.x86_64
NetworkManager-tui-1:1.36.0-3.el9_0.x86_64
jansson-2.14-1.el9.x86_64
Installed:
NetworkManager-initscripts-updown-1:1.36.0-3.el9_0.noarch
Complete!
```

To update everything, just run dnf update without specifying the package:

```
[root@rhel9 ~]# dnf update
Updating Subscription Management repositories.
Last metadata expiration check: 3:47:38 ago on Wed Apr 13 08:37:34 2022.
Dependencies resolved.
================================================================================
 Package         Arch       Version              Repository                  Size
================================================================================
Installing:
 kernel          x86_64     5.14.0-70.5.1.el9_0  rhel-9-for-x86_64-baseos-beta-rpms   580 k
Installing dependencies:
 kernel-core     x86_64     5.14.0-70.5.1.el9_0  rhel-9-for-x86_64-baseos-beta-rpms    34 M
 kernel-modules  x86_64     5.14.0-70.5.1.el9_0  rhel-9-for-x86_64-baseos-beta-rpms    23 M

Transaction Summary
================================================================================
Install  3 Packages

Total download size: 58 M
Installed size: 94 M
Is this ok [y/N]:
```

Figure 7.13 – Partial capture of RHEL updating with dnf/yum

The result of running `dnf update` in the system is the following:

```
Installed:
kernel-5.14.0-70.5.1.el9_0.x86_64
kernel-core-5.14.0-70.5.1.el9_0.x86_64
kernel-modules-5.14.0-70.5.1.el9_0.x86_64
Complete!
```

These were examples of packages upgraded in a system. Your system, depending on the time you last upgraded it and the newly released packages, may have a different output.

> **Important Note**
> The kernel is the most important part of the system. It enables hardware access and all the basic functions of the operating system. That's why, instead of upgrading it, a new version is installed. The system keeps the previous two versions just in case the system becomes unbootable, and one of them can be selected to run easily.

We can search the packages available with the `dnf search` command:

```
[root@rhel-instance ~]# dnf search wget
Updating Subscription Management repositories.
Last metadata expiration check: 3:52:23 ago on Wed Apr 13
08:37:34 2022.
=============== Name Exactly Matched: wget =========
wget.x86_64 : A utility for retrieving files using the HTTP or
FTP protocols
```

We can obtain extended information about a package, installed or not, with `dnf info`:

```
[root@rhel-instance ~]# dnf info wget
Updating Subscription Management repositories.
Last metadata expiration check: 3:53:13 ago on Wed Apr 13
08:37:34 2022.
Installed Packages
Name         : wget
Version      : 1.21.1
Release      : 6.el9
Architecture : x86_64
Size         : 3.1 M
Source       : wget-1.21.1-6.el9.src.rpm
```

```
Repository    : @System
From repo     : AppStream
Summary       : A utility for retrieving files using the HTTP or
FTP protocols
URL           : http://www.gnu.org/software/wget/
License       : GPLv3+
Description   : GNU Wget is a file retrieval utility which can
use either the HTTP or
              : FTP protocols. Wget features include the ability
to work in the
              : background while you are logged out, recursive
retrieval of
              : directories, file name wildcard matching, remote
file timestamp
              : storage and comparison, use of Rest with FTP
servers and Range with
              : HTTP servers to retrieve files over slow or
unstable connections,
              : support for Proxy servers, and configurability.
```

We can also remove an installed package with dnf remove:

```
[root@rhel-instance ~]# dnf remove screen -y
[omitted]
Removed:  screen-4.8.0-6.el9.x86_64
Complete!
```

Sometimes you want to install some packages that come together to perform a specific task, and that's what **package groups** are for. Let's get a list of the groups first with dnf grouplist:

```
[root@rhel-instance ~]# dnf grouplist | grep Tools
    RPM Development Tools
    Console Internet Tools
    Security Tools
    System Tools
    Development Tools
    Graphical Administration Tools
```

You may run it without | grep Tools to see the full list.

Let's install the `System Tools` group with `dnf groupinstall`:

```
[root@rhel-instance ~]# dnf groupinstall "System Tools"
Updating Subscription Management repositories.
Last metadata expiration check: 3:56:12 ago on Wed Apr 13
08:37:34 2022.
Dependencies resolved.
```

The entire output of the preceding command is shown in the following screenshot:

```
[root@rhel9 ~]# dnf groupinstall "System Tools"
Updating Subscription Management repositories.
Last metadata expiration check: 3:56:12 ago on Wed Apr 13 08:37:34 2022.
Dependencies resolved.
================================================================================
 Package              Arch       Version              Repository          Size
================================================================================
Upgrading:
 libipa_hbac          x86_64     2.6.2-2.el9          rhel-9-for-x86_64-baseos-beta-rpms     38 k
 libsmbclient         x86_64     4.15.5-105.el9_0     rhel-9-for-x86_64-baseos-beta-rpms     79 k
 libsss_certmap       x86_64     2.6.2-2.el9          rhel-9-for-x86_64-baseos-beta-rpms     81 k
 libsss_idmap         x86_64     2.6.2-2.el9          rhel-9-for-x86_64-baseos-beta-rpms     43 k
 libsss_nss_idmap     x86_64     2.6.2-2.el9          rhel-9-for-x86_64-baseos-beta-rpms     45 k
 libsss_sudo          x86_64     2.6.2-2.el9          rhel-9-for-x86_64-baseos-beta-rpms     36 k
 libwbclient          x86_64     4.15.5-105.el9_0     rhel-9-for-x86_64-baseos-beta-rpms     50 k
 openldap             x86_64     2.4.59-4.el9_0       rhel-9-for-x86_64-baseos-beta-rpms    287 k
 samba-client-libs    x86_64     4.15.5-105.el9_0     rhel-9-for-x86_64-baseos-beta-rpms    5.6 M
 samba-common         noarch     4.15.5-105.el9_0     rhel-9-for-x86_64-baseos-beta-rpms    151 k
 samba-common-libs    x86_64     4.15.5-105.el9_0     rhel-9-for-x86_64-baseos-beta-rpms    105 k
 sssd                 x86_64     2.6.2-2.el9          rhel-9-for-x86_64-baseos-beta-rpms     28 k
 sssd-ad              x86_64     2.6.2-2.el9          rhel-9-for-x86_64-baseos-beta-rpms    210 k
 sssd-client          x86_64     2.6.2-2.el9          rhel-9-for-x86_64-baseos-beta-rpms    156 k
 sssd-common          x86_64     2.6.2-2.el9          rhel-9-for-x86_64-baseos-beta-rpms    1.6 M
 sssd-common-pac      x86_64     2.6.2-2.el9          rhel-9-for-x86_64-baseos-beta-rpms     94 k
 sssd-ipa             x86_64     2.6.2-2.el9          rhel-9-for-x86_64-baseos-beta-rpms    275 k
 sssd-kcm             x86_64     2.6.2-2.el9          rhel-9-for-x86_64-baseos-beta-rpms    110 k
 sssd-krb5            x86_64     2.6.2-2.el9          rhel-9-for-x86_64-baseos-beta-rpms     82 k
 sssd-krb5-common     x86_64     2.6.2-2.el9          rhel-9-for-x86_64-baseos-beta-rpms     87 k
 sssd-ldap            x86_64     2.6.2-2.el9          rhel-9-for-x86_64-baseos-beta-rpms    158 k
 sssd-nfs-idmap       x86_64     2.6.2-2.el9          rhel-9-for-x86_64-baseos-beta-rpms     41 k
 sssd-proxy           x86_64     2.6.2-2.el9          rhel-9-for-x86_64-baseos-beta-rpms     72 k
Installing group/module packages:
 NetworkManager-libreswan x86_64 1.2.14-1.el9.3      rhel-9-for-x86_64-appstream-beta-rpms  141 k
 cifs-utils           x86_64     6.14-1.el9           rhel-9-for-x86_64-baseos-beta-rpms    102 k
 libreswan            x86_64     4.6-3.el9            rhel-9-for-x86_64-appstream-beta-rpms  1.3 M
 nmap                 x86_64     3:7.91-10.el9        rhel-9-for-x86_64-appstream-beta-rpms  5.6 M
 openldap-clients     x86_64     2.4.59-4.el9_0       rhel-9-for-x86_64-baseos-beta-rpms    168 k
 samba-client         x86_64     4.15.5-105.el9_0     rhel-9-for-x86_64-appstream-beta-rpms  674 k
 setserial            x86_64     2.17-54.el9          rhel-9-for-x86_64-baseos-beta-rpms     29 k
 tigervnc             x86_64     1.11.0-21.el9        rhel-9-for-x86_64-appstream-beta-rpms  296 k
 tmux                 x86_64     3.2a-4.el9           rhel-9-for-x86_64-baseos-beta-rpms    476 k
 zsh                  x86_64     5.8-9.el9            rhel-9-for-x86_64-baseos-beta-rpms    3.2 M
```

Figure 7.14 – Partial capture of RHEL installing a group dnf/yum

Once the preinstallation completes, we can see that we will install 31 packages:

```
Install  31 Packages
Total download size: 24 M
Is this ok [y/N]:y
```

Replying with y will perform the installation (note that the –y option works here too, assuming yes to all questions).

We can check the history of all the installation transactions with dnf history:

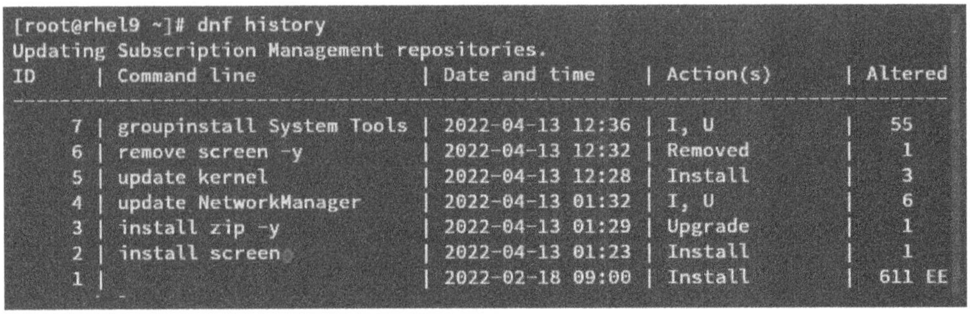

Figure 7.15 – Partial capture of RHEL dnf/yum history

It's easy to obtain specific info from each transaction specifying the number of it with dnf history:

```
[root@rhel-instance ~]# dnf history info 3
Updating Subscription Management repositories.
Transaction ID : 3
Begin time     : Wed Apr 13 01:29:25 2022
Begin rpmdb    : 604:ec03dbdf42251e063bf93ed3743b6cfd23622988
End time       : Wed Apr 13 01:29:25 2022 (0 seconds)
End rpmdb      : 604:97bafed2c0622d1ca789b1b6314e6353d512c339
User           : root <root>
Return-Code    : Success
Releasever     : 9
Command Line   : install zip -y
Comment        :
Packages Altered:
    Upgrade  zip-3.0-33.el9.x86_64 @rhel-9-for-x86_64-baseos-
rpms
    Upgraded zip-3.0-30.el9.x86_64 @@System
```

More interestingly, we can roll back to one of the previous points marked with `dnf history rollback`. To make it faster, *install* the wget package (remove it if it already exists) and then *roll back* to the previous number:

```
[root@rhel-instance ~]# dnf install wget
[omitted]
[root@rhel-instance ~]# dnf history
Updating Subscription Management repositories.
ID| Command line| Date and time   | Action(s)       | Altered
------------------------------------------------------------
10 | install wget | 2022-04-13 12:44 | Install        |    1
9  | remove wget  | 2022-04-13 12:44 | Removed        |    1
8  | install vim  | 2022-04-13 12:40 | Upgrade        |    2
[omitted]
```

The preceding command rolls back everything after transaction number 8 and will leave the OS as it was after transaction 8 was complete:

```
[root@rhel-instance ~]# dnf history rollback 8
[omitted]
Installed:  wget-1.21.1-7.el9.x86_64
Complete!
```

We can also undo a single transaction with `yum history undo`. Let's see it with this transaction:

```
[root@rhel-instance ~]# dnf history undo 10 -y
[omitted]
Removed:
    wget-1.21.1-7.el9.x86_64
Complete!
```

Let's review the most important transactions done with dnf:

Command	Usage
dnf search	Search for a package with a given name.
dnf info	Shows information of a package whether installed or not.
dnf install	Installs the packages mentioned and their dependencies.
dnf install -y	Installs packages and dependencies assuming yes to all questions.
dnf remove	Removes a specified package and the dependencies that would become orphans.
dnf check-update	Shows the packages that have an update available.
dnf update	Updates all packages that have a newer version available.
dnf grouplist	Shows a list of the groups available, whether installed or not.

Table 7.2 – dnf commands to manage transactions and packages

The following table shows a summary of the most important dnf parameters to manage the history actions:

Command	Usage
dnf groupinstall	Installs a group of packages.
dnf history	Shows the list of transactions done with dnf.
dnf history info	Provides info on a specific transaction.
dnf history rollback	Undoes all transactions done up to a specified point.
dnf history undo	Undoes a specific transaction.

Table 7.3 – dnf commands to list and manage the history

Modularity enables the possibility to have different versions of the same package on the system. It is all managed with dnf so there is no need to install extra software:

> **Important Note**
>
> The modules are enabled by RHEL 9.1. If you are using RHEL 9.0, the following command won't return any module.

```
[root@rhel-instance ~]# dnf module list
 Updating Subscription Management repositories.


Last metadata expiration check: 3:32:45 ago on Wed 31 Aug 2022
12:46:57 PM CEST.
Red Hat Enterprise Linux 9.1 AppStream BETA for x86_64 - (RPMs)
Name    Stream Profiles                   Summary
maven   3.8     common [d]                Java project management
and project comp rehension tool
nodejs 18       common [d], development Javascript runtime,
minimal, s2i
php     8.1     common [d], devel, mini PHP scripting language
mal
ruby    3.1     common [d]                An interpreter of object-
oriented scripting language
Hint: [d]efault, [e]nabled, [x]disabled, [i]nstalled
```

> **Tip**
>
> The `dnf module list` command, without specifying any package, will show the full list of modules. Try it!

As you can see, we have three different versions of the PostgreSQL database available in RHEL 9, which are 9.6, 10, and 12. None of them are enabled and the default one is version 10.

Let's enable version 8.1 of PHP using `dnf module`:

```
[root@rhel-instance ~]# dnf module enable php
[omitted]
Enabling module streams: php                     8.1
[omitted]
Is this ok [y/N]: y
Complete!
[root@rhel-instance ~]# dnf module list php
```

The output of the preceding command can be seen in the following screenshot:

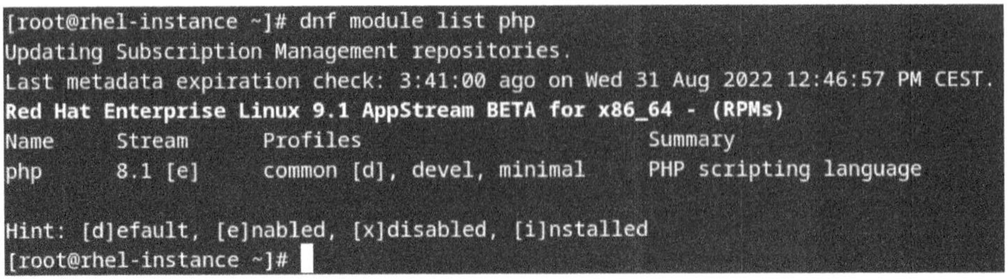

```
[root@rhel-instance ~]# dnf module list php
Updating Subscription Management repositories.
Last metadata expiration check: 3:41:00 ago on Wed 31 Aug 2022 12:46:57 PM CEST.
Red Hat Enterprise Linux 9.1 AppStream BETA for x86_64 - (RPMs)
Name      Stream      Profiles                       Summary
php       8.1 [e]     common [d], devel, minimal     PHP scripting language

Hint: [d]efault, [e]nabled, [x]disabled, [i]nstalled
[root@rhel-instance ~]#
```

Figure 7.16 – Capture of the PostgreSQL module list

From now on, Yum will install, update, and maintain version 8.1 of PHP on this system. Let's install it:

```
[root@rhel-instance ~]# dnf install php -y
[omitted]
Installed:

  apr-1.7.0-11.el9.x86_64
  apr-util-1.6.1-20.el9.x86_64
apr-util-bdb-1.6.1-20.el9.x86_64
  apr-util-openssl-1.6.1-20.el9.x86_64
Complete!
```

In the previous example, version 8.1 was installed.

We can remove the PostgreSQL package and reset the module status to go back to the initial point:

```
[root@rhel-instance ~]# dnf remove php -y
[omitted]
Removing:
  apr-1.7.0-11.el9.x86_64
  apr-util-1.6.1-20.el9.x86_64

[omitted]
Complete!
[root@rhel-instance ~]# dnf module reset php

[omitted]
Complete!
```

```
[root@rhel-instance ~]# dnf module list php
Updating Subscription Management repositories.
Last metadata expiration check: 1:23:21 ago on dom 14 feb 2021
19:25:32 CET.
Red Hat Enterprise Linux 9.1 AppStream for x86_64 - AppStream
(RPMs)
Name               Stream          Profiles                 Summary
php             8.1 [e]           common [d], devel, minimal PHP
scripting language

Hint: [d]efault, [e]nabled, [x]disabled, [i]nstalled
```

Let's review the commands shown in this section for modularity:

Command	Usage
dnf module list	Lists all modules available in the system
dnf module enable	Enables one specific module to work within the system
dnf module reset	Clears the configuration for a module

Table 7.4 – dnf commands and parameters

> **Tip**
> For more info on modularity, go to the system's manual page by running man dnf.modularity.

Now that we have learned how to handle software transactions in RHEL, let's go ahead with how to create and handle local repositories.

Creating and syncing repositories with createrepo and reposync

It's common that we receive an RPM file and keep it in a repository that we can use on our machine (and sometimes share it with other machines with a web server of an NFS share). It's also common that when we start building our own RPMs, we distribute them and, to do so, we need to create a repository. To do that, we can use the **createrepo** tool.

First, let's create a folder in /var/tmp for repos:

```
[root@rhel-instance ~]# cd /var/tmp/
[root@rhel-instance tmp]# mkdir repos
[root@rhel-instance tmp]# cd repos/
```

Then, let's create a folder for Slack, a common tool to communicate with your team, and download the RPM package:

```
[root@rhel-instance repos]# mkdir slack
[root@rhel-instance repos]# cd slack/
[root@rhel-instance repos]# curl -s -O https://downloads.slack-
edge.com/releases/linux/4.25.0/prod/x64/slack-4.25.0-0.1.fc21.
x86_64.rpm
[root@rhel-instance slack]# ls -l
total 70620
-rw-r--r--. 1 root root 72314252 Apr 13 19:03 slack-4.25.0-
0.1.fc21.x86_64.rpm
```

Now we have a repository with an RPM file. We could have one with as many RPMs as we want, but we will continue with only this single package.

Let's install the createrepo tool:

```
[root@rhel-instance slack]# dnf install -y createrepo
[omitted]
Installed:
    createrepo_c-0.17.7-4.el9_0.x86_64    createrepo_c-
libs-0.17.7-4.el9_0.x86_64
Complete!
```

And now we could simply run it to create a repository in the current folder with the following command:

```
[root@rhel-instance slack]# createrepo .
Directory walk started
Directory walk done - 1 packages
Temporary output repo path: ./.repodata/
Preparing sqlite DBs
Pool started (with 5 workers)
Pool finished
[root@rhel-instance slack]# ls -l
```

```
total 62656
total 70624
drwxr-xr-x. 2 root root     4096 Apr 13 19:04 repodata
-rw-r--r--. 1 root root 72314252 Apr 13 19:03 slack-4.25.0-
0.1.fc21.x86_64.rpm
```

We see that the `repodata` folder has been created. In it, we can find the `repomd.xml` file that defines the repository content and also the recently created index files:

```
[root@rhel-instance slack]# ls repodata/
14288f14d67dd571bc34ce3ecd735a68ed8231a3954672eecbd59accb
b3c2608-filelists.xml.gz
62ba8d6a9a1f41181f57bda494a8da7c80774b2be5abb64
5dd63376879276955-primary.sqlite.bz2
875605b939be5476644ce5d0ddd814b015f7c733ba7d9d9dd5432558b53d
b1d7-other.sqlite.bz2
b034107c628a1af32ce6f138ce06c2507a38407915825fa3300fbaa0ce095
ccc-filelists.sqlite.bz2
e381ab772c9a99fb9064d76eca11746d82ea76c46293612c9f206b39e7
e6c921-other.xml.gz
fc5d85aa39dc97d7153ba206d7502c1966cae0d88c15d9ce24015d69f59cf
77f-primary.xml.gz
repomd.xml
```

Now we can add the repository to the system. We could do it without gpg signatures, setting the gpgcheck variable to 0 but, to have better security, let's do it with the gpg signature. By searching the slack page, we find the signature and download it to the /etc/pki/rpm-gpg directory:

```
[root@rhel-instance slack]# curl https://slack.com/gpg/slack_
pubkey_2019.gpg -o /etc/pki/rpm-gpg/RPM-GPG-KEY-SLACK
```

Then, we add the repository to the system by creating the file /etc/yum.repos.d/local-slack.repo with the following content:

```
[local-slack-repo]
name=Local Slack Repository
baseurl=file:///var/tmp/repos/slack
enabled=1
gpgcheck=1
gpgkey=file:///etc/pki/rpm-gpg/RPM-GPG-KEY-SLACK
```

And now we can try searching `slack`. To install Slack, it would require that the package group *Server with a GUI* is installed and also that the EPEL repository is installed and enabled, however, for the purpose of this exercise, we can continue without its installation. We can run the search by running `dnf search slack`:

```
[root@rhel-instance slack]# # dnf search slack
Updating Subscription Management repositories.
Last metadata expiration check: 0:02:36 ago on Wed Apr 13
13:10:07 2022.
====== Name & Summary Matched: slack ============
slack.x86_64 : Slack Desktop
```

Once a new version of Slack appears, we can download it to the same folder, and regenerate the repository index by running `createrepo` again. This way, all the systems using this repository will update `slack` when they run `yum update`. It's a good way to keep all systems standardized and in the same version. For advanced features when managing RPM repositories, please check Red Hat Satellite.

Sometimes we want to have a local replica of the repositories in our system. To do that, we can use the **reposync** tool.

First, we install `reposync`, which comes in the `yum-utils` package:

```
[root@rhel-instance ~]# dnf install yum-utils -y
[omitted]
Installed:
   yum-utils-4.0.24-4.el9_0.noarch
Complete!
```

> **Tip**
> If you try to install the `dnf-utils` package, the same `yum-utils` package will be installed.

Now it's time to disable all repos provided by Red Hat except `rhel-9-for-x86_64-baseos-rpms`, which can be done with the following command:

```
[root@rhel-instance ~]# subscription-manager repos
--disable="*"
[root@rhel-instance ~]# subscription-manager repos
--enable="rhel-9-for-x86_64-baseos-rpms"
```

Time to check the change:

```
[root@rhel-instance ~]# dnf repolist
Updating Subscription Management repositories.
repo id                                      repo name
rhel-9-for-x86_64-baseos-rpms Red Hat Enterprise Linux 9 for
x86_64 - BaseOS (RPMs)
local-slack-repo                                    Local
Slack Repository
mirror.uv.es_mirror_fedora-epel_9_Everything_x86_64_ created
by dnf config-manager from http://mirror.uv.es/mirror/fedora-
epel/9/Everything/x86_64/
```

We can also disable the other repos, but this time we will do it in a different way, renaming them to something that doesn't end with .repo:

```
[root@rhel-instance ~]# mv /etc/yum.repos.d/local-slack.repo  /
etc/yum.repos.d/local-slack.repo_disabled
[root@rhel-instance ~]# mv /etc/yum.repos.d/mirror.uv.es_
mirror_fedora-epel_9_Everything_x86_64_.repo  /etc/yum.repos.d/
mirror.uv.es_mirror_fedora-epel_9_Everything_x86_64_.repo_
disabled
[root@rhel-instance ~]# dnf repolist
Updating Subscription Management repositories.
repo id                              repo name
rhel-9-for-x86_64-baseos-rpms Red Hat Enterprise Linux 9 for
x86_64 - BaseOS(RPMs)
```

Now we can run reposync with some options:

```
[root@rhel-instance ~]# cd /var/tmp/repos
[root@rhel-instance repos]# reposync --newest-only --download-
metadata --destdir /var/tmp/repos
Updating Subscription Management repositories.
[omitted]
(1137/1139): samba-libs-4.15.5-105.el9_0.x86_64.rpm
352 kB/s | 107 kB      00:00
(1138/1139): grub2-efi-x64-modules-2.06-27.el9_0.noarch.rpm
805 kB/s | 1.1 MB      00:01
(1139/1139): glibc-langpack-ro-2.34-28.el9_0.x86_64.rpm
533 kB/s | 483 kB      00:00
```

```
[root@rhel-instance repos]# ls
rhel-9-for-x86_64-baseos-rpms  slack
[root@rhel-instance repos]# ls rhel-9-for-x86_64-baseos-rpms/
Packages  repodata
[root@rhel-instance repos]# ls rhel-9-for-x86_64-baseos-rpms/
repodata/
0c625572996074ed6222993172444ac83ad60a6132d1fdd5f0978c342
c8b0935-filelists.xml.gz
11f262d81fac509a4ac8d8ddaccb3fdc924445a769ff6b7dad977caf1
12d2301-updateinfo.xml.gz
1d8ae930-bdbc-443a-857c-bbf32dc44512
1f6c95265723ea0480714d4e5005a5b7330adca989a1fb40704235e0
e0c29880-filelists.sqlite.bz2
52420e103dc1497dd802721146871a2e4700761b8bc8a308eaa48938fce5
56f8-primary.xml.gz
bceb83ebd1751392966aef17b680651d165467a16141d0c1923f8e9fa121
abe5-other.sqlite.bz2
c3e9411464f827d2bbad2d3e517ad1eedb8598bf2e7aedf02b46cb87c06a
c8d3-other.xml.gz
dc9cdc0c049addd83d60ac5e0edf3b20be010c1c3ac609b4e8b5e6fec3ee2
73d-comps.xml
e23e942c6db0f83fa07023cc435692e9582fcd53943200a457de0aedc8
d5b699-primary.sqlite.bz2
repomd.xml
```

This will download the latest packages for the enabled channels. Let's review the options:

- --newest-only: Red Hat repositories keep all the versions of the packages since the first release. This will download only the latest version.

- --download-metadata: To be sure that we download a fully functional repo, and we do not need to run createrepo on it, we can use this option, which will retrieve all metadata in a source repository.

- --destdir /var/tmp/repos: Sets the destination directory for the downloaded files. It will also create a directory for each repo configured so the specified directory will be the parent of them all.

With this replicated repository, we can also work in isolated environments. It could be very convenient to prepare test environments. For advanced repo management features, please remember to try Red Hat Satellite.

After learning the basics of repositories and how to use them to manage software, let's dive into the technology behind it, the **Red Hat Package Manager**, or **RPM**.

Understanding RPM internals

Linux distributions tend to have their own package manager, from Debian with .deb to Pacman in Arch Linux and other more exotic mechanisms. The intention of package managers is to keep software installed on the system, update it, patch it, keep dependencies, and maintain an internal database of what is installed on the system. RPM is used by distributions such as Fedora, openSUSE, CentOS, Oracle Linux, and, of course, RHEL.

To handle RPMs, the rpm command is available on the system, however, since the introduction of yum/dnf, it is hardly ever used in system administration, and is not included in RHCSA.

RPMs contain the following:

- The files to be installed on the system, stored in CPIO format and compressed
- Information on permissions and the assigned owner and group for each file
- The dependencies required and provided by each package, along with, conflicts with other packages
- Install, uninstall, and upgrade scripts to be applied in any of those phases
- A signature to ensure the package was not modified

To learn a bit about it, we will show some simple useful commands.

Commands to check packages include the following:

- rpm -qa: Lists all the installed packages on the system
- rpm -qf <filename>: Shows which package installed the mentioned filename
- rpm -ql <packagefile>: Lists the files included in a downloaded package (interesting to check previously downloaded packages)

Commands to install, upgrade, and remove include the following:

- rpm -i <packagefile>: Installs the list of provided packages, not fetching dependencies
- rpm -U <packagefile>: Upgrades a package to the downloaded one. Checks dependencies but doesn't manage them
- rpm -e <packagename>: Removes the packages specified, although it won't remove dependencies

If you want to understand how the dependency management system works in yum/dnf, try installing packages with rpm -i.

> **Tip**
> You can also check the following Red Hat Customer Portal solution to learn about how to download RPM packages manually from the Customer Portal: https://access.redhat.com/solutions/6996

It is important to know that all the databases of installed packages are located in /var/lib/rpm and can be managed with the rpmdb command.

Nowadays, having to work with the rpm command usually means having a low-level issue, so it's better to try to break a test system before having to use it in real life.

With this, we've completed our look at software management in RHEL systems.

Summary

In this chapter, we have gone through the admin parts of software management in a RHEL 9 system, from subscriptions to installation, to modularity, and other miscellaneous tips.

All the system patching, updating, and management in RHEL relies on yum/dnf and simplifies managing dependencies, installing the right versions of software, and distributing it in isolated environments. This is one of the tasks more common for system administrators and should be understood completely.

For the Red Hat Certified Engineer level, a more in-depth look will be required, including creating RPM packages, which are very useful to manage, maintain, and distribute internally produced software in your own environments leveraging the experience and tools that Red Hat provides.

Now that our systems are up to date, let's move on to learn how to manage them remotely in the upcoming chapter.

Part 2 –
Security with SSH, SELinux, a Firewall, and System Permissions

Security in production systems is a direct responsibility of the systems administrator. To handle that, RHEL includes capabilities such as SELinux, an integrated firewall, and of course, the standard system permissions. In this part, a good overview and understanding of the security mechanisms in RHEL are provided so that you can perform everyday maintenance tasks.

The following chapters are included in this part:

- *Chapter 8, Administering Systems Remotely*
- *Chapter 9, Securing Network Connectivity with firewalld*
- *Chapter 10, Keeping Your System Hardened with SELinux*
- *Chapter 11, System Security Profiles with OpenSCAP*

8

Administering Systems Remotely

When working with systems, once the server has been installed, and many times, even during the installation itself, administration can be performed remotely. Once a machine has been installed, the tasks that need to be performed during its life cycle are not that different from the ones that have already been performed.

In this chapter, we will cover, from a connection point of view, how to connect to remote systems, transfer files, and automate the connection so that it can be scripted and make it resilient if issues arise with the network link. Administration tasks that can be performed on the system are the same as the ones we described in previous chapters, such as installing software, configuring additional networking settings, and even managing users.

Since administering a system requires privileged credentials, we will focus on the available tools that are considered to be secure to perform such connections, as well as how to use them to encapsulate other traffic.

In addition, we will add a brief introduction to Ansible as an automation tool for system administration to present some of the basics of it and show how can be used for managing at scale.

We will cover the following topics:

- SSH and OpenSSH overview and base configuration
- Enabling root access via SSH (and when not to do it)
- Accessing remote systems with SSH
- Key-based authentication with SSH
- Remote file management with SCP/rsync
- Advanced remote management—SSH tunnels and SSH redirections
- Remote terminals with tmux
- Introduction to Ansible automation

By covering these topics, we will be able to master remote system access and bring our administration skills to the next level.

Let's start by talking about the SSH protocol and the OpenSSH client and server in the next section.

Technical requirements

You can continue using the **virtual machine** (**VM**) that we created at the beginning of this book in *Chapter 1*, *Getting RHEL Up and Running*. Any additional packages that are required will be indicated in the text. Any additional files that are required for this chapter can be downloaded from `https://github.com/PacktPublishing/Red-Hat-Enterprise-Linux-RHEL-9-Administration`.

SSH and OpenSSH overview and base configuration

SSH is an acronym for **Secure Shell Host**. It started to spread by replacing traditional Telnet usage, which was a remote login protocol that used no encryption for connecting to hosts, so the credentials that were used for logging in were transmitted in plain text. This means that anyone who had a system between the user terminal and the remote server could intercept the username and password and use that information to connect to remote systems. This is similar to what happens when credentials are transmitted to a web server via **Hypertext Transfer Protocol** (**HTTP**) and not **HTTP Secure** (**HTTPS**).

With SSH, a secure channel is created between the client and the target host, even if the connection is performed over untrusted or insecure networks. Here, the SSH channel that's created is secure and no information is leaked.

OpenSSH provides both a server and a client (the `openssh-server` and `openssh-clients` packages) in **Red Hat Enterprise Linux** (**RHEL**) that can be used to connect to and allow connections from remote hosts.

> Tip
> Knowing everything is not possible, so it is really important for **Red Hat Certified System Administrator** (**RHCSA**)-certified individuals (and even later certifications, if you followed that path) to be resourceful. We already know how to install packages and how to check the **manual** (**man**) pages that are installed by them, but we can also use those packages to find the necessary configuration files. This skill can be used to find the possible configuration files we need to edit to configure a service or a client. Remember to use `rpm -ql package` to review the list of files provided by a package if you cannot remember which one to use.

The default configuration for both the client and server allows connections, but there are many options that can be tuned.

OpenSSH server

OpenSSH is a free implementation based on the last free SSH version that was created by OpenBSD members and updated with all the relevant security and features. It has become a standard in many operating systems, both as a server and as a client, to make secure connections between them.

The main configuration file for the OpenSSH server is located at `/etc/ssh/sshd_config` (and you can use `man sshd_config` to get detailed information about the different options). Some of the most widely used options are listed here:

- `AcceptEnv`: Defines which environment variables that have been set by the client will be used on the remote host (for example, locale, terminal type, and so on).

- `AllowGroups`: A list of groups a user should be a member of to get access to the system.

- `AllowTcpForwarding`: Allows us to forward ports using the SSH connection (we will discuss this later in this chapter, in the *Advanced remote management – SSH tunnels and SSH redirections* section).

- `DisableForwarding`: This takes precedence over other forwarding options, making it easier to restrict the service.

- `AuthenticationMethods`: Defines which authentication methods can be used, such as disabling password-based access.

- `Banner`: Files to send to the connecting user before authentication is allowed. This defaults to no banner, which might also reveal who is running the service that might be providing too much data to possible attackers.

- `Ciphers`: A list of valid ciphers to use when you're interacting with the server. You can use + or – to enable or disable them.

- `ListenAddress`: The hostname or address and port where the `sshd` daemon should be listening for incoming connections.

- `PasswordAuthentication`: This defaults to `yes` and can be disabled to block users from interactively connecting to the system unless a public/private key pair is used.

- `PermitEmptyPasswords`: Allows accounts with no password to access the system (the default is `no`).

- `PermitRootLogin`: Defines how login works for the root user—for example, to avoid the root user from connecting remotely with a password. The default setting is `prohibit-password`—that is, only SSH keys can be used.

- `Port`: Related to `ListenAddress`, this defaults to `22`. This is the port number where the `sshd` daemon listens for incoming connections.

- `Subsystem`: Configures the command for the external subsystem. For example, it is used with `sftp` for file transfers.

- `X11Forwarding`: This defines whether `X11` forwarding is permitted so that remote users can open graphical programs on their local display by tunneling the connection.

The following screenshot shows the options that are installed by our system while we're removing comments (note—there are other folders that can keep configurations; make sure to check `/etc/ssh/sshd_config.d/`, `/etc/ssh/sshd_config`, and `/etc/crypto-policies/back-ends/opensshserver.config` for additional settings):

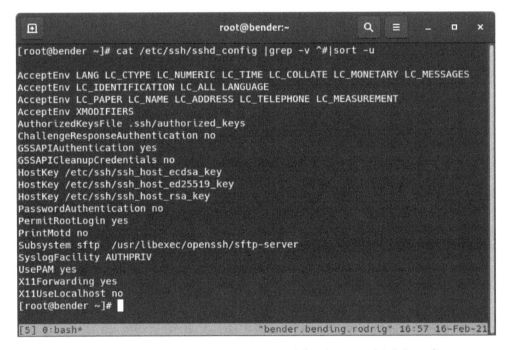

Figure 8.1 – Default values at installation time defined in /etc/ssh/sshd_config

We'll check the client part of the configuration in the next section.

OpenSSH client

The client part of OpenSSH is configured system-wide via the `/etc/ssh/ssh_config` file and the files in the `/etc/ssh/ssh_config.d/` folder. They are also configured via each user `~/.ssh/config` file.

Usually, the system-wide file just contains some comments, not actual settings, so we will be focusing on the per-user configuration file and command-line parameters.

One example entry in our `~/.ssh/config` file could look like this:

```
Host jump
    Hostname jump.example.com
    User root
    Compression yes
    StrictHostKeyChecking no
    GSSAPIAuthentication yes
    GSSAPIDelegateCredentials yes
    GSSAPIKeyExchange yes
    ProxyCommand connect-proxy -H squid.example.com:3128 %h %p
    ControlPath ~/.ssh/master-%r@%h:%p
    ControlMaster auto
```

In the previous example, we defined an entry named `jump` (that we can use with `ssh jump`) that will connect the `root` username to the `jump.example.com` host.

This is a basic setting, but we're also defining that we'll be using a helper program in `ProxyCommand` that will make use of a proxy server on `squid.example.com` on port 3128 to connect to the `%h` host and `%p` port to reach our target system. Additionally, we're making use of `Compression` and using `ControlMaster` with additional `GSSAPI` authentication.

One feature that has security implications is `StrictHostKeyChecking`. When we connect to a host for the first time, keys are exchanged between the client and the host, and the server identifies itself with the keys that are used. If they're accepted, they will be stored in the `.ssh/known_hosts` file at the user's home.

If the remote host key is changed, a warning will be printed on the `ssh` client's terminal and the connection will be refused, but when we set `StrictHostKeyChecking` to no, we will accept any key that's sent by the server, which might be useful if we're using a test system that gets redeployed frequently (and thus, generating a new host key). It is not recommended to be used in general, since it protects us from a server being replaced and also someone impersonating the server we want to connect to with a server that—for example—logs usernames and passwords to access our system later.

In the next section, we will learn about accessing remote systems with SSH.

Enabling root access via SSH (and when not to do it)

As we introduced in the previous section, there are some options that directly affect the access as root to the system, as shown in the following example:

```
PermitRootLogin
PasswordAuthentication
```

Depending on your security requirements, there might be better options or combinations to use, so let's discuss some scenarios.

It's common to set some requirements such as not allowing root login, which can be set via `PermitRootLogin no` in the `sshd` configuration file (`/etc/ssh/sshd_config`). This enforces that every user willing to become root will need to first SSH into the system with another user, and once connected, use `su` or `sudo` to switch to the root user.

Alternatively, you can use `PermitRootLogin prohibit-password`, which will only allow connections to the system as root when using public key authentication (ahead in this chapter).

Probably, the worst case is to set `PermitRootLogin yes` as this will allow access to the root account via `ssh` using passwords.

In general, allowing access with SSH empowers remote attackers to use password-testing tools to try getting into the system, but also allowing them to use root makes it even easier, as they can attempt to log in directly as a privileged user.

If you still need to use passwords for your users or root, there are third-party tools such as `fail2ban` that can help you secure your system a bit more, as those utilities can be configured to block the **Internet Protocol** (IP) addresses of users trying to brute-force the password after some failed attempts (but still, they might use proxies or other ways to get into the system).

My recommendation, whenever possible, would be to only allow root login when using private keys (`prohibit-password`), and in general, consider disabling any kind of passwords with `PasswordAuthenticaton no` as this will also protect other users, allowing them only to use their generated private key to access the system, especially to servers.

Let's learn a bit more about the different ways to reach systems in the next sections.

Accessing remote systems with SSH

SSH, as we mentioned earlier in this chapter, is a protocol that's used to connect to remote systems. In general, the syntax, in its most basic form, is just executing `ssh host` within a terminal.

The `ssh` client will then initiate a connection to the `ssh` server on the target host, using the username of the currently logged-in user by default, and will try to reach the remote server on port `22/tcp`, which is the default for the SSH service.

In the following screenshot, we can see the closest server to our `localhost` system, which means we will be connecting to our own server:

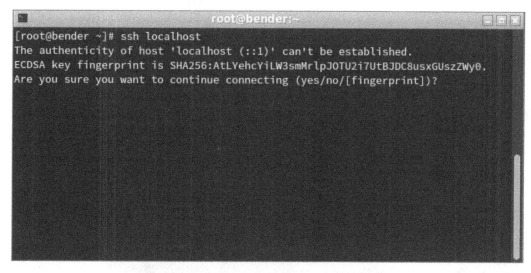

Figure 8.2 – Initiating an SSH connection to localhost

In the preceding screenshot, we can see how the first interaction with the server prints the fingerprint of the server to authenticate it. This is what was discussed in the *OpenSSH client* section; that is, `StrictHostKeyChecking`. Once accepted, if the host key changes, the connection will be denied until we manually remove the older key from the `.ssh/known_hosts` file to confirm that we're aware of the server change.

Let's add the key and try again. The output is shown in the following screenshot:

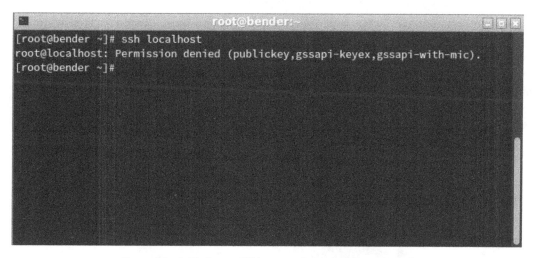

Figure 8.3 – Initiating an SSH connection to localhost denied

On our second attempt, the connection failed, but let's examine the output; that is, `Permission denied (publickey,gssapi-keyex,gssapi-with-mic)`. What does it mean? If we pay attention, `password` isn't listed, which means that we cannot connect to this host via a password prompt (which comes from setting `PasswordAuthentication` to `no`, which we defined in the `/etc/ssh/sshd_config` file).

In the following screenshot, we can see that once we set `PasswordAuthentication` to `yes`, the system asks for the password, which is not echoed on the screen. Once validated, we get a shell prompt so that we can start typing in commands:

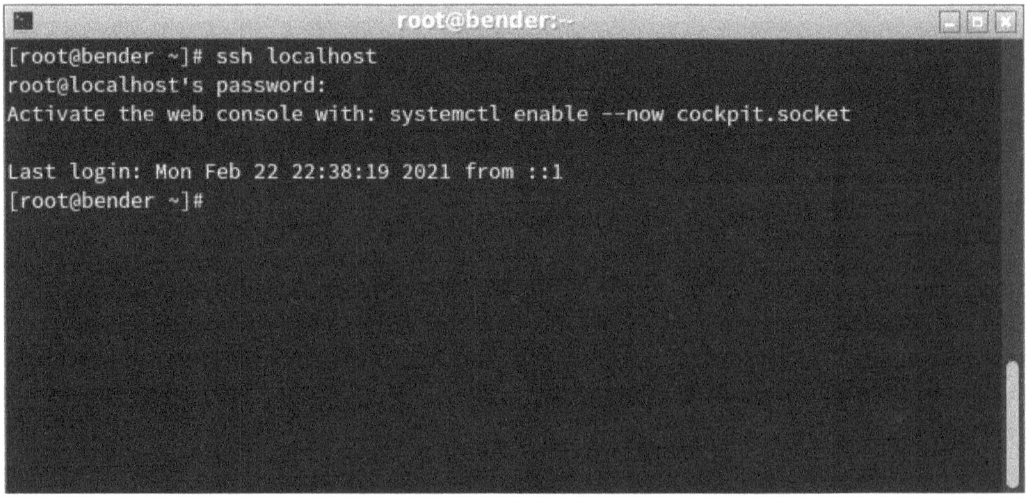

```
[root@bender ~]# ssh localhost
root@localhost's password:
Activate the web console with: systemctl enable --now cockpit.socket

Last login: Mon Feb 22 22:38:19 2021 from ::1
[root@bender ~]#
```

Figure 8.4 – SSH connection completed

In general, password authentication can be a security risk as the keyboard might be intercepted, someone might be looking over your shoulder, a brute-force attack might be used against the accounts, and so on. Due to this, it's common practice to at least disable it for the `root` user, meaning that someone trying to log in to the system should know the username and password of a user, and from there, use the system tools to become `root`.

Let's learn how to log in to remote systems that have passwords disabled by using authentication keys.

Key-based authentication with SSH

One big advantage of SSH connections is that commands can be given to be executed on remote hosts—for example, to grab updated data that can be used for monitoring without requiring a specific agent on the host.

Having to provide login details on each connection is not something that we could consider an improvement to the **user experience (UX)**, but SSH also allows us to create a key pair that can be used for authentication to remote systems so that no password or credential input is required.

The keys contain two parts: one that is public and must be configured in each host we want to connect to, and one that is private and must be secured as it will be used to identify us while we're trying to connect to remote hosts.

It is not necessary to say that this entire process happens over the encrypted connection created by SSH. So, using SSH and compression will also make our connections faster versus other legacy methods such as Telnet, which is unencrypted.

First of all, let's create one key pair for authentication.

> **Tip**
>
> It is recommended to have at least one key pair per user so that each user can have keys based on the roles when they're connecting to servers. Even if the keys can be shared for users in a role, it's better to have each user have their own set so that keys can be revoked individually. For example, we can keep several ssh key pairs to be used in different roles, such as personal systems, production systems, lab systems, and so on. Having to specify the key pair for connecting is also an extra security measure: we cannot connect to production systems unless we use the production key pair.

To create a key pair, we can use the `ssh-keygen` tool, which has several options for the key we are creating, as shown in the following screenshot:

```
 ⊞                            root@bender:~                      Q  ≡   _  ▫  ✕
usage: ssh-keygen [-q] [-b bits] [-t dsa | ecdsa | ed25519 | rsa] [-m format]
                  [-N new_passphrase] [-C comment] [-f output_keyfile]
       ssh-keygen -p [-P old_passphrase] [-N new_passphrase] [-m format]
                  [-f keyfile]
       ssh-keygen -i [-m key_format] [-f input_keyfile]
       ssh-keygen -e [-m key_format] [-f input_keyfile]
       ssh-keygen -y [-f input_keyfile]
       ssh-keygen -c [-P passphrase] [-C comment] [-f keyfile]
       ssh-keygen -l [-v] [-E fingerprint_hash] [-f input_keyfile]
       ssh-keygen -B [-f input_keyfile]
       ssh-keygen -D pkcs11
       ssh-keygen -F hostname [-f known_hosts_file] [-l]
       ssh-keygen -H [-f known_hosts_file]
       ssh-keygen -R hostname [-f known_hosts_file]
       ssh-keygen -r hostname [-f input_keyfile] [-g]
       ssh-keygen -G output_file [-v] [-b bits] [-M memory] [-S start_point]
       ssh-keygen -T output_file -f input_file [-v] [-a rounds] [-J num_lines]
                  [-j start_line] [-K checkpt] [-W generator]
       ssh-keygen -s ca_key -I certificate_identity [-h] [-U]
                  [-D pkcs11_provider] [-n principals] [-O option]
                  [-V validity_interval] [-z serial_number] file ...
       ssh-keygen -L [-f input_keyfile]
       ssh-keygen -A
       ssh-keygen -k -f krl_file [-u] [-s ca_public] [-z version_number]
                  file ...
       ssh-keygen -Q -f krl_file file ...
[root@bender ~]# ▮
```

Figure 8.5 – ssh-keygen options

When no arguments are provided, by default, it will create a key for the current user and ask for a password for the key. When we use the defaults and provide no values, we get an output similar to the one shown in the following screenshot:

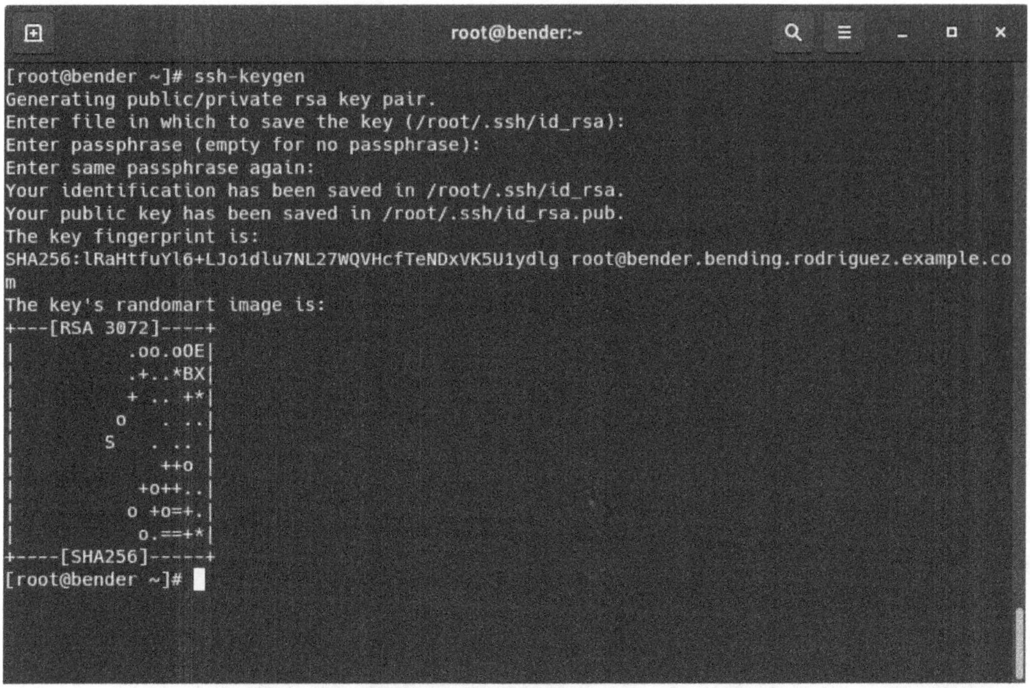

```
[root@bender ~]# ssh-keygen
Generating public/private rsa key pair.
Enter file in which to save the key (/root/.ssh/id_rsa):
Enter passphrase (empty for no passphrase):
Enter same passphrase again:
Your identification has been saved in /root/.ssh/id_rsa.
Your public key has been saved in /root/.ssh/id_rsa.pub.
The key fingerprint is:
SHA256:lRaHtfuYl6+LJo1dlu7NL27WQVHcfTeNDxVK5U1ydlg root@bender.bending.rodriguez.example.co
m
The key's randomart image is:
+---[RSA 3072]----+
|         .oo.oOE|
|         .+..*BX|
|         + .. +*|
|       o   . ..|
|      S   . ...  |
|         ++o |
|        +o++.. |
|       o +o=+. |
|        o.==+*|
+----[SHA256]-----+
[root@bender ~]# █
```

Figure 8.6 – ssh-keygen execution creating a Rivest-Shamir-Adleman
(RSA) key pair under ~/.ssh/{id_rsa,id_rsa.pub}

From this point on, this system has created a key pair for the root user, and it has stored the two pieces of it in the same folder, which is .ssh by default. The public one contains the .pub suffix, while the other contains the private key.

How do we use them? If we look inside the .ssh folder in our home directory, we can see several files: we have an authorized_keys file and a known_hosts file, in addition to the key pair we have just created. The authorized_keys file will contain one entry per line. This contains the public keys that can be used to log in to this system for this user.

> **Tip**
>
> The vast range of options that can be used with authorized_keys goes further than adding just regular keys—you can also define commands to execute, expiry times for keys, remote hosts that can be used to connect so that only those hosts will be able to use that key successfully, and much more. Again, man sshd is your friend, so check out the AUTHORIZED_KEYS FILE FORMAT section there to learn about more complex setups.

To simplify how keys are set up on remote systems, we have the `ssh-copy-id` utility, which connects via `ssh` to the remote host. This will ask for the `ssh` password and install the available public keys on our system. However, this requires the system to have password authentication enabled.

The alternate method consists of manually appending our public key to that file (`.ssh/authorized_keys`), as shown in the following screenshot:

```
[root@bender ~]# ssh-copy-id localhost
/usr/bin/ssh-copy-id: INFO: Source of key(s) to be installed: "/root/.ssh/id_rsa.pub"
/usr/bin/ssh-copy-id: INFO: attempting to log in with the new key(s), to filter out any tha
t are already installed
/usr/bin/ssh-copy-id: INFO: 1 key(s) remain to be installed -- if you are prompted now it
s to install the new keys
root@localhost: Permission denied (publickey,gssapi-keyex,gssapi-with-mic).
[root@bender ~]# cat .ssh/id_rsa.pub >> .ssh/authorized_keys
[root@bender ~]# ssh localhost uname -a
Linux bender.bending.rodriguez.example.com 4.18.0-240.1.1.el8_3.x86_64 #1 SMP Thu Nov 19 17
:20:08 UTC 2020 x86_64 x86_64 x86_64 GNU/Linux
[root@bender ~]#
```

Figure 8.7 – ssh-copy-id failure and manual authorization of the private key

The first line has attempted to use `ssh-copy-id`, but since we had password authentication enabled, it tried to copy our public key and failed. Then, we appended the public key using `>>` to the `authorized_keys` file. Finally, we demonstrated how to connect to `localhost` with `ssh` and execute a command without a password.

> **Important note**
>
> The permissions for the `.ssh` folder and the `authorized_keys` file must not be too wide open (for example, 777). If they are, the `ssh` daemon will reject them as someone could have appended new keys and tried to gain access without really being a legitimate user of the system. The `.ssh` folder should be set to 0700 permissions and the private keys such as `id_rsa` should be set to 0600 permissions.

What has just happened opens a new world of automation. Using the keys being exchanged between our system and the remote hosts, we can now connect remotely to them to run commands interactively or to script commands to be executed on remote hosts. We can check the results in our terminal. Let's consider this simple script, for a system load average check, which is available at `https://github.com/PacktPublishing/Red-Hat-Enterprise-Linux-RHEL-9-Administration/`

`blob/main/chapter-08-remote-systems-administration/loadaverage-check.sh`:

```
#!/usr/bin/bash
for system in host1 host2 host3 host4;
do
     echo "${system}: $(ssh ${system} cat /proc/loadavg)"
done
```

In this example, we're running a loop to connect to four systems and then outputting the name and the load average of that system, as shown in the following screenshot:

Figure 8.8 – Passwordless login to four hosts to check their load average

As we can see, we quickly grabbed the information from four hosts over `ssh`. If you want to test this in your environment, you might want to put into practice what we learned about creating entries in the `/etc/hosts` file, which points to `127.0.0.1` for the hostnames we want to try, so that the connection goes to your own practice system, as we explained in *Chapter 6, Enabling Network Connectivity*.

Now, think about the different options we have for administering our systems remotely, as follows:

- Check IPs for a range of hosts
- Install updates or add/remove one package
- Check the local time in case the system has drifted
- Restart one service after adding a new user to the system

Many more options exist, but these are the main ones.

Of course, there are more suitable tools for remotely administering systems and ensuring that errors are detected and handled properly, such as using Ansible, but in this case, for simple tasks, we are good to go.

Previously, we created a key and replied with <ENTER> when we were asked for a password. What if we had typed one in? We'll look at this in the next section.

SSH agent

If we have decided to create an SSH key with a password to protect it (good choice), we will need to input the passphrase each time we want to use the key, so in the end, it might be as insecure as having to type in the password, as someone might be checking over our shoulder. To overcome this, we can use a program called ssh-agent that temporarily keeps the passphrase in memory. This is convenient and reduces the chances of someone watching while you type in your key.

When you're using a graphical desktop, such as **GNU Network Object Model Environment** (**GNOME**), as provided by RHEL, the agent might be already set up to start at session login. In the event that you're using a console (local or remote), the agent must be started manually by executing ssh-agent.

When ssh-agent is executed, it will output some variables that must be set in our environment so that we can make use of it, as shown in the following screenshot:

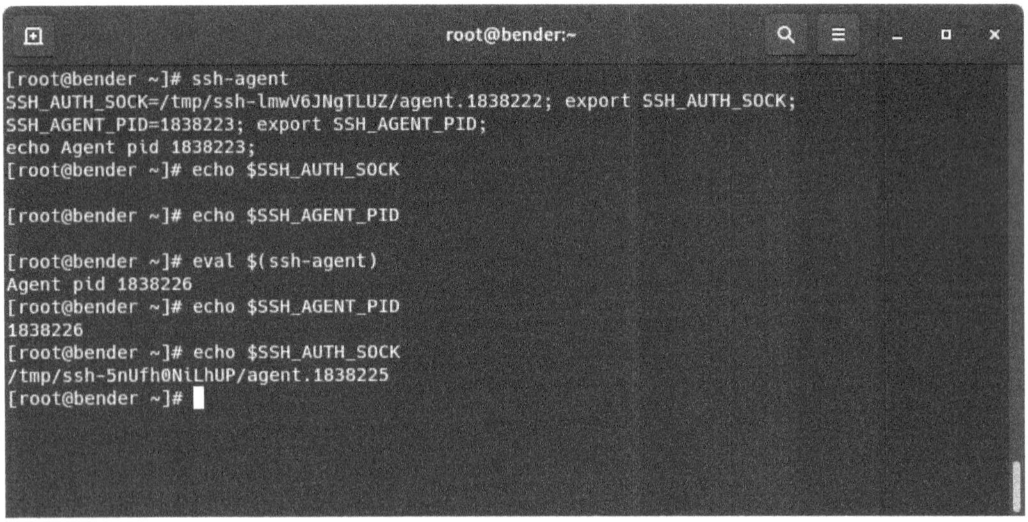

```
[root@bender ~]# ssh-agent
SSH_AUTH_SOCK=/tmp/ssh-lmwV6JNgTLUZ/agent.1838222; export SSH_AUTH_SOCK;
SSH_AGENT_PID=1838223; export SSH_AGENT_PID;
echo Agent pid 1838223;
[root@bender ~]# echo $SSH_AUTH_SOCK

[root@bender ~]# echo $SSH_AGENT_PID

[root@bender ~]# eval $(ssh-agent)
Agent pid 1838226
[root@bender ~]# echo $SSH_AGENT_PID
1838226
[root@bender ~]# echo $SSH_AUTH_SOCK
/tmp/ssh-5nUfh0NiLhUP/agent.1838225
[root@bender ~]# 
```

Figure 8.9 – ssh-agent being used to set the required variables

As shown in the preceding screenshot, before being executed, or just while we're executing the agent, the variables are undefined. However, if we were to execute `eval $(ssh-agent)`, we would accomplish the goal of having the variables defined and ready to use.

The next step is to add keys to the agent. This can be accomplished with the `ssh-add` command, which can be used without parameters or by specifying the key to be added. If the key requires a password, it will prompt you for it. Once we're done, we might be able to use that key to log in to the systems with the passphrase that's being cached until we exit the session that executed the agent, thus clearing the passphrase from memory.

The following screenshot shows the command that was used to generate a new key pair with a password. Here, we can see that the only difference is that we're storing it in a file named `withpass` versus what we did earlier in this chapter:

```
[root@bender ~]# ssh-keygen -f withpass
Generating public/private rsa key pair.
Enter passphrase (empty for no passphrase):
Enter same passphrase again:
Your identification has been saved in withpass.
Your public key has been saved in withpass.pub.
The key fingerprint is:
SHA256:ONGQ/GyYRQ1JYPCUKxCBycH/BSF44FNLEXWWahOWzFc root@bender.bending.rodriguez.example.co
m
The key's randomart image is:
+---[RSA 3072]----+
|o=*OB=BO*E       |
|+++.o@*=o .      |
| ooo. BB.        |
| ...++++         |
|    o.=.S         |
|     . .         |
|                 |
|                 |
|     [SHA256]    |
+-----------------+
[root@bender ~]# 
```

Figure 8.10 – Creating an additional ssh key pair with a password

We can see how to connect to our `localhost` (using the key that has a password defined). In the screenshot we can see how connection fails and once we use the key interactively or we load it into the agent (after the logout) it allows to connect without asking for the password as it's cached byt the agent:

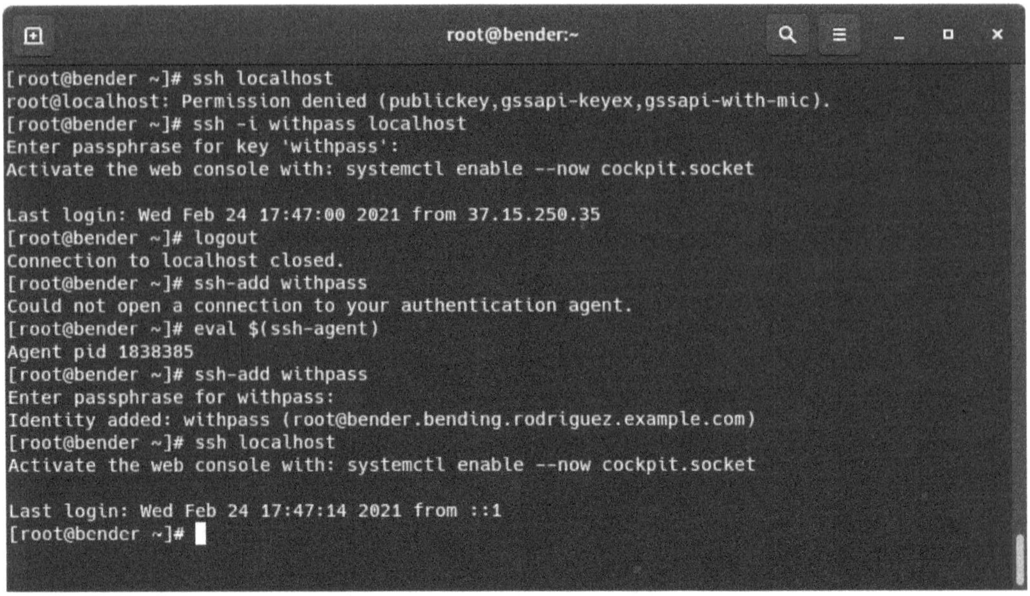

Figure 8.11 – Using ssh-agent to remember our passphrase

To make this clearer, let's analyze what's happening, as follows:

1. First, we `ssh` to the host. Permission is denied as the default key we used was removed from `authorized_keys`.

2. We `ssh` again, but while defining the identity file (the key pair) to connect to, as we can see, we're asked for the passphrase for the key, not for logging in to the system.

3. Then, we log out and the connection is closed.

4. Next, we try to add the key, and we get an error because we have not set the environment variables for the agent.

5. As instructed when we introduced the agent, we execute the command for loading the environment variables for the agent in the current shell.

6. When we retry adding the key with `ssh-add withpass`, the agent asks for our passphrase.

7. When we finally `ssh` to the host, we can connect without a password as the key is in memory for our key pair.

Here, we have achieved two things: we now have an automated/disattended method to connect to systems and have ensured that only authorized users will know the passphrase to unlock them.

We'll learn how to do remote file management in the next section!

Remote file management with SCP/rsync

Similar to `telnet`, which was replaced with `ssh` on many devices and systems, using insecure solutions for file transfer is being reduced. By default, the **File Transfer Protocol** (**FTP**) uses **Transmission Control Protocol** (**TCP**) port `21`, but since communication happened in plain text, it was a perfect target for intercepting credentials. FTP is still used today, mostly for serving files on servers that only allow anonymous access and wish to move to more secure options.

SSH usually enables two interfaces for copying files: `scp` and `sftp`. The first one is used in a similar way to the regular `cp` command, but here, we're accepting remote hosts as our target or source, while `sftp` uses a client approach similar to the traditional `ftp` command that interacts with FTP servers. Just remember that in both cases, the connection is encrypted and happens over port `22/tcp` on the target host.

We'll dig into `scp` in the next section.

Transferring files with an OpenSSH secure file copy

The `scp` command, which is part of the `openssh-clients` package, allows us to copy files between systems using the `ssh` layer for the whole process. This allows us to securely transfer a file's contents, plus all the automation capabilities that were introduced by key-pair login, to various systems.

To set up this example, we will have to relax the `PasswordAuthentication` setting in `/etc/ssh/sshd_config` allowing us to use a password to copy the keys that we'll be creating for the users, as shown in the following screenshot:

```
[root@bender ~]# adduser kys
[root@bender ~]# echo letmein|passwd kys --stdin
Changing password for user kys.
passwd: all authentication tokens updated successfully.
[root@bender ~]# ssh-copy-id kys@localhost
/usr/bin/ssh-copy-id: INFO: Source of key(s) to be installed: "/root/.ssh/id_rsa.pub"
/usr/bin/ssh-copy-id: INFO: attempting to log in with the new key(s), to filter out any tha
t are already installed
/usr/bin/ssh-copy-id: INFO: 1 key(s) remain to be installed -- if you are prompted now it
s to install the new keys
kys@localhost's password:

Number of key(s) added: 1

Now try logging into the machine, with:   "ssh 'kys@localhost'"
and check to make sure that only the key(s) you wanted were added.

[root@bender ~]# 
```

Figure 8.12 – Preparing our system with an additional user to practice file transfers

You can find the preceding commands in a script available at `https://github.com/ PacktPublishing/Red-Hat-Enterprise-Linux-RHEL-9-Administration/blob/ main/chapter-08-remote-systems-administration/create-kys-user.sh`.

Once a user has been created and the key has been copied, we can start testing! (Note—remember to always secure created keys.)

Earlier in this chapter, we created a key named `withpass` with a public counterpart at `withpass. pub`. To provide the key to the newly created user, we can copy both files to the `kys` user via the following command:

```
scp withpass* kys@localhost:
```

Let's analyze each part of the command using this template:

```
scp source target
```

In our case, `source` is indicated with `withpass.*`, which means that it will select all files starting with the `withpass` string.

Our `target` value is a remote host. Here, the username is `kys`, the host is `localhost`, and the folder you should store the files in is the default one—usually, the home folder of the user indicated (the one with an empty path after the `:` symbol).

In the following screenshot, we can see the output of the command and the validation we can perform later via remote execution:

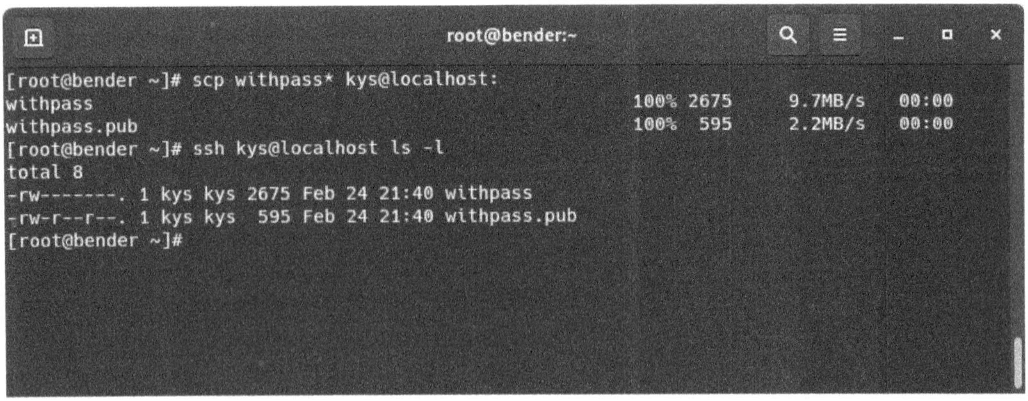

Figure 8.13 – Copying SCP files to a remote path and validating the files that have been copied

In the preceding screenshot, you can also check that the files that were owned by the root user are copied. The copied ones are owned by the `kys` user, so the file's contents are the same, but since the creator on the target is the `kys` user, files have their ownership.

We can also make more complex copies by indicating remote files first and local paths as targets so that we download files to our system, or even copy files across remote locations for both the origin and target (unless we specify the `-3` option, they will go directly from `origin` to `target`).

> **Tip**
>
> Time for a reminder! `man scp` will show you all the available options for the `scp` command, but since it is based on `ssh`, most of the options we use with `ssh` are available, as well as the host definitions we made in the `.ssh/config` file.

We'll explore the `sftp` client in the next section.

Transferring files with sftp

Compared to `scp`, which can be scripted in the same way we can script with the regular `cp` command, `sftp` has an interactive client for navigating a remote system. However, it can also automatically retrieve files when a path containing files is specified.

To learn about the different commands that are available, you can invoke the `help` command, which will list the available options, as shown in the following screenshot:

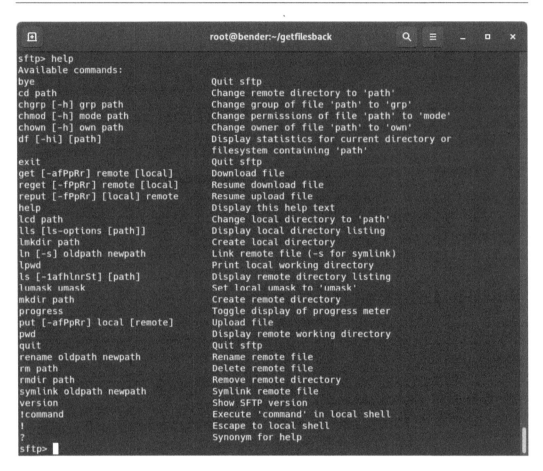

Figure 8.14 – Available sftp interactive mode commands

Let's look at an example of this with the help of the following screenshot:

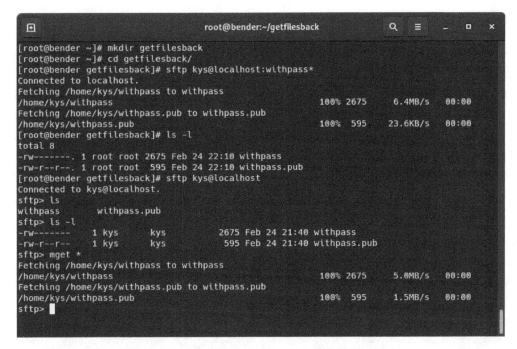

Figure 8.15 – Both modes of operation with sftp: automated transfer or interactive transfer

In this example, we've created a local folder to be our work folder, called getfilesback. First, we have invoked sftp with a remote path with the files we've identified. Here, sftp has automatically transferred the files and has stopped executing. The files we have received are now the property of our user.

In the second command, when we invoke sftp with the user and host and we enter interactive mode, we can execute several commands, similar to what we can do on a remote shell session. Finally, using the mget command with the * wildcard character, we transferred the files to our local system.

In both cases, the files have been transferred from the remote system to our local system, so our goal has been accomplished. However, using scp requires knowing the exact path of the files you want to transfer. On the other hand, it might be a bit more user-friendly to navigate the system using the ls and cd commands within the sftp interactive client until we reach the files we want to transfer if we can't remember the path.

Now, let's learn how to quickly transfers files and trees with rsync.

Transferring files with rsync

Although we can use the -r option of scp to transfer files recursively, scp only handles the full copy of the file, which is not ideal if we are just keeping some folders in sync across systems.

In 1996, `rsync` was launched, and many systems implemented it by using a dedicated server that was listening to client connections. This was to allow trees to be synchronized with files, which was done by copying over the differences between the files. Here, parts of the source and destination were compared to see whether there were differences that should be copied over.

With `ssh`, and with the `rsync` package installed on both the client and the server, we can take advantage of the secure channel that's created by `ssh` and the faster synchronization provided by `rsync`.

The difference between using the `rsync` daemon and using `ssh` is the syntax for the source or destination, which either uses the `rsync://` protocol or `::` after the hostname. In other cases, it will use `ssh`, or even the local filesystem.

The following screenshot shows us mentioning the schema for **Uniform Resource Locators** (**URLs**) via the `rsync -help` command:

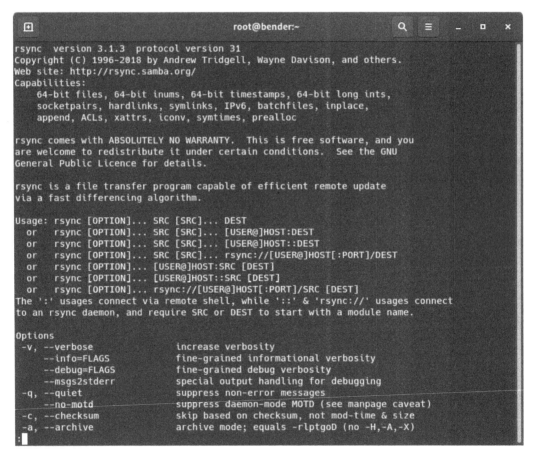

Figure 8.16 – The rsync command's help output

Now, let's review some of the useful options we can use with `rsync`, as follows:

- `-v`: Provides more verbose output during the transfer
- `-r`: Recurses into directories
- `-u`: Update—only copies files that are newer than the ones at the target
- `-a`: Archive (this includes several options, such as `-rlptgoD`)
- `-X`: Preserves extended attributes
- `-A`: Preserves **access control lists** (**ACLs**)
- `-S`: Sparse—sequences of nulls will be converted into sparse blocks
- `--preallocate`: Claims the space that's required for files before transferring them
- `--delete-during`: Deletes files on the target that are not hosted during the copy
- `--delete-before`: Deletes files on the target that are not hosted before the copy
- `--progress`: Shows progress information on the copy (copied files versus total files)

The `rsync` algorithm breaks the file into chunks and calculates checksums for each chunk that's transmitted to the source. They are then compared to the ones for local files. We are only allowed to share the differences between the source and the target. `rsync` doesn't check the modification file date and size by default, so if the file has changed without changes being left in both, the change might not be detected unless a checksum check is forced for each file candidate to be transferred.

Let's look at this basic example: `rsync -avr getfilesback/ newfolder/` will copy the files in the local `getfilesback/` folder to `newfolder/` by showing a progress update, but only for updated files, as shown in the following screenshot:

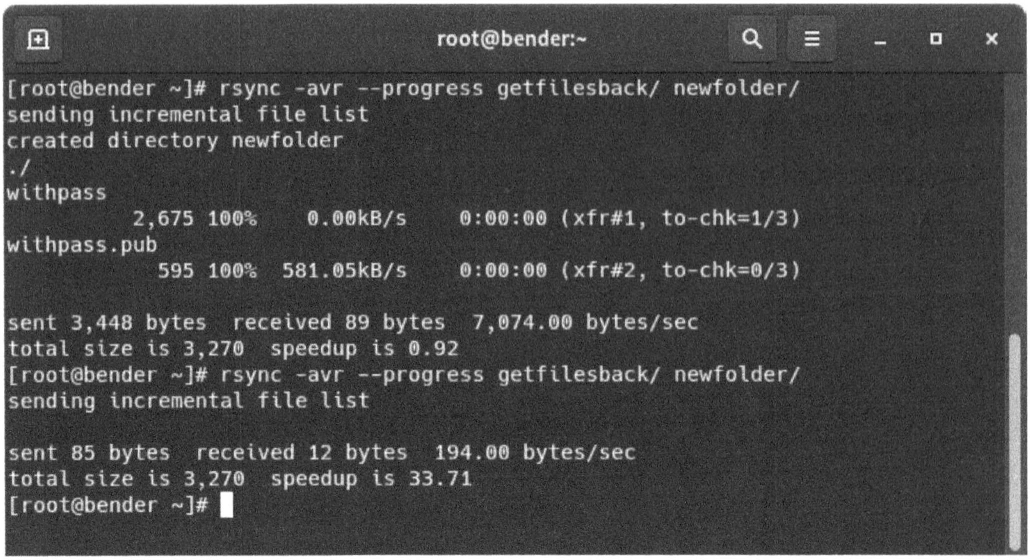

Figure 8.17 – The rsync operation being used on the same source/
destination, repeated to illustrate transfer optimization

As we can see, the second operation just sent 85 bytes and received 12 bytes. This is because there was a little checksum operation happening internally to validate across the folders as the files hadn't been changed. The same output can be obtained if we use the remote target approach with `rsync -avr --progress getfilesback/ root@localhost:newfolder/`, but in this case, `ssh` transport will be used.

Let's get some bigger sample files and compare them by checking out a Git repository at some point in time, transferring the files, then updating to the latest version to simulate work on the repository. Then, we will synchronize again.

First, let's install `git` if it's not installed and check out a sample repository by executing the following code (note that we should make sure that the required repositories must be available—for example, with `subscription-manager repos --enable="rhel-9-for-x86_64-baseos-rpms"`):

```
dnf -y install git    # install git in our system
git clone https://github.com/citellusorg/citellus.git   # clone
a repository over https
cd citellus # to enter into the repository folder
git reset HEAD~400   # to get back 400 commits in history
```

At this point, we have a folder with files ready for transfer. Once we've done this, we'll execute `git pull` to sync with the latest changes and use `rsync` again to copy the differences with `rsync -avr -progress /root/citellus/ root@localhost:newfolder/`. Later, we'll additionally use the `--delete` modifier to remove any files that no longer exist on the source.

Let's check out the sequence shown in the following screenshot:

```
tests/plugins-unit-tests/test_pacemaker_stonith_enabled.py
          2,876 100%   18.98kB/s     0:00:00 (xfr#334, to-chk=11/465)
tests/plugins-unit-tests/setup/
tests/plugins-unit-tests/setup/bugzilla/
tests/plugins-unit-tests/setup/bugzilla/httpd_bug_1406417.sh
          1,481 100%    9.77kB/s     0:00:00 (xfr#335, to-chk=6/465)
tests/plugins-unit-tests/setup/bugzilla/keystone_bug_1473713.sh
          1,771 100%   11.69kB/s     0:00:00 (xfr#336, to-chk=5/465)
tests/plugins-unit-tests/setup/openstack/
tests/plugins-unit-tests/setup/openstack/crontab/
tests/plugins-unit-tests/setup/openstack/crontab/keystone_cleanup.sh
          1,606 100%   10.60kB/s     0:00:00 (xfr#337, to-chk=3/465)
tests/plugins-unit-tests/setup/pacemaker/
tests/plugins-unit-tests/setup/pacemaker/stonith_enabled.sh
          1,827 100%   12.06kB/s     0:00:00 (xfr#338, to-chk=2/465)
tools/
tools/www/
tools/www/citellus.html
         35,136 100%  230.29kB/s     0:00:00 (xfr#339, to-chk=0/465)

sent 9,134,017 bytes  received 7,085 bytes  6,094,068.00 bytes/sec
total size is 9,103,471  speedup is 1.00
[root@bender citellus]#
```

Figure 8.18 – Synchronizing the git folder to a new folder with rsync

In the preceding screenshot, pay attention to the speedup that's reported in the latest line of the command.

Now, let's execute `git pull` to get the 400 changes we were missing and repeat `rsync`. We will get an output similar to this:

```
          1,120,131 100%     5.62MB/s     0:00:00 (xfr#357, to-chk=8/529)
tests/jsons/osp/
tests/jsons/osp/ctrl0/
tests/jsons/osp/ctrl0/citellus.json
        155,085 100%   788.80kB/s     0:00:00 (xfr#358, to-chk=6/529)
tests/other/
tests/other/dmidecode
          4,332 100%    22.03kB/s     0:00:00 (xfr#359, to-chk=5/529)
tests/plugins-unit-tests/
tests/plugins-unit-tests/test_all_plugins.py
          4,805 100%    24.44kB/s     0:00:00 (xfr#360, to-chk=4/529)
tests/plugins-unit-tests/test_all_plugins_for_cf_usage_without_sourcing.py
          2,373 100%    12.07kB/s     0:00:00 (xfr#361, to-chk=3/529)
tests/plugins-unit-tests/test_brackets-parenthesis.py
          1,315 100%     6.69kB/s     0:00:00 (xfr#362, to-chk=2/529)
tests/plugins-unit-tests/test_executable_bit.py
          1,158 100%     5.89kB/s     0:00:00 (xfr#363, to-chk=1/529)
tests/plugins-unit-tests/test_no_echo_rc.py
          1,882 100%     9.57kB/s     0:00:00 (xfr#364, to-chk=0/529)

sent 10,499,211 bytes  received 7,495 bytes  21,013,412.00 bytes/sec
total size is 18,132,086  speedup is 1.73
[root@bender citellus]#
```

Figure 8.19 – Using rsync again to copy over the differences

In the preceding screenshot, pay attention to the speedup reported in the last line so that you can compare it with the previous one.

From this sequence of screenshots, we can check the last numbers for the total bytes that were sent to see the improvement in transfer, along with some of the files that were received (because we added the –v modifier to get verbose output and --progress).

The biggest advantage comes when a copy is performed over slower network links and it's performed periodically—for example, as a way to copy to an offsite copy for backup purposes. This is because rsync will only copy the changes, update the newer files that have been modified on the source, and allow us to use compression over the ssh channel. For example, the Linux kernel at https://www.kernel.org/ can be mirrored using rsync.

In the next section, we will dig into a very interesting feature of SSH to make connecting to servers with no direct access easy.

Advanced remote management – SSH tunnels and SSH redirections

SSH has two really powerful features; that is, SSH tunnels and SSH redirections. When an SSH connection is established, it can not only be used to send commands to the remote host and let us work on them as if they were our local system, but we can also create tunnels that interconnect our systems.

Let's try to imagine a scenario that is common in many companies, where a **virtual private network** (**VPN**) is used to reach the internal network with all the services and servers, but with SSH instead of a regular VPN.

So, let's put some context into this imaginary scenario.

We can use a host that gets external traffic for `ssh` redirected from our internet router to the `ssh` service in that system. So, in brief, our router gets connections on port 22 via TCP, and the connection is forwarded to our server. We will be naming this server `bastion` in this exercise.

With this in place, our common sense tells us that we will be able to reach that bastion host via SSH, even if we can use other tools or even `ssh` it to connect to other systems later.

Can we connect directly to other hosts in the internal network? The answer is *yes*, because, by default, SSH allows us to use TCP forwarding (`sshd_config` setting `AllowTcpForwarding`), which empowers us, as remote login users, to create port redirections and even a **Socket Secure** (**SOCKS**) proxy to be used for our connections.

For example, we can create a tunnel using that bastion host to reach our internal mail server via the **Internet Message Access Protocol** (**IMAP**) and **Simple Mail Transfer Protocol** (**SMTP**) protocols by just executing the following code:

```
ssh -L 10993:imap.example.com:993 -L 10025:smtp.example.com:25
user@bastionhost
```

This command will listen on local ports `10993` and `10025`. All the connections that are performed there will be tunneled until `bastionhost` connects those to `imap.example.com` at port `993` and `smtp.example.com` at port `25`. This allows our local system to configure our email account using those custom ports and use `localhost` as the server, and still be able to reach those services.

Tip

Ports under `1024` are considered privileged ports, and usually, only the root user can bind services to those ports. That's why we use them for our redirection ports `10025` and `10093` so that those can be used by a regular user instead of requiring the root user to perform the `ssh` connection. Pay attention to `ssh` messages when you're trying to bind to local ports in case those are in use, as the connections might fail. Furthermore, you can use additional `-v`, `-vv`, and `-vvv` modifiers to increase verbosity.

Additionally, from the target server's point of view, the connections will appear as if they originated in the bastion server as this is the one effectively performing the connections.

When the list of open ports starts to grow, it is better to go back to what we explained at the beginning of this chapter: the `~/.ssh/config` file can hold the host definition, along with the redirections we want to create, as shown in this example:

```
Host bastion
     ProxyCommand none
     Compression yes
     User myuser
     HostName mybastion.example.com
     Port 330
     LocalForward 2224 mail.example.com:993
     LocalForward 2025 smtp.example.com:25
     LocalForward 2227 ldap.example.com:389
     DynamicForward 9999
```

In this example, when we are connecting to our bastion host (via `ssh bastion`), we are automatically enabling `Compression`, setting the host to connect to `mybastion.example.com` at port `330`, and defining port forwarding for our `imap`, `smtp`, and `ldap` servers and one dynamic forward (SOCKS proxy) at port `9999`. If we have different identities (key pairs), we can also define the one we wish to use via the `IdentityFile` configuration directive for each host, or even use wildcards such as `Host *.example.com` to automatically apply those options to hosts ending in that domain that have no specific configuration stanza.

> **Note**
>
> Sometimes, while using `ssh`, `scp`, or `sftp`, the goal is to reach a system that is accessible from a bastion host. Other port forwarding is not needed here—only reaching those systems is required. In this case, you can use the handy `-J` command-line option (equivalent to defining a `ProxyJump` directive) to use that host as a jump host to the final target you want to reach. For example, `ssh -J bastion mywebsiteserver.example.com` will transparently connect to `bastion` and jump from there to `mywebsiteserver.example.com`.

In the next section, we will learn how to protect ourselves from network issues with our remote connections and get the most out of our remote terminal connection.

Remote terminals with tmux

`tmux` is a terminal multiplexer, which means that it allows us to open and access several terminals within a single screen. A good analogy is a window manager in a graphical desktop that allows us to open several windows so that we can switch contexts while using only one monitor.

tmux also allows us to detach and reattach to the sessions, so it's the perfect tool in case our connection drops. Think, for example, about performing a software upgrade on a server. If for whatever reason, the connection drops, it will be equivalent to abruptly stopping the upgrade process in whichever status it was in at that moment, which can lead to bad consequences. However, if the upgrade was launched inside tmux, the command will continue executing, and once the connection is restored, the session can be reattached and the output will be available to be examined.

First of all, let's install it on our system via dnf -y install tmux. This line will download the package and make the tmux command available. Bear in mind that the goal of tmux is not to install it on our system (even if this is useful) but for it to be available on the servers we connect to, to get that extra layer of protection in case a disconnection happens. So, it's a good habit to get used to installing it on all the servers we connect to.

In the following screenshot, we can see what tmux looks like with the default configuration after executing tmux on a command line:

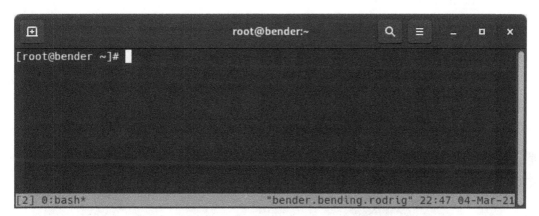

Figure 8.20 – tmux default layout after execution

As shown in the preceding screenshot, it hasn't changed a lot of the view of our terminal except for the status bar in the lower part of the window. This shows some information about the host, such as its name, time, date, and a list of open windows, with 0:bash being the active one, as denoted by the asterisk (*) symbol.

There are lots of combinations for using tmux, so let's get familiar with some of them that will cover the initial use case, as follows:

- Run tmux to create a new session
- Run tmux at to attach to a previous session (for example, after reconnecting to a host)
- Run tmux at -d to attach to a previous session and detach other connections from it

Once we're inside `tmux`, there is a whole world of commands we can use that are preceded by the *Ctrl + B* keys. Let's view some important ones here (remember that *Ctrl + B* must be pressed before you use the next item in the list):

- **?**: Displays inline help about the shortcuts to use
- **c**: Creates a new window
- **n/p**: Go to the next/previous window
- **d**: Detaches the `tmux` session
- **0-9**: Go to the window numbered with the pressed number
- **,**: Renames windows
- **"**: Splits the pane top/bottom
- **%**: Splits the pane right/left
- **space**: Switches to the next layout
- **&**: Kills the window
- **pg down/pg up**: Go higher or lower in the window history
- **Arrow keys**: Select the pane in the direction of the pressed key

Let's look at an example in action in the following screenshot:

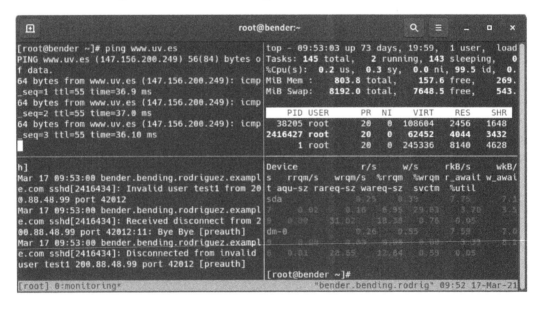

Figure 8.21 – tmux with four panes running different commands inside the same window

As we can see, there are several commands running at the same time—`top`, `journalctl -f`, `iostat -x`, and `ping`—so this is a good way to monitor a system while operations are being performed on it.

Additionally, one of the advantages is that `tmux` can be scripted, so if we are using one layout while administering systems, we can copy that script and execute it as soon as we connect to them so that we can enjoy the same layout and even the commands being executed.

You can find the following code with extra comments and descriptions at `https://github.com/PacktPublishing/Red-Hat-Enterprise-Linux-RHEL-9-Administration/blob/main/chapter-08-remote-systems-administration/term.sh` if you want to try it on your system:

```bash
#!/bin/bash
SESSION=$USER
tmux -2 new-session -d -s $SESSION # create new session
tmux select-window -t $SESSION:0  # select first window
tmux rename-window -t $SESSION "monitoring" #rename to
monitoring
tmux split-window -h #split horizontally
tmux split-window -v #split vertically
tmux split-window -h # split again horizontally
tmux select-layout tiled #tile panes
tmux selectp -t1 # select pane 1
tmux send-keys "top" C-m #run top by sending the letters +
RETURN
tmux selectp -t2 # select pane 2
tmux send-keys "journalctl -f" C-m # run journalctl
tmux selectp -t3 # select pane 3
tmux send-keys "iostat -x" C-m # run iostat
tmux selectp -t0 #select the pane without commands executed
```

Once the session with `tmux` has been set, we can attach the session we've just created and configured by executing `tmux`, which will show a layout similar to the one shown in the preceding screenshot.

Introduction to Ansible automation

As defined on its website at `https://www.ansible.com`: *Ansible is a foundation for building and operating automation across an organization.*

In this section, we will see at a very introductory level how Ansible can be used to automate some tasks across different hosts.

Ansible uses **YAML Ain't Markup Language** (**YAML**) files to define a set of tasks named playbooks that act over a set of hosts in an inventory. For each host, the tasks define the status desired and Ansible connects by default over SSH, executes the required actions, and then removes the files that were uploaded for the execution. It requires no daemon or agent to execute, and as the playbooks are text-based, they are the perfect candidate for a **version control system** (**VCS**) taking care of the changes, and so on.

For installing Ansible, we'll execute `dnf install ansible-core rhel-system-roles`, which will trigger the installation of the core and several dependencies required for it.

Inventory

The inventory is one of the fundamental pieces of Ansible. It defines a set of hosts that will be acted on, that can be grouped, and so on, and it could either be a text file or be dynamically generated out of some program execution, also allowing us to build dynamic sets based on the tasks, locations, and so on.

An example inventory could look like this:

```
[shellservers]
localhost
```

In this example, a group named `shellservers` that contains a host reachable by `localhost` is defined.

Playbook

A playbook uses a set of modules that perform different tasks. Some of them might have some other requirements—for example, a module to interact with `firewalld` will require the `firewalld` package and some libraries to be available—and others are part of the base set of included tools in any system.

Let's see an example here:

```
---
- hosts: all
  user: root
  tasks:
   - name: Print host name
     debug:
       msg: "{{ ansible_hostname }}"
```

> **Note**
>
> This file is available at the GitHub repository at `https://github.com/PacktPublishing/Red-Hat-Enterprise-Linux-RHEL-9-Administration/blob/main/chapter-08-remote-systems-administration/hostname.yaml`.

Now, let's execute it against our host using the `ansible-playbook` command with `ansible-playbook -i localhost, hostname.yaml`, as illustrated in the following screenshot:

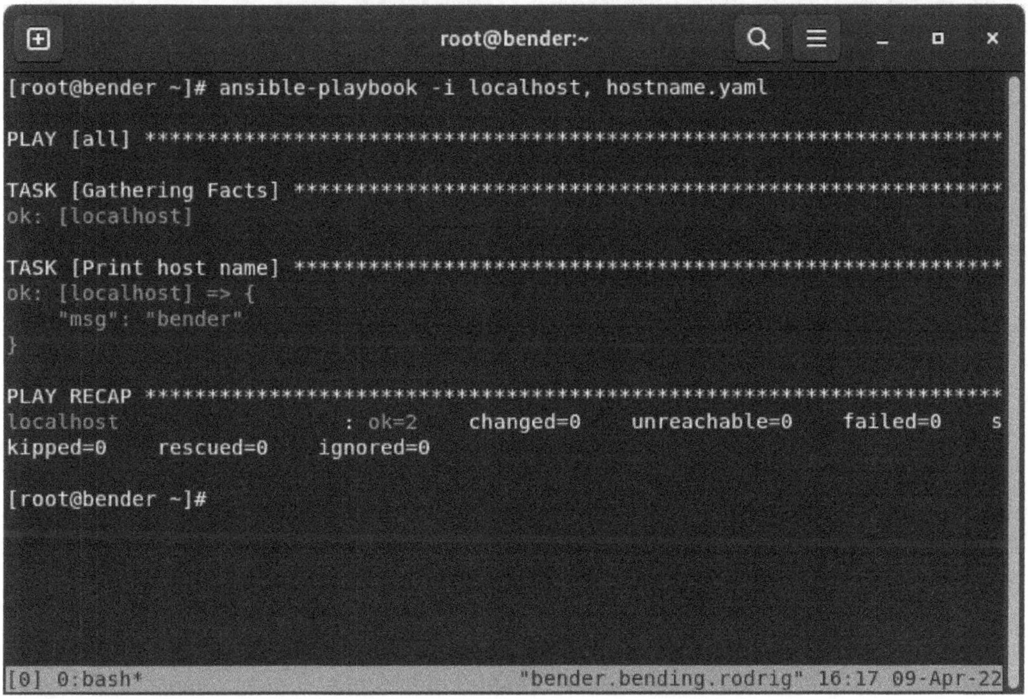

Figure 8.22 – Ansible playbook execution

In the preceding screenshot, we can see that first, we provided the inventory on the command line. Also, note that there's a comma after the hostname—this is on purpose as we're using an inventory not based on a file.

When Ansible execution started, several sections show the output, and one of the first is `Gathering Facts`. This is really important when Ansible connects to hosts and grabs variables from the remote system to perform its action (such as running some tasks based on the operating system of the host, amount of **random-access memory** (**RAM**), and so on). Later, we can see the actual task execution, where we get the `msg` line containing the hostname.

If we repeat in the same host by adding as inventory `-i localhost,127.0.0.1`—that is, two different ways to refer to the same host (as a trick for not requiring an additional one)—we get this output:

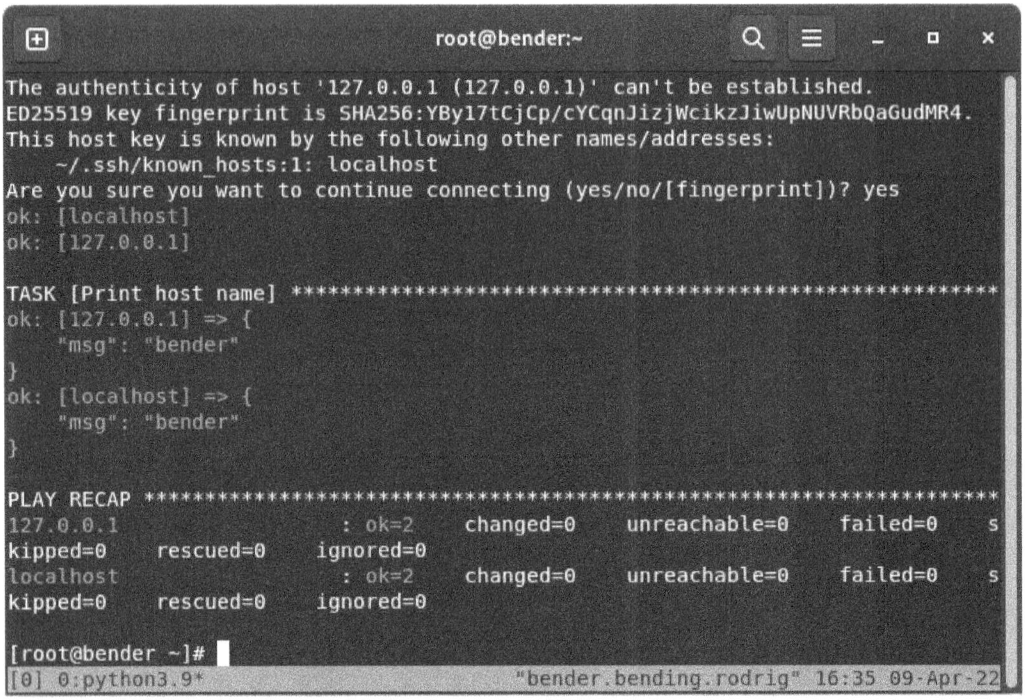

```
The authenticity of host '127.0.0.1 (127.0.0.1)' can't be established.
ED25519 key fingerprint is SHA256:YBy17tCjCp/cYCqnJizjWcikzJiwUpNUVRbQaGudMR4.
This host key is known by the following other names/addresses:
    ~/.ssh/known_hosts:1: localhost
Are you sure you want to continue connecting (yes/no/[fingerprint])? yes
ok: [localhost]
ok: [127.0.0.1]

TASK [Print host name] **************************************************
ok: [127.0.0.1] => {
    "msg": "bender"
}
ok: [localhost] => {
    "msg": "bender"
}

PLAY RECAP **************************************************************
127.0.0.1                  : ok=2    changed=0    unreachable=0    failed=0    s
kipped=0    rescued=0    ignored=0
localhost                  : ok=2    changed=0    unreachable=0    failed=0    s
kipped=0    rescued=0    ignored=0

[root@bender ~]#
[0] 0:python3.9*                          "bender.bending.rodrig" 16:35 09-Apr-22
```

Figure 8.23 – Ansible playbook execution with two hosts

The important piece here is that Ansible considers the hosts to be different and has executed the tasks twice, one for each host. We can see a summary, with the different hosts, tasks completed, changes performed, hosts unreachable, and so on.

You can find more documentation at the developer's site at `https://www.ansible.com/overview/how-ansible-works` and find some online interactive courses at `https://developers.redhat.com/products/ansible/getting-started` that will allow you to practice.

Let's work on creating a sample playbook that does several tasks, and let's keep explaining alongside (also available in the GitHub repository). Have a look at the following code snippet:

```
---
- hosts: all
  user: root
  vars:
```

```
        firewall:
          - service: https
            permanent: true
            state: enabled
            immediate: yes
          - service: http
            permanent: true
            state: enabled
            immediate: yes
      tasks:
      - name: Install httpd package
        package:
          name: httpd
          state: installed
      - name: Start http service
        service:
          name: httpd
          state: started
          enabled: yes
      - name: Open Firewall ports
        include_role:
          name: rhel-system-roles.firewall
```

The tasks defined will install the httpd package, later enable the service, and open the firewall ports for HTTP and HTTPS services. We can later use ansible-playbook -i myhost, httpd.yaml to get it applied to it.

Ansible is taking care of the error control. If we reapply a playbook over a server that has the changes defined, it will just verify it is in the desired state, so it is very useful to also verify the state or to even repurpose servers if we take care of disabling unused services, once we no longer require a specific usage for them.

The use of conditionals could help, together with data collection tasks to later switch the logic of the playbook based on some system specifics that help in better adapting the tasks and making more complex logic.

The scope of covering those use cases is beyond the scope of this chapter, but with the indicated URLs you can get familiar and start using them to learn a lot more! Also, always check the Ansible documentation at https://docs.ansible.com/ansible/latest/collections/ansible/index.html for syntax and examples of how to use the different modules for defining tasks.

Summary

In this chapter, we covered SSH and how to use it to connect to remote systems, how to use keys to authenticate with or without a password, and how to take advantage of it for automation, transferring files, and even making services accessible or reachable via port redirection. With tmux, we learned how to make our administration sessions survive network interruptions and, at the same time, show important information at a glance by automating the layouts for it, and with Ansible, we learned how to start automating at scale changes in our systems.

In the next chapter, we'll be digging into securing our system network via firewalld to only expose the services that are required for operation.

Securing Network Connectivity with firewalld

A great mentor and technologist working with military-restricted environments once told me that *"The only secure system is the one that is switched off, disconnected from any network, and buried in the middle of the desert."* He is right, of course, but we must provide a service to make the system useful. This means having it running and connected to a network.

One of the techniques that are used in security to reduce incidents, such as avoiding unexpected exposure to a vulnerability and enabling unauthorized remote access, is reducing the attack surface and applying **defense in depth** principles. When you do that in a network, step one is filtering connections using a **firewall**. The firewall management tool that's included in **Red Hat Enterprise Linux** (RHEL) is **firewalld** and it helps us manage zones, profiles, services, and ports. It also includes a command-line tool called `firewall-cmd` and a `systemd` service unit to simplify its management.

In this chapter, we will cover the following topics to get a good understanding of how to manage the default firewall in RHEL:

- Introduction to the RHEL firewall – firewalld
- Enabling firewalld on the system and reviewing the default zones
- Reviewing the different configuration items under firewalld
- Enabling and managing services and ports
- Creating and using service definitions for firewalld
- Configuring firewalld with the web interface

Introduction to the RHEL firewall – firewalld

RHEL comes with two low-level network traffic filtering mechanisms: **nftables**, for filtering IP-related traffic, and **ebtables**, for transparent filtering in bridges. These mechanisms are static and use a set of rules to accept or reject traffic, though they do provide a myriad of other capabilities. In RHEL, they are both handled and managed dynamically by **firewalld**. Unless there is a specific need to have a very low-level usage of these low-level filtering mechanisms, please use firewalld (or its main command, firewall-cmd) instead. In this section, we will take a look at the firewall defaults in RHEL.

firewalld is installed by default in the system, which we can check by using the rpm command, so there is no need to install it:

```
[root@rhel-instance ~]# rpm -qa | grep firewalld
firewalld-filesystem-1.0.0-3.el9_b.noarch
firewalld-1.0.0-3.el9_b.noarch
```

If we have an installation that doesn't include firewalld for some reason, we can install it by running dnf install firewalld.

firewalld includes a service called firewalld, which is configured to run by default at startup. We can check this by using the systemctl status firewalld command:

```
[root@rhel9 ~]# systemctl status firewalld
• firewalld.service - firewalld - dynamic firewall daemon
     Loaded: loaded (/usr/lib/systemd/system/firewalld.service; enabled; vendor preset: enabled)
     Active: active (running) since Mon 2022-05-09 08:36:41 CEST; 6min ago
       Docs: man:firewalld(1)
   Main PID: 725 (firewalld)
      Tasks: 2 (limit: 11113)
     Memory: 39.4M
        CPU: 530ms
     CGroup: /system.slice/firewalld.service
             └─725 /usr/bin/python3 -s /usr/sbin/firewalld --nofork --nopid

May 09 08:36:40 rhel9.example.com systemd[1]: Starting firewalld - dynamic firewall daemon...
May 09 08:36:41 rhel9.example.com systemd[1]: Started firewalld - dynamic firewall daemon.
```

Figure 9.1 – Output of "systemctl status firewalld"

As we can see, the firewalld service is enabled and running. This is the default status in an RHEL system.

The main way for a sysadmin to configure firewalld is by using the firewall-cmd command. However, you can also do the following:

- Add new files with service definitions in /etc/firewalld/ (as explained in the *Creating and using service definitions for firewalld* section of this chapter).

- Use the web interface, called **cockpit**, to configure firewalls (as explained in the *Configuring firewalld with the web interface* section of this chapter).

- Use the `firewall-config` graphical interface in your desktop environment.

In this chapter, we will review the main mechanism and the web interface.

Now that we know the defaults of the RHEL main firewall, let's learn how to enable it.

Enabling firewalld in the system and reviewing the default zones

We have already seen that **firewalld** is enabled by default in the system. However, we may need to disable (that is, check whether the firewall is interfering with a service), re-enable (that is, after restoring configuration files), and start and stop it (that is, to reload configuration or to do a quick check). These tasks are managed like any other service in the system, that is, using `systemctl`. Let's stop the `firewalld` service:

```
[root@rhel-instance ~]# systemctl stop firewalld
[root@rhel-instance ~]# systemctl status firewalld
○ firewalld.service - firewalld - dynamic firewall daemon
     Loaded: loaded (/usr/lib/systemd/system/firewalld.service;
enabled; vendor preset: enabled)
     Active: inactive (dead) since Mon 2022-05-09 08:45:54
CEST; 5s ago
       Docs: man:firewalld(1)
    Process: 725 ExecStart=/usr/sbin/firewalld --nofork --nopid
$FIREWALLD_ARGS (code=exited, status=0/SUCCESS)
   Main PID: 725 (code=exited, status=0/SUCCESS)
        CPU: 589ms
May 09 08:36:40 rhel9.example.com systemd[1]: Starting
firewalld - dynamic firewall daemon...
May 09 08:36:41 rhel9.example.com systemd[1]: Started firewalld
- dynamic firewall daemon.
May 09 08:45:54 rhel9.example.com systemd[1]: Stopping
firewalld - dynamic firewall daemon...
May 09 08:45:54 rhel9.example.com systemd[1]: firewalld.
service: Deactivated successfully.
May 09 08:45:54 rhel9.example.com systemd[1]: Stopped firewalld
- dynamic firewall daemon.
```

In the previous output, as shown in bold, the service is inactive. We can check this by using the `firewall-cmd --state` command:

```
[root@rhel-instance ~]# firewall-cmd --state
not running
```

At the moment, the firewall service has stopped and all the rules have been dropped. The configuration for the service, however, has not changed, so if we reboot the system, firewalld will be running again.

> **Tip**
> We can always see the underlying `netfilter` rules by running the `nft list tables` command. You may want to run it before and after stopping the service to see the difference.

Now, let's try and start the service again:

```
[root@rhel-instance ~]# systemctl start firewalld
[root@rhel-instance ~]# systemctl status firewalld
○ firewalld.service - firewalld - dynamic firewall daemon
     Loaded: loaded (/usr/lib/systemd/system/firewalld.service;
enabled; vendor preset: enabled)
     Active: active (running) since Mon 2022-05-09 08:47:34
CEST; 3s ago
       Docs: man:firewalld(1)
   Main PID: 4377 (firewalld)
      Tasks: 2 (limit: 11113)
     Memory: 21.5M
        CPU: 291ms
     CGroup: /system.slice/firewalld.service
             └─4377 /usr/bin/python3 -s /usr/sbin/firewalld
--nofork --nopid
```

Let's check that `firewalld` is running:

```
[root@rhel-instance ~]# firewall-cmd --state
running
```

To fully disable the service, we will need to run the following command:

```
[root@rhel-instance ~]# systemctl disable firewalld
Removed /etc/systemd/system/multi-user.target.wants/firewalld.
service.
Removed /etc/systemd/system/dbus-org.fedoraproject.FirewallD1.
service.
```

Let's see how the service is disabled but still running:

```
[root@rhel-instance ~]# systemctl status firewalld -n0
○ firewalld.service - firewalld - dynamic firewall daemon
     Loaded: loaded (/usr/lib/systemd/system/firewalld.service;
disabled; vendor preset: enabled)
     Active: active (running) since Mon 2022-05-09 08:47:34
CEST; 1min 29s ago
       Docs: man:firewalld(1)
   Main PID: 4377 (firewalld)
      Tasks: 2 (limit: 11113)
     Memory: 24.5M
        CPU: 395ms
     CGroup: /system.slice/firewalld.service
             └─4377 /usr/bin/python3 -s /usr/sbin/firewalld
--nofork --nopid
```

When you're managing services with systemd using systemctl, you need to understand that enabling and disabling the service only affects how it behaves during the startup sequence, while starting and stopping only affects the current status of the service.

> **Tip**
> To disable and stop in one command, we can use the --now option, for example, systemctl disable firewalld --now. This option can also be used to enable and start, for example, systemctl enable firewalld --now.

Let's reenable the service again and ensure it's running:

```
[root@rhel-instance ~]# systemctl enable firewalld --now
Created symlink /etc/systemd/system/dbus-org.fedoraproject.
FirewallD1.service → /usr/lib/systemd/system/firewalld.service.
Created symlink /etc/systemd/system/multi-user.target.wants/
firewalld.service → /usr/lib/systemd/system/firewalld.service.
[root@rhel-instance ~]# firewall-cmd --state
running
```

Now that we know how to start and stop, as well as enable and disable, the `firewalld` service, let's understand the configuration structure and learn how to interact with it by reviewing the default configuration.

Reviewing the different configuration items under firewalld

firewalld manages three concepts in its configuration:

- **Zones**: A firewalld zone is a group of rules that can be activated altogether and assigned to a network interface. It includes different services and rules but also settings that alter the behavior of network traffic filtering.

- **Services**: A firewalld service is a port or group of ports that must be configured together for a specific system service (hence the name) to work properly.

- **Ports**: A firewalld port includes a port number (that is, 80) and a type of traffic (that is, TCP) and can be used to manually enable network traffic to a custom system service.

firewalld manages two types of configurations:

- **Running**: The rules that have currently been applied to the system

- **Permanent**: The rules that have been saved and will be loaded when the service starts

> **Important Note**
>
> The concept behind running versus permanent is to try network filtering rules in a running system and, once ensured that they work well, save them as permanent ones. Remember to check that the rules you want in the system have been saved properly.

Now, let's check out our system and see which zones are available:

```
[root@rhel-instance ~]# firewall-cmd --get-zones
block dmz drop external home internal nm-shared public trusted
work
```

Let's also check which zone is applied by default:

```
[root@rhel-instance ~]# firewall-cmd --get-default-zone
public
```

Let's review the zones that are available in firewalld by looking at the following table:

Zone	Enabled services	Other characteristics
public	cockpit, ssh, and dhcpv6-client	Default zone in firewalld. Assigned by default to new network interfaces. Accepts incoming traffic related to outgoing connections.
work	cockpit, ssh, ipp-client, and dhcpv6-client	Accepts incoming traffic related to outgoing connections.
internal	cockpit, ssh, mdns, ipp-client, samba-client, and dhcpv6-client	Accepts incoming traffic related to outgoing connections.
home	cockpit, ssh, mdns, ipp-client, samba-client, and dhcpv6-client	Accepts incoming traffic related to outgoing connections. (Same as internal!)
trusted	N/A	Accepts all incoming traffic.
external	ssh	Accepts incoming traffic related to outgoing connections. Any IPv4 traffic that's forwarded through the interface with this zone assigned will be masqueraded, and look as though it originated from this machine.
dmz	ssh	Accepts incoming traffic related to outgoing connections.
block	N/A	Rejects all incoming traffic unless it is related to outgoing connections.
drop	N/A	Ignores all incoming traffic, not even notifying us that the connections haven't been accepted.

Table 9.1 – Services and characteristics of each zone

> **Important Note**
>
> You can always access the information about these zones, and more, by accessing the firewalld.zones manual pages available in the system by running man firewalld.zones. A good exercise is to review the aforementioned manual page.

The aforementioned services will be reviewed in more detail in the next section. For now, let's learn how to manage zones.

Let's change the default zone to home:

```
[root@rhel-instance ~]# firewall-cmd --set-default-zone=home
success
[root@rhel-instance ~]# firewall-cmd --get-default-zone
home
```

We can establish a public zone as the default and assign a home zone to our local network:

```
[root@rhel-instance ~]# firewall-cmd --set-default-zone=public
success
[root@rhel-instance ~]# firewall-cmd --permanent
--zone=internal \
--add-source=192.168.122.0/24
success
[root@rhel-instance ~]# firewall-cmd --reload
success
[root@rhel-instance ~]# firewall-cmd --get-active-zones
internal
  sources: 192.168.122.0/24
public
  interfaces: enp1s0
```

This configuration allows us to publish services to the local network only, which is defined as 192.168.122.0/24 and assigned to the internal zone. Any service or port that's assigned to the internal zone from now on will be only accessible if it's accessed from an IP address in the internal network. We avoid allowing access to these services from other networks.

Also, to enable services to be accessed from any other network, we only need to assign them to the public zone.

Let's review the main options that are used and a couple more that could be useful:

- --get-zones: Lists the zones that have been configured in the system.

- --get-default-zone: Shows the zone that's configured by default.

- --set-default-zone=<zone>: Sets the default zone. This is applied to the running and permanent configuration.

- `--get-active-zones`: Shows the zones being used and what networks/interfaces they apply to.

- `--zone=<zone>`: Used to specify a zone for another option.

- `--permanent`: Used to apply the changes to the saved configuration. When you use this option, the changes will not be applied to the running configuration.

- `--reload`: Loads the saved configuration as running.

- `--add-source=<network>`: Adds a source network, in CIDR format, to a specified zone. The default zone is used if one hasn't been specified. Changes are applied to the running config; use `--permanent` to save them.

- `--remove-source=<network>`: Removes a source network, in CIDR format, to a specified zone. The default zone is used if one hasn't been specified. Changes are applied to the running config; use `--permanent` to save them.

- `--add-interface=<interface>`: Routes traffic from an interface to a zone. The default zone is used if one hasn't been specified.

- `--change-interface=<interface>`: Changes the traffic that's being routed to an interface to a zone. The default zone is used if one hasn't been specified.

Although this list of options may be very helpful, the full list of options is available on the manual page for `firewall-cmd`. You should review this page as you will be using it often when you're reconfiguring your firewall options.

> **Tip**
> To view the `firewall-cmd` manual page, simply run `man firewall-cmd`.

Now that we are aware of what zones are and how they are selected, let's learn how to manage services and ports.

Enabling and managing services and ports

As we mentioned in the previous section, a `firewalld` service is a port or group of ports that are configured together for a specific system service (hence the name) to work properly. There are a set of services that are enabled by default in one or many of the available **firewalld zones**. Let's start by reviewing them:

- `ssh`: Provides access to the **Secure Shell** (**SSH**) service in the system, which also enables remote management. The traffic that's accepted goes to port 22 and is of the TCP type.

- `mdns`: Provides access to the **Multicast DNS (MDNS)** service that's used to announce services in the local network. Traffic is accepted to multicast address 224.0.0.251 (IPv4) or ff02::fb (IPv6), on port 5353, and is of the UDP type.

- `ipp-client`: Provides access to the **Internet Printing Protocol (IPP)** client, which goes to port 631 and uses the UDP protocol.

- `samba-client`: This is a file and print sharing client that's compatible with Microsoft Windows. It uses ports 137 and 138 and is of the UDP type.

- `dhcpv6-client`: A **Dynamic Host Configuration Protocol (DHCP)** for IPv6. Its destination is the special network fe80::/64, its port is 546, and it's of the UDP type.

- `cockpit`: The web management interface for RHEL. Its destination is port 9090 and it's of the TCP type.

As you can see, a `firewalld` service can specify more than one port, a target address, and even a target network.

Now, let's take a look at the services that have been configured in our firewall:

```
[root@rhel-instance ~]# firewall-cmd --list-services
cockpit dhcpv6-client ssh
[root@rhel-instance ~]# firewall-cmd --list-services
--zone=internal
cockpit dhcpv6-client mdns samba-client ssh
```

Please note that when you're not establishing a zone, the services that are displayed are those related to the default zone – in this case, `public`. However, consider that we have configured more than one zone.

Now, let's install a web server – in this case, the Apache `httpd` server:

```
[root@rhel-instance ~]# dnf install httpd -y
Updating Subscription Management repositories.
Last metadata expiration check: 0:00:10 ago on Mon May  9
09:05:10 2022.
Dependencies resolved.
================================================================
Package                     Architecture    Version
Repository                                   Size
================================================================
Installing:
```

```
httpd x86_64 2.4.51-7.el9_0 rhel-9-for-x86_64-appstream-rpms
1.5 M
Installing dependencies:
apr x86_64 1.7.0-11.el9 rhel-9-for-x86_64-appstream-rpms 127 k
apr-util x86_64 1.6.1-20.el9 rhel-9-for-x86_64-appstream-rpms
98 k
[omitted]
Complete!
```

Let's enable and start the httpd service:

```
[root@rhel-instance ~]# systemctl enable httpd --now
Created symlink /etc/systemd/system/multi-user.target.wants/
httpd.service → /usr/lib/systemd/system/httpd.service.
[root@rhel-instance ~]# systemctl status httpd -n0
○ httpd.service - The Apache HTTP Server
   Loaded: loaded (/usr/lib/systemd/system/httpd.service;
enabled; vendor preset: disabled)
     Active: active (running) since Mon 2022-05-09 09:08:51
CEST; 11s ago
     Docs: man:httpd.service(8)
Main PID: 32019(httpd)
  Status: "Total requests: 0; Idle/Busy workers 100/0;Requests/
sec: 0; Bytes served/sec:   0 B/sec"
    Tasks: 213 (limit: 8177)
   Memory: 25.0M
      CPU: 107
    CGroup: /system.slice/httpd.service
            ├─32019 /usr/sbin/httpd -DFOREGROUND
            ├─32020 /usr/sbin/httpd -DFOREGROUND
            ├─32021 /usr/sbin/httpd -DFOREGROUND
            ├─32022 /usr/sbin/httpd -DFOREGROUND
            └─32023 /usr/sbin/httpd -DFOREGROUND
```

Now, let's check that the service is listening on all the interfaces:

```
[root@rhel-instance ~]# ss -a -A "tcp" | grep http
LISTEN     0          511                     *:http
  *:*
```

Optionally, we can check whether the port is open by using an external machine (if we have one):

```
[root@external:~]# nmap 192.168.122.196 -Pn
Starting Nmap 7.80 ( https://nmap.org ) at 2022-05-09 19:12
CEST
Nmap scan report for 192.168.122.196
Host is up (0.55s latency).
Not shown: 998 filtered ports
PORT      STATE   SERVICE
22/tcp    open    ssh
9090/tcp closed zeus-admin
Nmap done: 1 IP address (1 host up) scanned in 70.44 seconds
```

Now, we can enable the http service on the firewall:

```
[root@rhel-instance ~]# firewall-cmd --add-service http \
--zone=public --permanent
success
[root@rhel-instance ~]# firewall-cmd --add-service http \
--zone=internal --permanent
success
[root@rhel-instance ~]# firewall-cmd --reload
success
[root@rhel-instance ~]# firewall-cmd --list-services
cockpit dhcpv6-client http ssh
[root@rhel-instance ~]# firewall-cmd --list-services
--zone=internal
cockpit dhcpv6-client http mdns samba-client ssh
```

With that, the service has been enabled and the port is open. We can verify this from an external machine, like so (this is optional):

```
[root@external:~]# nmap 192.168.122.8
Starting Nmap 7.80 ( https://nmap.org ) at 2022-05-09 19:13
CEST
Nmap scan report for rhel.redhat.lan (192.168.122.8)
Host is up (0.00032s latency).
Not shown: 997 filtered ports
PORT      STATE   SERVICE
```

```
22/tcp    open    ssh
80/tcp    open    http
9090/tcp closed zeus-admin
Nmap done: 1 IP address (1 host up) scanned in 12.64 seconds
```

We can see that port 80 is open now. We can also retrieve the main page from the web server and show the first line:

```
[root@external:~]# curl -s http://192.168.122.8 | head -n 1
<!DOCTYPE html PUBLIC "-//W3C//DTD XHTML 1.1//EN" "http://www.
w3.org/TR/xhtml11/DTD/xhtml11.dtd">
```

> **Important Note**
>
> The definitions of the services in firewalld are kept in independent files in the /usr/lib/
> firewalld/services directory. If you need to check out the details of a service, you can
> go there and inspect the file and its definition.

Now, let's try to remove the service from the public network since this will be an internal service:

```
[root@rhel-instance ~]# firewall-cmd --list-services
--zone=public
cockpit dhcpv6-client http ssh
[root@rhel-instance ~]# firewall-cmd --remove-service http \
--zone=public --permanent
success
[root@rhel-instance ~]# firewall-cmd --reload
success
[root@rhel-instance ~]# firewall-cmd --list-services
--zone=public
cockpit dhcpv6-client ssh
```

Let's assume we didn't have the service definition and we still wanted to open port 80 on TCP in the public interface:

```
[root@rhel-instance ~]# firewall-cmd --list-ports --zone=public
[root@rhel-instance ~]# firewall-cmd --add-port 80/tcp
--zone=public --permanent
success
[root@rhel-instance ~]# firewall-cmd --reload
success
```

```
[root@rhel-instance ~]# firewall-cmd --list-ports --zone=public
80/tcp
```

We can review the ports and services in one go, like so:

```
[root@rhel-instance ~]# firewall-cmd --list-all --zone=public
public (active)
  target: default
  icmp-block-inversion: no
  interfaces: enp1s0
  sources:
  services: cockpit dhcpv6-client ssh
  ports: 80/tcp
  protocols:
  masquerade: no
  forward-ports:
  source-ports:
  icmp-blocks:
  rich rules:
```

Now, we can remove the port:

```
[root@rhel-instance ~]# firewall-cmd --list-ports --zone=public
80/tcp
[root@rhel-instance ~]# firewall-cmd --remove-port 80/tcp
--zone=public --permanent
success
[root@rhel-instance ~]# firewall-cmd --reload
success
[root@rhel-instance ~]# firewall-cmd --list-ports --zone=public
[root@rhel-instance ~]#
```

With this, we know how to add and remove services and ports to/from a firewall and check their statuses. Let's review the options that we can use for `firewall-cmd` to do so:

- `--zone=<zone>`: Used to specify a zone. When no zone is specified, the default one is used.
- `--list-services`: Displays a list of services for the specified zone.
- `--add-service`: Adds a service to the specified zone.

- `--remove-service`: Removes a service from the specified zone.

- `--list-ports`: Lists the open ports in the specified zone.

- `--add-port`: Adds a port to the specified zone.

- `--remove-port`: Removes a port from the specified zone.

- `--list-all`: Lists the ports, services, and all configuration items associated with the specified zone.

- `--permanent`: Rules will be applied to the saved configuration instead of the running one.

- `--reload`: Reloads the rules from the saved configuration.

Now that we know how to assign services and ports to different zones in the firewall, let's take a look at how they are defined.

Creating and using service definitions for firewalld

Service definitions for firewalld are stored in the `/usr/lib/firewalld/services` directory. Let's take a look at a simple service, such as the `ssh` service stored in the `ssh.xml` file, which has the following content:

```
<?xml version="1.0" encoding="utf-8"?>
<service>
  <short>SSH</short>
  <description>Secure Shell (SSH) is a protocol for logging
into and executing commands on remote machines. It provides
secure encrypted communications. If you plan on accessing your
machine remotely via SSH over a firewalled interface, enable
this option. You need the openssh-server package installed for
this option to be useful.</description>
  <port protocol="tcp" port="22"/>
</service>
```

Here, we can see that we only need an XML file with three sections to describe a basic service:

- `short`: The short name for the service

- `description`: A long description of what the service does

- `port`: The port to be opened for this service

Let's say we want to install an Oracle database on our server. We must have port 1521 open, and it must be of the TCP type. Let's create the /etc/firewalld/services/oracledb.xml file with the following content:

```
<?xml version="1.0" encoding="utf-8"?>
<service>
  <short>OracleDB</short>
  <description>Oracle Database firewalld service. It allows
connections to the Oracle Database service. You will need to
deploy Oracle Database in this machine and enable it for this
option to be useful.</description>
  <port protocol="tcp" port="1521"/>
</service>
```

We can enable it by using the following code:

```
[root@rhel-instance ~]# firewall-cmd --reload
success
[root@rhel-instance ~]# firewall-cmd --add-service oracledb
success
[root@rhel-instance ~]# firewall-cmd --list-services
cockpit dhcpv6-client oracledb ssh
```

Now, it's ready to be used in the running configuration. We can add it to the permanent configuration like so:

```
[root@rhel-instance ~]# firewall-cmd --add-service oracledb
--permanent
success
```

> **Tip**
>
> It would be unusual to have to open more complex services. In any case, the manual page that describes how to create firewalld services is firewalld.service and can be opened by running man firewalld.service.

With this, we have an easy way to standardize the services to be opened in the firewalls of our systems. We can include these files in our configuration repositories so that they can be shared with the whole team.

Now that we can create a service, let's take a look at an easier way to configure the firewall in RHEL, that is, using the web interface.

Configuring firewalld with the web interface

To use the RHEL web administrative interface of RHEL 9, we must install it. The package and service running it are both called `cockpit`. We can install it by running the following code:

```
[root@rhel-instance ~]# dnf install cockpit -y
Updating Subscription Management repositories.
[omitted]
Installing:
cockpit                          x86_64 264.1-1.el9 rhel-9-for-
x86_64-baseos-rpms        46 k
[omitted]
    cockpit-264.1-1.el9.x86_64
Complete!
```

Now, let's enable it:

```
[root@rhel-instance ~]# systemctl enable --now cockpit.socket
Created symlink /etc/systemd/system/sockets.target.wants/
cockpit.socket → /usr/lib/systemd/system/cockpit.socket.
```

> **Tip**
> `cockpit` uses a clever trick to save resources. The interface is stopped but a socket is enabled to listen on port `9090`. When it receives a connection, `cockpit` is started. This way, it will only consume resources in your machine when it is in use.

Now, let's learn how to add the `dns` service to the `public` zone.

Let's access `cockpit` by pointing a browser to the IP of the machine and port `9090` – in this case, `https://192.168.122.196:9090`. Let's log in as `root` with the password that was provided during installation:

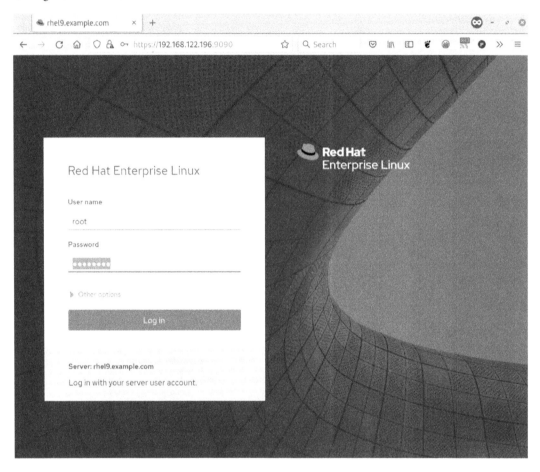

Figure 9.2 – cockpit login screen

Now, we can access the `cockpit` dashboard, which contains information about the system:

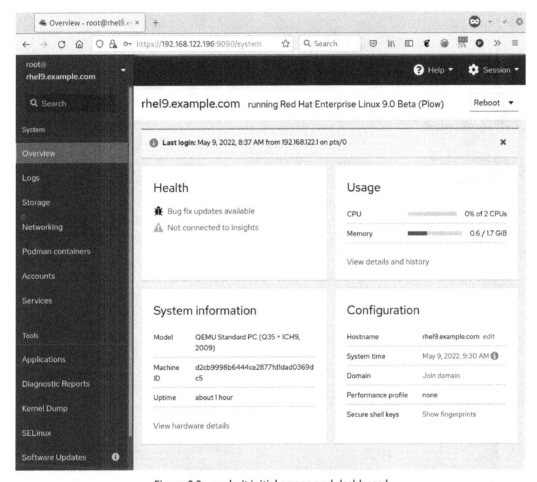

Figure 9.3 – cockpit initial screen and dashboard

Now, let's go to **Networking** and then click on **Edit rules and zones**, as shown in the following screenshot:

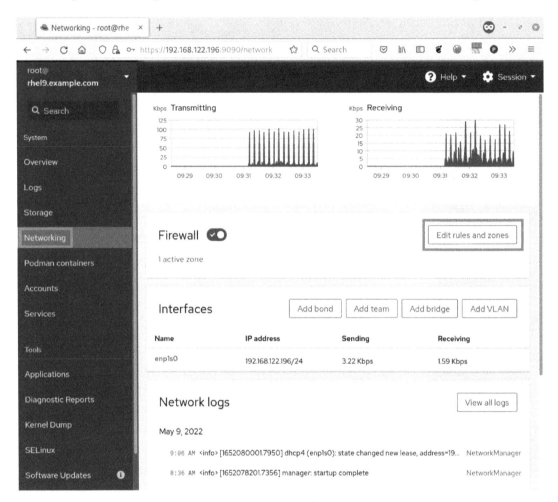

Figure 9.4 – cockpit accessing the firewall configuration

At this point, we can click on **Add Services** in the **public zone** section to modify it and add one more service:

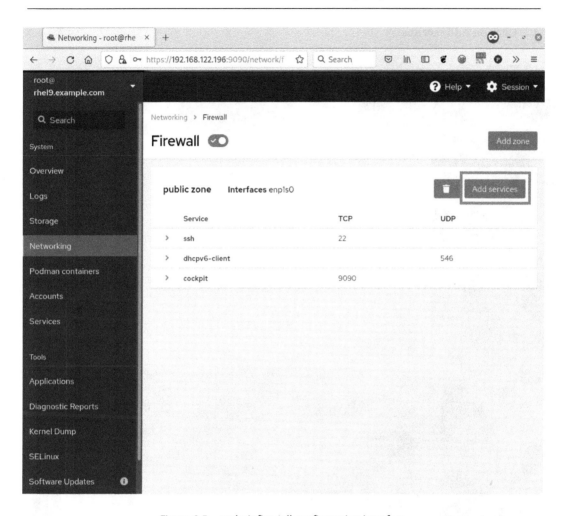

Figure 9.5 – cockpit firewall configuration interface

The steps to add the dns service to the **public zone** section of the firewall are simple:

1. Click on **Services**.
2. Filter the services by typing dns into the **Filter services** input field.
3. Select the dns service with TCP:53 and UDP:53.

4.   Click on **Add services**:

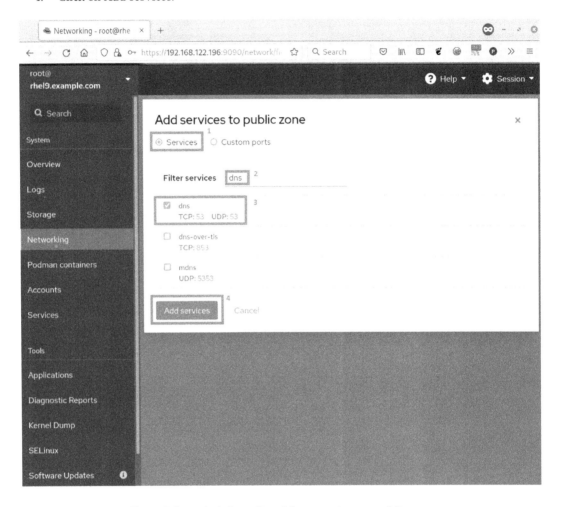

Figure 9.6 – cockpit firewall – adding a service to a public zone

Once you've done this, the service will be added to the running and permanent configurations. It will be displayed in the **public zone** section of `cockpit`:

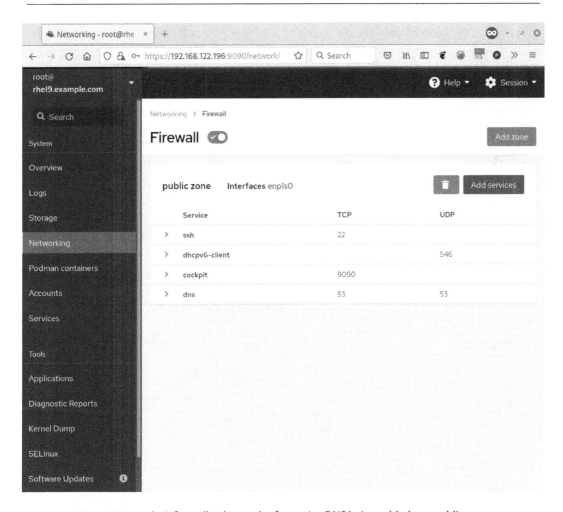

Figure 9.7 – cockpit firewall – the result of a service DNS being added to a public zone

With this, we know how to make modifications to the firewall in RHEL 9 using a web interface. We'll leave it as an exercise for you to remove and redo the configuration we did with the command line at the beginning of this chapter, but with the web interface instead.

Summary

Security is a very important part of system administration. Disabling the security measures on a system just because it's in an isolated network goes against the **defense in depth** principle, so this is heavily discouraged.

In this chapter, we saw how simple and easy it is to configure a firewall using firewalld in RHEL 9, thus providing us with another tool to manage, filter, and secure the network connections in our system. We also worked with `cockpit`, a web administration tool that makes this task more visual and easier to perform.

We can now take control of the network connectivity of our systems, provide access to the services we want to provide, and add a layer of security to them. We also know how to manage zones and how to use them, depending on our system's use case. We are now ready to define our own custom services so that we can always filter network connectivity for them. We can now also deploy more secure systems by using the firewall included in RHEL.

Now, we are ready to learn more about security in RHEL, which is what we will do in the next chapter. Remember, security is a team sport, and the system administrators are key.

10

Keeping Your System Hardened with SELinux

In this chapter, we are going to familiarize ourselves with SELinux. SELinux has been around for a while, but a lack of understanding regarding how it works leads many people to suggest disabling it.

This is not something we want, as it would be similar to telling a user to forego a password because it is hard to remember.

We will introduce the origins of SELinux, and what the default modes and policies are. Then, we will understand how SELinux applies to our files, folders, and processes, and how to restore them to the system defaults.

Additionally, we will explore how to fine-tune the policies using Booleans and troubleshoot common issues with the help of the following sections:

- SELinux usage in enforcing and permissive modes
- Reviewing the SELinux context for files and processes
- Tweaking the policy with semanage
- Restoring changed file contexts to the default policy
- Using SELinux Boolean settings to enable services
- SELinux troubleshooting and common fixes
- Integrity Measurement Architecture, digital hashes, and signatures for enhancing security

In the last section, we'll be learning about how we can secure our system by using signatures and digital hashes to validate the integrity of our system and make sure that no unauthorized changes to the files are performed.

By the end, we will better understand how to use SELinux properly and how to benefit from the additional protection that it provides to our system.

During the chapter, there will be detailed explanations of how SELinux works to aid our understanding of the way it operates, even if using it, in reality, is a lot simpler. We will also use these examples to illustrate cases where SELinux prevents attacks or misconfigurations.

Let's get hands-on with SELinux!

Technical requirements

It is possible to continue the practice of using the virtual machine created at the beginning of this book in *Chapter 1*, *Getting RHEL Up and Running*. Any additional packages required for this chapter will be indicated alongside the text and can be downloaded from `https://github.com/PacktPublishing/Red-Hat-Enterprise-Linux-RHEL-9-Administration`.

SELinux usage in enforcing and permissive modes

Security-Enhanced Linux (**SELinux**) was introduced in December 2000 via the Linux-Kernel mailing list as a product started by the **National Security Agency** (**NSA**) to improve the security of the operating system by means of mandatory access controls and role-based access control, as opposed to the traditional discretionary access controls that were available in the system.

Before SELinux was introduced in the Linux kernel, discussions took place regarding the proper way to do it, and finally, a kernel framework named **Linux Security Modules** (**LSM**) was introduced and SELinux was implemented using it so that other approaches could use LSM too, and not just SELinux.

SELinux provides security improvements to Linux as access to files made by users, processes, or even other resources can be controlled in a very granular way.

Let's take one example to make it clearer when SELinux comes into play: when a web server is serving pages from a user's account, it reads files from the user's home directory inside the `public_html` or www folders (the most standard ones). Being able to read files from the user's home directory can reveal the contents in the event that the web server process is hijacked by an attacker, and this precise moment is when SELinux comes into play, as it will automatically block access to files that should not be accessible for a web server.

SELinux then confines the processes and services to only perform what they are supposed to and only use the resources that are authorized. This is a really important feature that keeps things under control, even in the event of software bugs that may lead to unexpected files or resources being accessed. SELinux will block it if it has not been authorized by the active policy.

> **Important Tip**
> SELinux permissions always arise following regular **discretionary access control** (**DAC**) if a user has no access to a file because of improper file permissions. SELinux has nothing to do there.

By default, the system installation should deploy it in the `enforcing` mode using the `targeted` policy. It is possible to check your current system status via the execution of `sestatus`, as shown in the following screenshot:

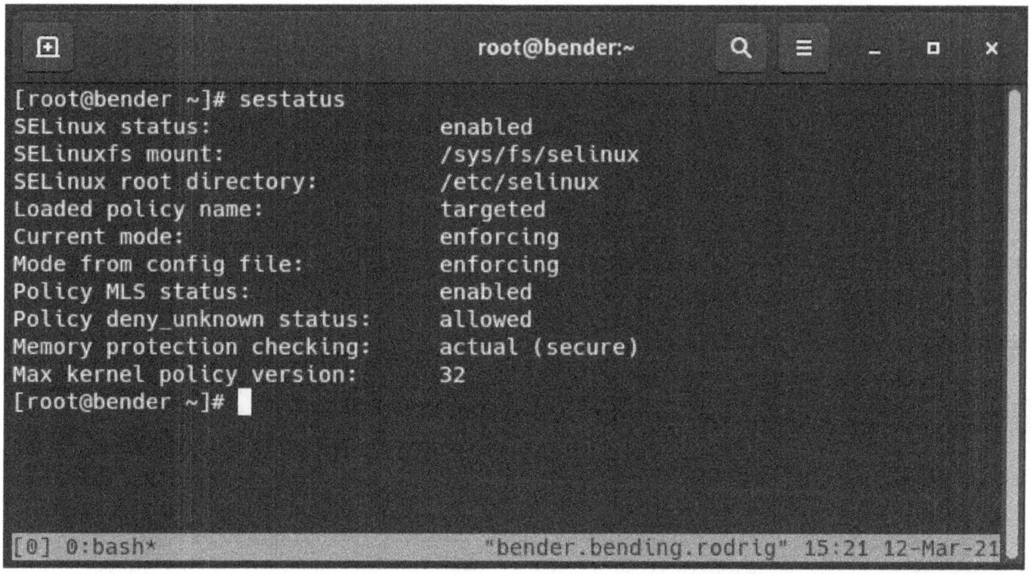

Figure 10.1 – Output of sestatus for our system

As we can see, our system has SELinux enabled using the `targeted` policy and is currently `enforcing`. Let's learn what this means.

SELinux works by defining a **policy**, that is, a set of predefined rules for granting or denying access to resources. The ones available can be listed via `dnf list selinux-policy-*` in your system, with `targeted` and `mls` being the most common ones.

We will focus on the `targeted` policy, but to make an analogy regarding `mls`, the **multi-level security** (**MLS**) policy, it is about allowing users to interact based on their security clearance level, similar to what we can see in movies where someone has clearance to know some information, but not other people. How does this apply to a system? Well, the root user might have access to perform certain actions but not others, and if the user became root via `su` or `sudo`, they would still have the original SELinux label attached, so permissions could be reduced if the root login happened over a local terminal or a remote connection and `sudo` execution.

The mode, listed as `enforcing`, means that the policy is currently being enforced, which is the opposite of `permissive`. We can consider this as being active and offering protection, while `permissive` entails being active but only providing a warning, and not offering protection.

Why do we have `permissive` instead of just disabling it? This question is a bit tricky, so let's explain a bit more about how it works to provide a better answer.

SELinux uses extended attributes in the filesystem to store the labels. Each time a file is created, a label is assigned based on the policy, but this only happens while SELinux is active, so this makes SELinux `disabled` different from SELinux `permissive`, as the first one will not create those labels for the new files created.

Additionally, SELinux in the `permissive` mode allows us to see the errors that will be raised if a program has not received a good policy for it or if a file has no proper labels.

It is really easy to temporarily switch from `enforcing` to `permissive` and vice versa, and always via the `setenforce` command, while we can use `getenforce` to retrieve the current status, as we can see in the following screenshot:

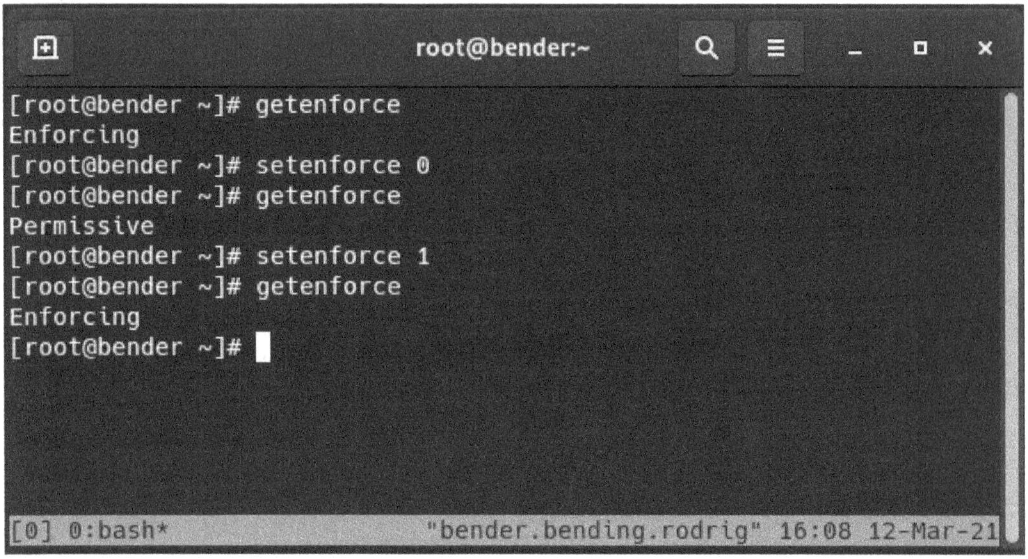

```
[root@bender ~]# getenforce
Enforcing
[root@bender ~]# setenforce 0
[root@bender ~]# getenforce
Permissive
[root@bender ~]# setenforce 1
[root@bender ~]# getenforce
Enforcing
[root@bender ~]#
```
```
[0] 0:bash*                    "bender.bending.rodrig" 16:08 12-Mar-21
```

Figure 10.2 – Changing SELinux enforcing status

It might look basic, but it really is as easy as that – a matter of running a command. However, if the status was disabled, it would be a completely different story.

The SELinux status is configured by editing the `/etc/selinux/config` file, but changes only take effect after a system reboot; that is, we can switch from `enforcing` to `permissive` in real time or from `permissive` to `enforcing`, but when changing the policy from `disabling` to `enabling`, or vice versa, SELinux will require us to reboot the system.

The general advice is to leave SELinux in the enforcing mode, but if, for whatever reason, it was disabled, the recommendation is to switch SELinux to permissive as the first step when moving from disabled. This will allow us to check that the system actually works without being locked out of it because of a kernel blocking access to files and resources.

> **Note**
>
> During the reboot after switching from disabled to permissive or enforcing, the system will force a relabeling of the filesystem based on the policy. This is accomplished by the creation of a file in the root folder of our filesystem named /.autorelabel, which will trigger the process and reboot again afterward.

But why opt for disabling instead of permissive? For example, some software might require to set it in the disabled mode even if later, it can be re-enabled for operations or other reasons, but bear in mind that SELinux is a security feature that protects your system and should be kept.

Keep in mind that SELinux uses **Access Vector Cache** (**AVC**) messages that are logged to the /var/log/audit/audit.log file as well as system journals, and yes, it's a cache, so rules are not checked as frequently so as to speed up operations.

Let's go back to the idea of the filesystem storing labels and let's jump into the next section to see how they relate to processes, files, and the RBAC provided by SELinux.

Reviewing the SELinux context for files and processes

SELinux uses labels, also referred to as the security context attached to each file, and defines several aspects. Let's check one example in our home folder with the ls -l command, but with a special modifier, Z, that will show SELinux attributes as well, as we can see in the following screenshot:

```
                                    root@bender:~
[root@bender ~]# ls -lZ
total 1316
drwxr-xr-x. 10 root root unconfined_u:object_r:admin_home_t:s0    4096 Mar  2 22:46 citellus
-rw-r--r--.  1 root root unconfined_u:object_r:admin_home_t:s0 1254424 Mar 12 22:52 datecron
drwxr-xr-x.  2 root root unconfined_u:object_r:admin_home_t:s0      42 Feb 24 22:10 getfilesback
-rw-r--r--.  1 root root unconfined_u:object_r:admin_home_t:s0   64955 Feb  7 09:01 index.html
drwxr-xr-x. 10 root root unconfined_u:object_r:admin_home_t:s0    4096 Mar  2 22:46 mynewrepo
drwxr-xr-x.  2 root root unconfined_u:object_r:admin_home_t:s0      42 Feb 24 22:10 newfolder
-rw-r--r--.  1 root root unconfined_u:object_r:admin_home_t:s0     540 Mar  6 19:33 term.sh
-rw-r--r--.  1 root root unconfined_u:object_r:admin_home_t:s0     496 Mar  6 19:32 term.sh~
-rw-------.  1 root root unconfined_u:object_r:admin_home_t:s0    2675 Feb 24 17:45 withpass
-rw-r--r--.  1 root root unconfined_u:object_r:admin_home_t:s0     595 Feb 24 17:45 withpass.pub
[root@bender ~]#



[1] 0:bash*                                     "bender.bending.rodrig" 22:54 12-Mar-21
```

Figure 10.3 – File listing showing SELinux attributes

Let's focus on the output for one of the files:

```
-rw-r--r--.  1 root unconfined_u:object_r:admin_
home_t:s0      540 Mar  6 19:33 term.sh
```

The SELinux attributes are the ones listed as `unconfined_u:object_r:admin_home_t:s0`:

- The first part is the user mapping: `unconfined_u`

- The second part is the role: `object_r`

- The third part is the type: `admin_home_t`

- The fourth part is used for the level: `s0` in multi-level security and multi-category security

Something similar happens with processes, and similarly, we can append `Z` to many of the common commands to get the contexts, for example, with `ps  Z`, as we can see in the following screenshot:

```
[root@bender ~]# ps Z
LABEL                                   PID TTY      STAT   TIME COMMAND
system_u:system_r:getty_t:s0-s0:c0.c1023 1651 tty1  Ss+    0:00 /sbin/agetty -o -p -- \u --noclear tty1 li
system_u:system_r:getty_t:s0-s0:c0.c1023 1652 ttyS0 Ss+    0:00 /sbin/agetty -o -p -- \u --keep-baud 1152
unconfined_u:unconfined_r:unconfined_t:s0-s0:c0.c1023 2279084 pts/1 Ss+  0:00 -bash
unconfined_u:unconfined_r:unconfined_t:s0-s0:c0.c1023 2287639 pts/0 Ss   0:00 -bash
unconfined_u:unconfined_r:unconfined_t:s0-s0:c0.c1023 2287661 pts/0 S+   0:00 tmux
unconfined_u:unconfined_r:unconfined_t:s0-s0:c0.c1023 2287662 pts/2 Ss   0:00 -bash
unconfined_u:unconfined_r:unconfined_t:s0-s0:c0.c1023 2288150 pts/2 R+   0:00 ps Z
[root@bender ~]#

[1] 0:bash*                                                   "bender.bending.rodrig" 23:12 12-Mar-21
```

Figure 10.4 – ps output with SELinux contexts

Again, let's examine one of the lines:

```
unconfined_u:unconfined_r:unconfined_t:s0-s0:c0.c1023 2287661
pts/0 S+    0:00 tmux
```

Again, we can see the same approach: user, role, type, and level for multi-level security and multi-category security.

Now that we've introduced what the context looks like, let's focus on how it works in the `targeted` policy.

The `targeted` policy allows everything to run as if SELinux was not enabled in the system, except for the services targeted by it. This makes a good compromise between security and usability.

During the development of the policy, new services are added, while others are refined, and many of the most common services have policies written for protecting them.

SELinux also features something named **transitions**. A transition allows one process started by a user, with a binary with some specific role, to transition via the execution into some other role, which is used later to define what the permissions are for it.

As you might imagine, our user also has an SELinux context, and similarly, we can use the `id -Z` command for checking it:

```
unconfined_u:unconfined_r:unconfined_t:s0-s0:c0.c1023
```

So, going back to the first example, Apache Web Server is provided by the `httpd` package, which can be installed via `dnf -y install httpd`. Once installed, let's start it with `systemctl start httpd` and enable it with `systemctl enable httpd`, and then open the firewall with `firewall-cmd --add-service=http` and `firewall-cmd --add-service=https`, as we've done with other services in previous chapters.

Previous commands can be found in the following script: `https://github.com/PacktPublishing/Red-Hat-Enterprise-Linux-RHEL-9-Administration/blob/main/chapter-10-selinux/apache.sh`.

Let's see how all that comes into play in the following screenshot:

```
[root@bender ~]# which httpd|xargs ls -Z
system_u:object_r:httpd_exec_t:s0 /usr/sbin/httpd
[root@bender ~]# ps axZ|grep http
system_u:system_r:httpd_t:s0    1347331 ?        Ss       2:15 /usr/sbin/httpd -DFOREGROUND
system_u:system_r:httpd_t:s0    2320645 ?        S        0:00 /usr/sbin/httpd -DFOREGROUND
system_u:system_r:httpd_t:s0    2320646 ?        Sl       0:31 /usr/sbin/httpd -DFOREGROUND
system_u:system_r:httpd_t:s0    2320647 ?        Sl       0:32 /usr/sbin/httpd -DFOREGROUND
system_u:system_r:httpd_t:s0    2320648 ?        Sl       0:31 /usr/sbin/httpd -DFOREGROUND
unconfined_u:unconfined_r:unconfined_t:s0-s0:c0.c1023 2358694 pts/3 R+   0:00 grep --color=auto
http
[root@bender ~]# ls -dZ /var/www/html/
system_u:object_r:httpd_sys_content_t:s0 /var/www/html/
[root@bender ~]# 
```

Figure 10.5 – Web server SELinux contexts

Here, we can see how the executable on disk has the `httpd_exec_t` context, the process is `httpd_t`, and the files/folder served by it is `httpd_sys_content_t`, and it works!

Let's now create an `index.htm` file in our `home` folder and move it to the `Apache Web Root` folder as follows:

```
# echo '<html><head><title>Our test</title></head><body>This is
our test html</body></html>' > index.htm
# cp index.htm /var/www/html/index2.htm
# mv index.htm /var/www/html/index1.htm
```

Let's see what happens when we try to access the files as shown in the following screenshot:

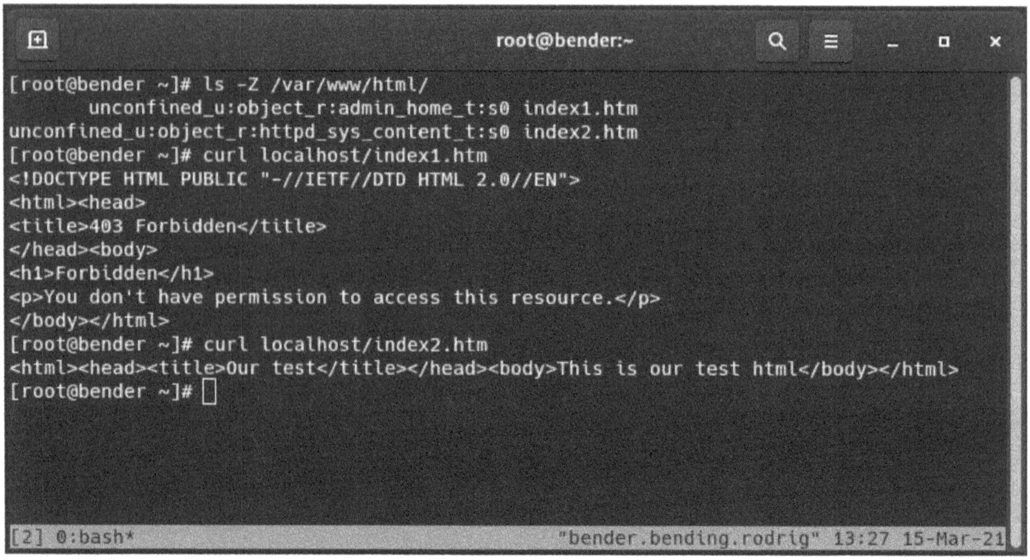

Figure 10.6 – Apache behavior with the generated files

As we can see, each file has one SELinux context, but on top of that, Apache is denying access to the one we moved (`index1.htm`), but showing the contents for the one we copied (`index2.htm`).

What has happened here? We copied one file and moved the other, out of the same source, but they got two different SELinux contexts, as we can see in the following figure:

Figure 10.7 – Different SELinux contexts

Let's extend the test as shown in the following screenshot:

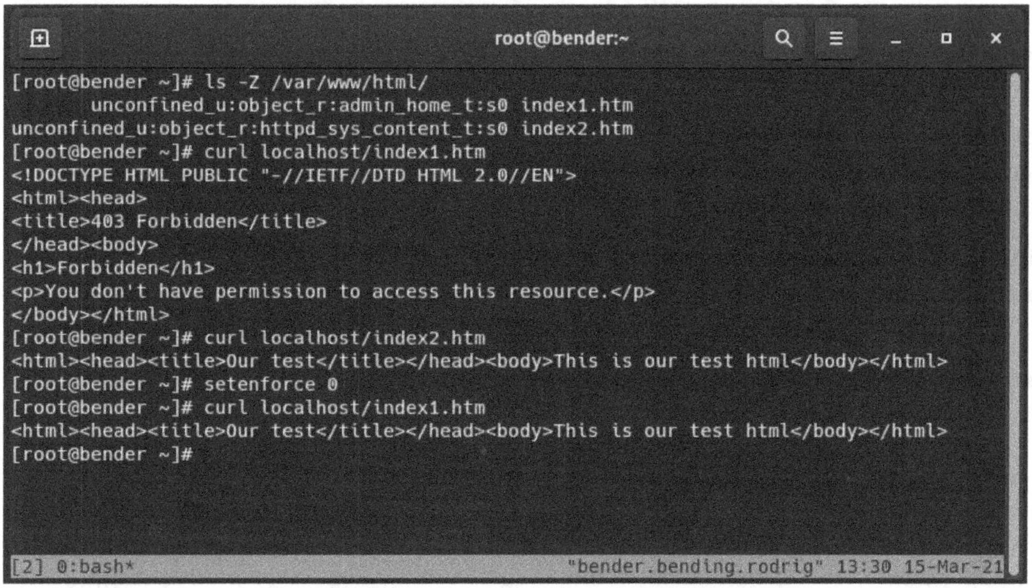
root@bender:~
[root@bender ~]# ls -Z /var/www/html/
unconfined_u:object_r:admin_home_t:s0 index1.htm
unconfined_u:object_r:httpd_sys_content_t:s0 index2.htm
[root@bender ~]# curl localhost/index1.htm
<!DOCTYPE HTML PUBLIC "-//IETF//DTD HTML 2.0//EN">
<html><head>
<title>403 Forbidden</title>
</head><body>
<h1>Forbidden</h1>
<p>You don't have permission to access this resource.</p>
</body></html>
[root@bender ~]# curl localhost/index2.htm
<html><head><title>Our test</title></head><body>This is our test html</body></html>
[root@bender ~]# setenforce 0
[root@bender ~]# curl localhost/index1.htm
<html><head><title>Our test</title></head><body>This is our test html</body></html>
[root@bender ~]#
[2] 0:bash*
"bender.bending.rodrig" 13:30 15-Mar-21

Figure 10.8 – Retrying with SELinux in permissive mode

As we can see in the preceding screenshot, we are now able to access file contents, so you could say *"What is wrong with SELinux that does not allow my site to work?"* but the right way to express it would be *"Look how SELinux has protected us from disclosing a personal file on a website."*

Note that as we've temporarily disabled SELinux enforcing via `setenforce 0`, we need to enable it back with `setenforce 1` to keep our system protected.

Instead of directly moving a file into Apache's **DocumentRoot** (`/var/www/html`), if it was an attacker trying to reach our `home` folder files, SELinux would have denied those accesses by default. The `httpd_t` process cannot access the `admin_home_t` context.

A similar thing happens when we try to get Apache (or any other service under the `targeted` policy) to listen on a port that is not the one configured by default, and the best way to get familiar with what we can or cannot do is to learn about the `semanage` utility.

Using `semanage`, we can list, edit, add, or delete the different values in the policy, and even export and import our customizations, so let's use it to learn a bit more about it using our example with `httpd`.

Let's learn about `semanage` in the following section.

Tweaking the policy with semanage

As we introduced earlier, the `targeted` policy contains some configurations that are enforced for the services it has defined, allowing the protection of those services while not interfering with the ones it does not know about.

Still, sometimes we need to tweak several settings, such as allowing `http` or the `ssh` daemon to listen on alternate ports or accessing some other file types, but without losing the additional layer of protection provided by SELinux.

First, let's ensure that `policycoreutils` and `policycoreutils-python-utils` are installed in our system with `dnf -y install policycoreutils-python-utils policycoreutils`, as they provide the tools we will use in this and the next sections of this chapter.

Let's learn with the help of an example. Let's see which ports `httpd_t` can access with `semanage port -l|grep http`:

```
http_cache_port_t              tcp      8080, 8118, 8123,
10001-10010
http_cache_port_t              udp      3130
http_port_t                    tcp      80, 81, 443, 488, 8008,
8009, 8443, 9000
pegasus_http_port_t            tcp      5988
pegasus_https_port_t           tcp      5989
```

As we can see, `http_port_t`, used by Apache Daemon, is allowed, by default, to use ports 80, 81, 443, 488, 8008, 9009, 8443, and 9000 via `tcp`.

That means that if we want to run Apache on any of those ports, no changes to the policy will be required.

If we repeat the command but for `ssh`, we only see port 22 opened (executing `semanage port -l|grep ssh`):

```
ssh_port_t                     tcp      22
```

For example, we might want to add another port, let's say 2222, to the list of possible ports, so that we hide the standard one being tested by port scanners. We will be able to do it via `semanage port -a -p tcp -t ssh_port_t 2222` and then validate with the prior command, `semanage port -l|grep ssh`, which now shows the following:

```
ssh_port_t                     tcp      2222, 22
```

As we can see, port 2222 has been added to the list of available ports for the `ssh_port_t` type, and that enables the `ssh` daemon to start listening on it (this, of course, requires additional configuration for the `ssh` daemon configuration and to the firewall before we get a working service).

In the same way, for example, some web services require writing to specific folders for storing configurations, but by default, the context on /var/www/html is httpd_sys_content_t, which does not allow writing to disk.

We can check the available file contexts with semanage fcontext -l in a similar way to what we did with the ports, but the list of files is huge, as a web server might use common locations such as logs and cgi-bin, as well as filesystem files for certificates, configuration, and home directories, and extensions such as PHP and others. When you check the contexts with the preceding command, pay attention to the different types that are available and what the structure is for one listing, as in the following example:

```
/var/www/html(/.*)?/wp-content(/.*)?                    all
files           system_u:object_r:httpd_sys_rw_content_t:s0
```

As we can see, there is a regular expression that matches the files in the wp-content folder inside the /var/www/html path applying to all files and sets an SELinux context of httpd_sys_rw_content_t, which allows read-write access. This folder is used by the popular blog software **WordPress**, so the policy is already prepared for covering some of the most popular services, folders, and requirements without requiring system administrators to write them ad hoc.

When invoking semanage, it will output that it has some subcommands we can use, such as the following:

- import: This allows the importing of local modifications.
- export: This allows the exporting of local changes.
- login: This allows the login and SELinux user associations to be managed.
- user: This manages SELinux users with roles and levels.
- port: This manages port definitions and types.
- ibpkey: This manages InfiniBand ibpkey definitions.
- ibendport: This manages end port InfiniBand definitions.
- interface: This defines network interface definitions.
- module: This manages policy modules for SELinux.
- node: This manages definitions of network nodes.
- fcontext: This manages file context definitions.
- boolean: This manages Booleans for tweaking policies.
- permissive: This manages the enforcing mode.
- dontaudit: This manages the dontaudit rules in the policy.

For each one of the preceding commands, we can use the `-h` argument to list, get help about, and learn about the extra arguments that can be used for each one.

For the day-to-day use case, most of the time we'll be using `port` and `fcontext` as those will cover extending or tuning the available services that come with Red Hat Enterprise Linux, like the example we have showcased with `ssh` listening on an additional port.

> **Important Tip**
>
> Traditionally, **Red Hat Certified System Administrator** (**RHCSA**) and **Red Hat Certified Engineer** (**RHCE**) courses used to have a reboot validation. This meant that for each service that was installed and started, it was also mandatory to remember to enable it to be active on the next reboot. A similar thing happens with SELinux. If we are adding a piece of software that will stay in our system, the best approach is to define, via `semanage`, `regexp` for the path that will be used. When following this approach, if the filesystem is relabeled or the context restored, the application will continue to work.

Let's see how to manually set the context for files and how to restore the defaults in the next section.

Restoring changed file contexts to the default policy

In the previous section, we mentioned how `semanage` enables us to perform changes to the policy, which is the recommended way to perform changes and to persist them for future files and folders, but that is not the only way we can perform operations.

From the command line, we can use the `chcon` utility to change the context for a file. This will allow us to define the user, the role, and the type for the file we want to alter, and similar to other filesystem utilities such as `chmod` or `chown`, we can also affect files recursively, so it's easy to set a full folder hierarchy to the desired context.

One feature that I always found very interesting is the ability to copy the context of a file via the `--reference` flag so that the same context as the referenced file is applied to the target one.

When we were introducing the example of `httpd` earlier in this chapter, we did a test with two files, `index1.htm` and `index2.htm`, that were moved and copied to the `/var/www/html` folder. To go deeper into this example, we will make additional copies of `index1.htm` (such as `index3.htm`) to demonstrate in the next screenshot the usage of `chcon`. Bear in mind that creating the files directly in the `/var/www/html` folder will set the files to have the proper context, so we need to create them at `/root` and then move them to the target folder, as we can see in the following screenshot:

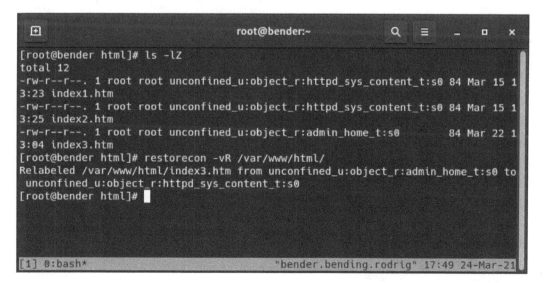

Figure 10.9 – Demonstrating chcon usage

As we can see, both the `index1.htm` and `index3.htm` files now have the proper context – in the first case, using the reference, and in the second, defining the type to use.

Of course, this is not the only method. As we indicated earlier, the recommended way for setting the context for applications is to define the `regexps` path via `semanage`, and this empowers us to use the `restorecon` command to apply the right context, according to the configuration, to the files. Let's see how it operates in the following screenshot:

```
[root@bender html]# ls -lZ
total 12
-rw-r--r--. 1 root root unconfined_u:object_r:httpd_sys_content_t:s0 84 Mar 15 1
3:23 index1.htm
-rw-r--r--. 1 root root unconfined_u:object_r:httpd_sys_content_t:s0 84 Mar 15 1
3:25 index2.htm
-rw-r--r--. 1 root root unconfined_u:object_r:admin_home_t:s0          84 Mar 22 1
3:04 index3.htm
[root@bender html]# restorecon -vR /var/www/html/
Relabeled /var/www/html/index3.htm from unconfined_u:object_r:admin_home_t:s0 to
 unconfined_u:object_r:httpd_sys_content_t:s0
[root@bender html]#
```

Figure 10.10 – Using restorecon to restore context

As we can see, we used `restorecon -vR /var/www/html/` and it automatically changed the `index3.htm` file into `httpd_sys_content_t`, which is defined for that folder as we saw when we were testing `semanage` to list the contexts. The arguments used, v and R, make the utility report the changes (verbose) and work recursively on the paths provided.

Let's say we have messed the system up by running `chcon` over the root filesystem. What would be the way to fix it? In this case, as we mentioned earlier, we should do the following:

1. Set the operation mode to permissive to not block further access via `setenforce 0`.
2. Put the marker to have the filesystem relabeled via touch `/.autorelabel`.
3. Modify the `/etc/selinux/config` file to set the boot mode to permissive.
4. Reboot the system to let relabeling happen.
5. Once the system reboots, edit `/etc/selinux/config` again to define the operation mode as enforcing.

By operating in this way, instead of just running `restorecon -R /`, we are making sure that the system is operational and will continue to operate after reboot and a full relabel is applied to the filesystem, so it is left ready to re-enable the `enforcing` mode safely.

In the next section, let's see how to tune the policy within itself, using Booleans to tune how it works.

Using SELinux Boolean settings to enable services

Many services have a wide range of configuration options for many common cases, but not always the same. For example, the `http` server should not access user files, but at the same time, it's a common operation to enable personal websites from the www or `public_html` folders in each user's home directory.

To overcome that use case and at the same time, provide enhanced security, the SELinux policy makes use of Booleans.

A Boolean is a tunable that can be set by the administrator that can enable or disable conditionals in the policy code. Let's see, for example, a list of Booleans available for `httpd` by executing `getsebool -a|grep ^http` (list reduced):

```
httpd_can_network_connect --> off
httpd_can_network_connect_db --> off
httpd_can_sendmail --> off
httpd_enable_homedirs --> off
httpd_use_nfs --> off
```

This list is a reduced subset of the Booleans available, but it does give us an idea of what it can accomplish; for example, `http`, by default, cannot use the network to connect to other hosts, or send an email (usually done in PHP scripts), or even access home folders for users.

For example, if we want to enable users in a system to publish their personal web pages from the www folder in their home directory, that is /home/user/www/, we will have to enable the httpd_enable_homedirs Boolean by running the following command:

```
setsebool -P httpd_enable_homedirs=1
```

This will tweak the policy to enable http to access the user's home directory to serve the pages there. If additionally, the servers will be stored on a **Network FileSystem** (**NFS**) or **Common Internet FileSystem** (**CIFS**) mount, additional Booleans will be required. We're still using the same targeted policy, but we've enabled the internal conditionals to permit that the accesses are not to be blocked by SELinux.

> **Important Tip**
>
> The -P parameter to setsebool is required to make the change *permanent*. That means writing the change so that it is persisted; without it, the change will be lost once we restart our server.

As we've seen, getsebool and setsebool allow us to query and set the values for the Booleans that tune the policy, but also, semanage boolean -l can help here, as we can see in the following screenshot:

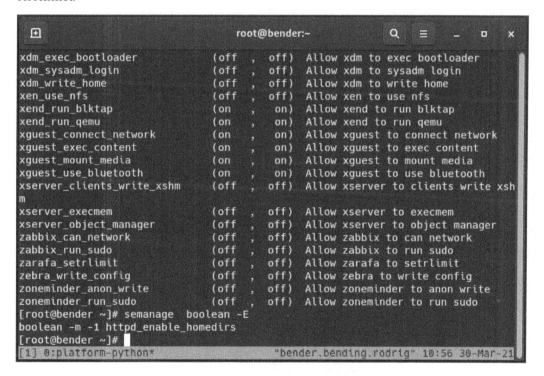

Figure 10.11 – Using semanage to manage Booleans

In the previous screenshot, we can see not only the Boolean we edited using `setsebool` but also a description of the intended behavior.

One of the benefits is that `semanage`, as we introduced, allows us to export and import the local changes to the policy, so any customization made can be exported and imported to another system to ease the setup of similar server profiles.

All the possible Booleans in the policy can be checked with `semanage boolean -l`, similar to what we did to list the binding ports for applications in our `http` example.

We have learned about using Booleans to tune how the policy is adapting to some specific but pretty common cases. Next, we will explore probably the most frequently used part for administrators, that is, troubleshooting, but with a focus on SELinux.

SELinux troubleshooting and common fixes

One of the main problems in getting used to SELinux is that many people who are not familiar with it blame it for things not working; however, this argument is getting a bit outdated, as SELinux was introduced in Red Hat Enterprise Linux 4, which was back in 2005.

Most of the time, issues with SELinux and our systems are related to changed file contexts and changing ports for services, and fewer time issues with the policy itself.

First of all, there are several places where we can check for errors, but in our list, we should start with the audit log or the system messages. For example, we can start with the `/var/log/audit/audit.log` file that we introduced earlier in this chapter.

Also bear in mind that SELinux **mandatory access control** (**MAC**) only plays once we have cleared access from regular **discretionary access control** (**DAC**), that is, if we've no permission to check a file (for example, mode `400` and our user not being the owner). In this case, it's highly unlikely that SELinux is blocking access.

Most of the time, our system will have installed the `setroubleshoot-server` and `setroubleshoot-plugins` packages that provide several tools, including `sealert`, to query the received SELinux messages and many times also, to suggest changes. Those packages provided additional tools for troubleshooting problems related to SELinux in an easier way.

Let's cover some of the basics that we should always validate:

- Review all the other controls (user and group ownership and permissions are properly set).

- Do not disable SELinux (`https://stopdisablingselinux.com/`).

  If a program is not working properly and it was shipped with the OS, it might be a bug and should be reported via a support case or Bugzilla at `https://bugzilla.redhat.com`.

Only if a program is not working properly might it be made to run unconfined, but leaving all remaining system services protected via the `targeted` policy.

- Think about what was done before the error happened if this is an existing program.

  Perhaps files were moved instead of copied or created upon reaching their destination, or perhaps the ports or folders for the software were changed.

Having arrived at this point, we should check `audit.log` for relevant messages. For example, as regards the example we mentioned regarding the wrong context with the files in `/var/www/html/`, an example audit entry would be as follows:

```
type=AVC msg=audit(1617210395.481:1603680):
avc:  denied  { getattr } for  pid=2826802 comm="httpd"
path="/var/www/html/index3.htm" dev="dm-0"
ino=101881472 scontext=system_u:system_r:httpd_t:s0
tcontext=unconfined_u:object_r:admin_home_t:s0 tclass=file
permissive=0
```

It looks strange, but if we check the parameters, we see the path of the affected file, the PID, the source context (`scontext`), and the target context (`tcontext`), so in brief, we can see that `httpd_t` tried to access (get attributes) for a target context `admin_home_t` and that was denied.

At the same time, if we're using `setroubleshoot`, we will get a message like this in the system journal:

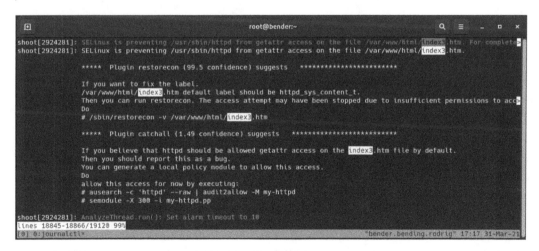

Figure 10.12 – setroubleshoot logging in the system journal

As we can see in the preceding screenshot, `setroubleshoot` already identified that one of the plugins suggests applying the `restorecon` command over the file, as it's not matching the one for the folder it's in, and even suggests the exact command to use for restoring the label.

Another plugin suggests generating a custom policy using the following two commands:

```
# ausearch -c 'httpd' --raw | audit2allow -M my-httpd
# semodule -X 300 -i my-httpd.pp
```

However, this kind of recommendation should be taken with knowledge of what is being done, which means that the preceding commands will fix httpd_t in terms of getting access to the home_admin_t file. We can learn about what would happen by only running the first command, together with the audit2allow pipe.

Running ausearch -c 'httpd' --raw | audit2allow -M my-httpd creates several files named my-httpd in the current folder, one named my-httpd.te, and another named my-httpd.pp. The second command, which we will *not* use, installs the modified policy, but please, don't ever do that until you have an understanding of what's going on, as we will see in the following lines.

The interesting file for us now is the my-httpd.te one (where *te* means *type enforcement*):

```
module my-httpd 1.0;
require {
        type httpd_t;
        type admin_home_t;
        class file getattr;
}
#============= httpd_t ==============
allow httpd_t admin_home_t:file getattr;
```

From there, we can see that it uses a requirements session for the types involved, and later, the rule itself, which allows httpd_t access to admin_home_t files for using the getattr function, nothing else, nothing more.

As has been said previously, will this fix our issue? It will effectively allow httpd_t to obtain access to the index3.html file, so there will no longer be any errors, but this comes with a significant cost. From that point, httpd_t could also read home directory files without any complaints.

> **Important Note**
> I don't know how many times this fact needs to be reinforced but think twice before acting on a system. SELinux is a protection mechanism for increasing the safety of your system; do not disable it, and do not blindly accept audit2allow-created policies without some initial investigation and understanding of what the issue might be and what the proposed resolution does, as it may almost be equivalent to disabling SELinux.

At this point, if we have installed that module, we can use `semodule` to do the following:

- List: `semodule -l`
- Install: `semodule -i $MODULE_NAME`
- Remove: `semodule -r $MODULE_NAME`

With the preceding commands, we can check or alter the current status of the policy-loaded modules.

Going back to reviewing system logs, we may realize that something is actually failing some time after it began, but not from the very beginning, so using `ausearch` or passing the full logs to `audit2allow` might not prove helpful; however, we can use the command suggested by `setroubleshoot` to list them:

```
Mar 31 17:06:41 bender setroubleshoot[2924281]: SELinux is
preventing /usr/sbin/httpd from getattr access on the file /
var/www/html/index3.htm. For complete SELinux messages run:
sealert -l 1b4d549b-f566-409f-90eb-7a825471aca8
```

If we execute `sealert -l <ID>`, we will receive the output provided by the different plugins to fix the issue as well as context information similar to what is shown in *Figure 10.11*.

In the case of new software being deployed that has no SELinux support, we can do the following checks the other way around in a test system:

- Set SELinux to the `permissive` mode.
- Deploy the software.
- Analyze all the alerts received to see whether anything is unexpected.
- Contact software vendors and initiate a support case with Red Hat to work on a policy.

If we're getting locked out of our system because SELinux is enforcing and we have badly messed the labels up, for example, by running a bad `chcon` command recursively against our root folder (for example, scripting a context change depending on a variable and that variable being empty), we still have the following ways to get out of trouble:

- Use `setenforce 0` to put SELinux in the `permissive` mode.
- Run `touch /.autorelabel`.
- Reboot the host so that at the next boot, SELinux restores the appropriate labels

If we are in a really bad situation and, for example, are unable to use `setenforce 0` or the system cannot even boot or perform relabeling correctly, there is still hope, but some additional steps are required.

When a system is rebooting, we can see the list of installed kernels at the grub prompt and use it to edit the kernel boot parameters.

Using the `selinux=0` parameter, we completely disable SELinux, and this is something we don't want, but we can use `enforcing=0` to accomplish having SELinux enabled, but in the `permissive` mode.

Once we have our system booting into the `permissive` mode, we can repeat the previous procedure to get back to the previous behavior and continue debugging the situation within the system itself with the prior indications given (checking system logs, and others).

Let's move to the next section to learn about how we can increase our system security with another kernel function named Integrity Measurement Architecture that will monitor file contents to be validated before execution, access, and so on.

Integrity Measurement Architecture, digital hashes, and signatures for enhancing security

SELinux, as aforementioned, can increase your system security by confining processes to what is intended for them to do, but in order to improve the protection, you can make use of another feature of the kernel, the integrity subsystem, provided by two components (`http://linux-ima.sourceforge.net/`):

- **Integrity Measurement Architecture (IMA)**, which maintains a runtime list of measures
- **Extended Verification Module (EVM)**, which detects alterations to measurements

The first component, IMA, checks the file contents when opened or executed, stores the obtained values, and allows the validation while permitting tunning via custom policies.

The second component, EVM, hashes the values from IMA, and SELinux attributes via cryptographic functions.

Of course, if the system has support for the **Trusted Platform Module (TPM)**, it can be used to increase the level of hardening of the solution. TPM can provide some services to improve security such as creating and storing encryption keys, for example.

Note that IMA is somewhat similar to what software such as AIDE can do; there's a comparison on the documentation at `https://access.redhat.com/documentation/en-us/red_hat_enterprise_linux/9/html/security_hardening/checking-integrity-with-aide_security-hardening#file-integrity-tools-aide-and-ima_checking-integrity-with-aide` that might help you decide what fits better with your requirements.

In the following subsections, we will introduce some of the commands and how data is stored so that you can get familiar with this technology.

EVM and trusted keys

Red Hat Enterprise Linux 9 Kernel uses the `evm-key` keyring, which accepts encrypted keys.

Encrypted keys are generated by the kernel and make use of the keyring kernel service. Those can be used by the EVM to confirm that the running system has integrity.

Used in conjunction with TPM, the keys will get additional values used to seal them with a set of integrity management values that will only allow decrypting the keys on the same system they were generated while allowing to update the keys to support new versions of the software (for example, updated kernel versions will be allowed to boot).

For using trusted keys, the system must have the *trusted* kernel module loaded. Check the `keyctl` man page to learn about the different options for creating keys.

Just for following this section, ensure that the required packages are installed with the following:

```
dnf install ima-evm-utils attr keyutils
```

Let's see how to enable this feature in the next section.

Enabling IMA and EVM in the system kernel

Both IMA end EVM are part of the **kernel integrity system**. To enable it on our system, we need to first check that `securityfs` is mounted by running the following commands:

```
mount|grep security
dmesg |grep -iE 'EVM|IMA'
```

The commands can be seen in action in the following screenshot:

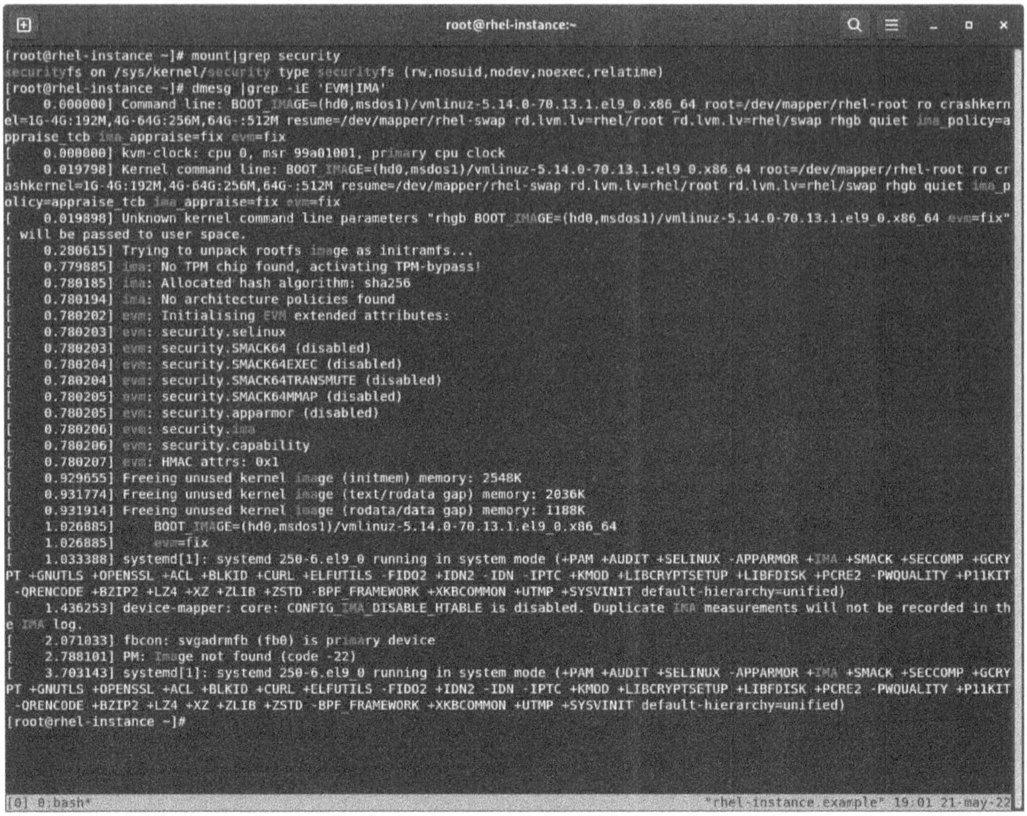

Figure 10.13 – Security kernel filesystem mounted

The first command should show that the filesystem is mounted, and the second, the entries of IMA. Note that in the preceding figure, the kernel command line contains some parameters, and to enable them, if you don't have them in your system, you can add them via the following:

```
grubby --update-kernel=/boot/vmlinuz-$(uname -r) --args="ima_
policy=appraise_tcb ima_appraise=fix evm=fix"
```

And once added, reboot the system to apply the changes. You can always verify the current kernel parameters via the `cat /proc/cmdline` command. In *Figure 10.12*, you can see it on the fourth line (coming from the kernel boot log).

Okay, now that we have our system ready, we can start adding keys. Previously, we indicated to check the manual page for `keyctl` but now we'll see some examples while we continue setting IMA.

When following the documentation at `https://sourceforge.net/p/linux-ima/wiki/Home/,` we can create the keys that we'll be later using.

The first step is to create the **kernel master key** (kmk). In the following examples, we'll be using the commands for systems without TPM enabled to allow testing of this inside an RHEL 9 VM:

```
keyctl add user kmk "$(dd if=/dev/urandom bs=1 count=32 2> /
dev/null)" @u
keyctl add encrypted evm-key "new user:kmk 64" @u
```

This gives an output like the one shown in the following figure:

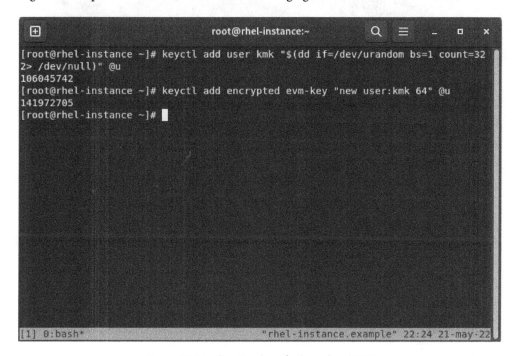

Figure 10.14 – Creating keys for kernel and EVM

We can also save the keys with the following commands so that an unencrypted copy of them will be stored on disk:

```
mkdir -p /etc/keys
keyctl pipe $(keyctl search @u encrypted evm-key) > /etc/keys/
evm-key
keyctl pipe $(keyctl search @u user kmk) > /etc/keys/kmk
```

We can check the new keys in the kernel and enable EVM, as shown in the next screenshot:

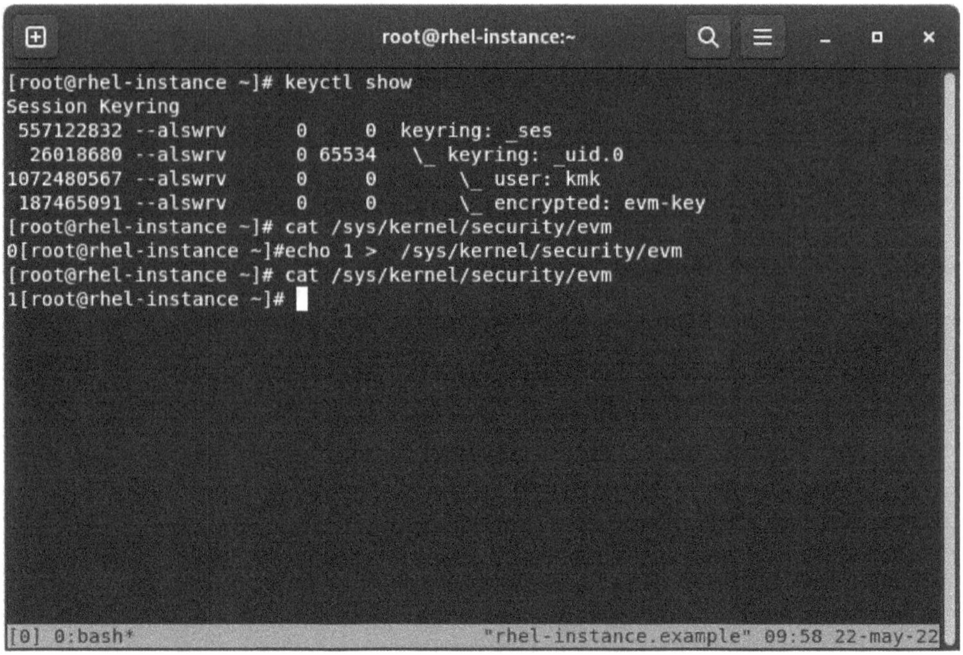

Figure 10.15 – Checking created keys and enabling EVM

As you can see, `keyctl show` is listing a tree-like display of the keys, and by querying and writing to the special `/sys/kernel/security/evm` file, we can enable it.

Now that IMA is enabled, the files on our system will get some extended attributes; for example, let's check the extended attributes of the `.bash_history` file with the following:

```
getfattr -m . -d .bash_history
```

It will return an output similar to this:

```
# file: .bash_history
security.ima=0sBAT4/qlLGUMDFUL8AmKNg5BJ9YFS8vU/Z0vi1wezgWGveA==
security.selinux="unconfined_u:object_r:admin_home_t:s0"
```

As you can see, the file doesn't show the evm attribute, because this was done before evm was created, so let's use the following command to update the file and get the attributes again:

```
touch .bash_history
[root@rhel-instance ~]# getfattr -m . -d .bash_history
# file: .bash_history
security.evm=0sAi194Jg8KrZA2Wy1tzAcTX9nO7mJ
security.ima=0sBAT4/qlLGUMDFUL8AmKNg5BJ9YFS8vU/Z0vi1wezgWGveA==
security.selinux="unconfined_u:object_r:admin_home_t:s0"
```

Now, we can see that `security.ima` keeps the same value, but the new `security.evm` key has been listed.

As aforementioned, check the provided links for more information and examples of how to use IMA and EVM to protect and go further than this introduction.

Summary

This chapter has introduced SELinux, how it works, how we can check the processes, files, and ports, and how to fine-tune them either by adding new options or using Booleans. We also covered several initial troubleshooting skills that we should explore further to enhance our knowledge and experience.

SELinux, as we've seen, is a powerful tool for keeping our system secured with an extra layer that protects our system even from unknown issues that might come from defects in the software itself.

We have covered how to find the SELinux context in files and processes, how those are applied via the policy, and how to tune it so that our system is protected and still able to provide the expected service.

Troubleshooting SELinux is a skill that will help us in adapting the software that doesn't come with Red Hat Enterprise Linux to still perform properly. Last but not least, we introduced IMA and EVM so that they can be leveraged to increase the security of the system via the usage of hashing algorithms to detect changes in our system

In the next chapter, we will learn about security profiles with OpenSCAP to continue keeping our system safe.

11

System Security Profiles with OpenSCAP

SCAP stands for **Security Content Automation Protocol**, a standardized way to check, verify, and report vulnerability assessment and policy assessment. **Red Hat Enterprise Linux** (RHEL) 9 includes the **OpenSCAP** tool and profiles to audit and manage the security of systems. This helps ensure the systems you are managing comply with standard security policies such as the **Payment Card Industry Data Security Standard (PCI DSS)** or the **Protection Profile for General Purpose Operating Systems—** or **Operating System Protection Profile** (OSPP) for short—as well as discovering vulnerabilities. New security profiles, such as the **Health Insurance Portability and Accountability Act (HIPAA)** security profile, have been added to RHEL 9 to cover systems containing personal health information.

RHEL 9 includes this tool to review security profiles in order to discover possible attack vectors (misconfigurations or vulnerabilities) and can obtain guidance on how to better harden the system. We will learn how to perform a scan on a system and discover what needs to be changed to prepare it, in order to ensure it is completely aligned with the regulatory requirements. We will also learn how this tool can be used to improve the security of a system for general use by reviewing it and applying the recommended changes.

To review how to use OpenSCAP, in this chapter, we will go through the following topics:

- Getting started with OpenSCAP and discovering system vulnerabilities
- Using OpenSCAP with security profiles for OSPP and PCI DSS

Getting started with OpenSCAP and discovering system vulnerabilities

Let's get started in OpenSCAP in a practical way by first reviewing the `Security Tools` software group, which has some tools that are good to know, and then proceeding to run some scans.

Our initial step will be to get information on `Security Tools`. Here's how we can do this:

```
[root@rhel-instance ~]# dnf group info "Security Tools"
Updating Subscription Management repositories.
Last metadata expiration check: 0:00:20 ago on Fri Jun  3
08:17:27 2022.
Group: Security Tools
Description: Security tools for integrity and trust
verification.
Default Packages:
   scap-security-guide
Optional Packages:
   aide
   hmaccalc
   openscap
   openscap-engine-sce
   openscap-utils
   scap-security-guide-doc
   scap-workbench
   tpm2-tools
   tss2
   udica
```

This group includes several security tools, such as `aide`, to ensure file integrity in the system; `tpm2-tools`, to manage the **Trusted Platform Module** (**TPM**) to store encryption keys; and `openscap-utils`, to review the security policies in the system.

We can get more information on those tools by using `dnf`. Let's review the one that is more relevant for this chapter—`openscap-utils`—as follows:

```
[root@rhel-instance ~]# dnf info openscap-utils
Updating Subscription Management repositories.
Last metadata expiration check: 0:01:41 ago on Fri Jun  3
08:17:27 2022.
Available Packages
Name        : openscap-utils
Epoch       : 1
Version     : 1.3.6
Release     : 3.el9
```

```
Architecture  : x86_64
Size          : 32 k
Source        : openscap-1.3.6-3.el9.src.rpm
Repository    : rhel-9-for-x86_64-appstream-rpms
Summary       : OpenSCAP Utilities
URL           : http://www.open-scap.org/
License       : LGPLv2+
Description   : The openscap-utils package contains command-line
tools build on top
              : of OpenSCAP library. Historically, openscap-
utils included oscap
              : tool which is now separated to openscap-scanner
sub-package.
```

We can see in the output of the previous command what the openscap-utils package is about, with a brief description and a link to the main web page, which includes more extensive information.

> **Tip**
>
> It would be useful to run the dnf info command for each of the tools mentioned and visit their web pages. This way, you will be able to gain a better understanding of the capabilities these tools provide and be able to use them.

Let's now install openscap-utils, as follows:

```
[root@rhel-instance ~]# dnf install openscap-utils -y
Updating Subscription Management repositories.
Last metadata expiration check: 0:01:41 ago on Fri Jun  3
08:17:27 2022.
Dependencies resolved.
================================================================
Package              Arch    Version       Repository      Size
================================================================
Installing:
openscap-utils x86_64 1:1.3.6-3.el9 rhel-9-for-x86_64-
appstream-rpms 32 k
Installing dependencies:
binutils x86_64 2.35.2-17.el9 rhel-9-for-x86_64-baseos-rpms 5.1
M
[omitted]
```

```
xmlsec1-1.2.29-9.el9.x86_64
xmlsec1-openssl-1.2.29-9.el9.x86_64
zip-3.0-33.el9.x86_64
Complete!
```

Now, let's install `scap-security-guide`, which includes the RHEL-specific SCAP profiles. To do so, we'll run the following command:

```
[root@rhel-instance ~]# dnf install scap-security-guide -y
Updating Subscription Management repositories.
Last metadata expiration check: 0:08:34 ago on Fri Jun  3
08:17:27 2022.
Dependencies resolved.
================================================================
Package
Arch    Version         Repository                      Size
================================================================
Installing:
scap-security-guide
noarch 0.1.60-6.el9_0 rhel-9-for-x86_64-appstream-rpms 683 k
Installing dependencies:
xml-common noarch 0.6.3-58.el9   rhel-9-for-x86_64-appstream-
rpms   36 k
[omitted]
Installed:
  scap-security-guide-0.1.60-6.el9_0.noarch
  xml-common-0.6.3-58.el9.noarch
Complete!
```

With this package comes the SCAP Security Guides, including the one related to vulnerabilities for RHEL 9, which is located at `/usr/share/xml/scap/ssg/content/ssg-rhel9-ds.xml`.

The `oscap info` sub-command can display information about a selected data stream. **Extensible Markup Language** (**XML**) files containing data streams are indicated by the `-ds` string in their filenames. You can find a list of available profiles and their **identifiers** (**IDs**) in the `Profile` section, as illustrated in the following code snippet:

```
[root@rhel-instance ~]# oscap info /usr/share/xml/scap/ssg/
content/ssg-rhel9-ds.xml | grep Title -A 1
WARNING: Datastream component 'scap_org.open-scap_cref_
```

```
security-data-oval-com.redhat.rhsa-RHEL9.xml.bz2' points out
to the remote 'https://access.redhat.com/security/data/oval/
com.redhat.rhsa-RHEL9.xml.bz2'. Use '--fetch-remote-resources'
option to download it.
WARNING: Skipping 'https://access.redhat.com/security/data/
oval/com.redhat.rhsa-RHEL9.xml.bz2' file which is referenced
from datastream
Title: ANSSI-BP-028 (enhanced)
Id: xccdf_org.ssgproject.content_profile_anssi_bp28_enhanced
Title: ANSSI-BP-028 (high)
Id: xccdf_org.ssgproject.content_profile_anssi_bp28_high

[omitted]
Title: [DRAFT] DISA STIG with GUI for Red Hat Enterprise Linux
9
Id: xccdf_org.ssgproject.content_profile_stig_gui
```

We can check the additional details about any of the profiles using the `--profile` parameter and the profile ID. For example, to check the details for the HIPAA profile, we can run the following command:

```
[root@rhel-instance ~]# oscap info --profile xccdf_org.
ssgproject.content_profile_hipaa /usr/share/xml/scap/ssg/
content/ssg-rhel9-ds.xml
Document type: Source Data Stream
Imported: 2022-03-28T15:05:12
[skipped]
Profile
     Title: Health Insurance Portability and Accountability Act
(HIPAA)
     Id: xccdf_org.ssgproject.content_profile_hipaa
     Description: The HIPAA Security Rule establishes U.S.
national standards to protect individuals' electronic personal
health information that is created, received, used, or
maintained by a covered entity.
[skipped]
```

> **Tip**
> The profiles available in a data stream XML file can be listed using the `--profiles` subparameter from the `info` parameter—for example, `oscap info --profiles /usr/share/xml/scap/ssg/content/ssg-rhel9-ds.xml`.

We can now run an initial scan using all the checks included in the profile. Note that this will include 2,323 tests and that this will be done as an exercise to learn about possible vulnerabilities and actions to harden a system. So, let's run it, as follows:

```
[root@rhel-instance ~]# oscap oval eval --report \
/tmp/vulnerability.html \
/usr/share/xml/scap/ssg/content/ssg-rhel9-ds.xml
Definition oval:ssg-zipl_vsyscall_argument:def:1: false
Definition oval:ssg-zipl_slub_debug_argument:def:1: false
Definition oval:ssg-zipl_page_poison_argument:def:1: false
Definition oval:ssg-zipl_bootmap_is_up_to_date:def:1: false
[omitted]
Definition oval:ssg-accounts_logon_fail_delay:def:1: false
Definition oval:ssg-accounts_have_homedir_login_defs:def:1:
true
Definition oval:ssg-account_unique_name:def:1: true
Definition oval:ssg-account_disable_post_pw_expiration:def:1:
false
Evaluation done.
```

A file called `vulnerability.html` in the `/tmp/` folder will be generated with the output of the scan. The results will look like this:

OVAL Results Generator Information				
Schema Version	Product Name	Product Version	Date	Time
5.11	cpe:/a:open-scap:oscap	1.3.6	2022-06-03	08:28:27
#✗	#✓	#Error	#Unknown	#Other
574	323	358	21	73

OVAL Definition Generator Information				
Schema Version	Product Name	Product Version	Date	Time
5.11	combine_ovals.py from SCAP Security Guide	ssg: [0, 1, 60], python: 3.9.10	2022-03-24	00:00:00
#Definitions	#Tests	#Objects	#States	#Variables
1349 Total 1276 73 0 0	2499	2598	1147	840

System Information	
Host Name	rhel9.example.com
Operating System	Red Hat Enterprise Linux
Operating System Version	9.0 (Plow)
Architecture	x86_64
Interfaces	Interface Name: lo
	IP Address: 127.0.0.1
	MAC Address: 00:00:00:00:00:00
	Interface Name: eth0
	IP Address: 192.168.122.106
	MAC Address: 52:54:00:A5:EF:41
	Interface Name: lo
	IP Address: ::1
	MAC Address: 00:00:00:00:00:00
	Interface Name: eth0
	IP Address: fe80::3475:8455:93b4:b272
	MAC Address: 52:54:00:A5:EF:41

OVAL System Characteristics Generator Information				
Schema Version	Product Name	Product Version	Date	Time
5.11	cpe:/a:open-scap:oscap	ssg: [0, 1, 60], python: 3.9.10	2022-06-03	08:28:27

OVAL Definition Results

| | ✗ | | ✓ | | Error | | Unknown | | Other |

ID	Result	Class	Reference ID
oval:ssg-zipl_vsyscall_argument:def:1	false	compliance	[CCE-84100-7], [zipl_vsyscall_ar
oval:ssg-zipl_slub_debug_argument:def:1	false	compliance	[CCE-84094-2], [zipl_slub_debug_
oval:ssg-zipl_page_poison_argument:def:1	false	compliance	[CCE-84101-5], [zipl_page_poison_
oval:ssg-zipl_page_alloc_shuffle_argument:def:1	false	compliance	[CCE-85880-3], [zipl_page_alloc_shuf

Figure 11.1 – Initial results of an OpenSCAP test scan

Let's check some of the details of the report. In the top-left corner of the following screenshot, we will find **OVAL Results Generator Information**, with details of the run and a summary of the results:

OVAL Results Generator Information				
Schema Version	Product Name	Product Version	Date	Time
5.11	cpe:/a:open-scap:oscap	1.3.6	2022-06-03	08:28:27
#✗	#✓	#Error	#Unknown	#Other
574	323	358	21	73

Figure 11.2 – OpenSCAP test scan summary

In the top-right corner of the following screenshot, we can see **OVAL Definition Generator Information**, with a summary of the definitions used for checks:

OVAL Definition Generator Information				
Schema Version	**Product Name**	**Product Version**	**Date**	**Time**
5.11	combine_ovals.py from SCAP Security Guide	ssg: [0, 1, 60], python: 3.9.10	2022-03-24	00:00:00
#Definitions	**#Tests**	**#Objects**	**#States**	**#Variables**
1349 Total 1276 73 0 0 0	2499	2598	1147	840

Figure 11.3 – OpenSCAP test scan definitions summary

Right below those tokens of information, we can see a basic summary of the system, which is useful if we have a long list of scans and we want to assign this scan to the proper system. You can see a representation of this in the following screenshot:

System Information		
Host Name	rhel9.example.com	
Operating System	Red Hat Enterprise Linux	
Operating System Version	9.0 (Plow)	
Architecture	x86_64	
Interfaces	**Interface Name**	lo
	IP Address	127.0.0.1
	MAC Address	00:00:00:00:00:00
	Interface Name	eth0
	IP Address	192.168.122.106
	MAC Address	52:54:00:A5:EF:41
	Interface Name	lo
	IP Address	::1
	MAC Address	00:00:00:00:00:00
	Interface Name	eth0
	IP Address	fe80::3475:8455:93b4:b272
	MAC Address	52:54:00:A5:EF:41

Figure 11.4 – OpenSCAP test scan system summary

Underneath it, we have information on the generator, as shown here:

OVAL System Characteristics Generator Information				
Schema Version	Product Name	Product Version	Date	Time
5.11	cpe:/a:open-scap:oscap	ssg: [0, 1, 60], python: 3.9.10	2022-06-03	08:28:27

Figure 11.5 – OpenSCAP test scan generator information

And finally, here are the results of the checks. In a full report, you can easily differentiate the checks passed from the rest just by looking at the colors:

OVAL Definition Results				
☐☒ ☐✓ ☐ Error ☐ Unknown ☐ Other				
ID	Result	Class	Reference ID	Title
oval:ssg-zipl_vsyscall_argument:def:1	false	compliance	[CCE-84100-7], [zipl_vsyscall_argument]	Disable vsyscalls in zIPL
oval:ssg-zipl_slub_debug_argument:def:1	false	compliance	[CCE-84094-2], [zipl_slub_debug_argument]	Enable SLUB/SLAB allocator poisoning in zIPL
oval:ssg-zipl_page_poison_argument:def:1	false	compliance	[CCE-84101-5], [zipl_page_poison_argument]	Enable page allocator poisoning in zIPL
oval:ssg-zipl_page_alloc_shuffle_argument:def:1	false	compliance	[CCE-85880-3], [zipl_page_alloc_shuffle_argument]	Enable randomization of the page allocator in zIPL
oval:ssg-zipl_init_on_alloc_argument:def:1	false	compliance	[CCE-85868-8], [zipl_init_on_alloc_argument]	Configure kernel to zero out memory before allocation in zIPL
oval:ssg-zipl_bootmap_is_up_to_date:def:1	false	compliance	[CCE-84098-3], [zipl_bootmap_is_up_to_date]	Ensure zIPL bootmap is up to date
oval:ssg-zipl_audit_backlog_limit_argument:def:1	false	compliance	[CCE-84099-1], [zipl_audit_backlog_limit_argument]	Extend Audit Backlog Limit for the Audit Daemon in zIPL
oval:ssg-zipl_audit_argument:def:1	false	compliance	[CCE-84096-7], [zipl_audit_argument]	Enable Auditing to Start Prior to the Audit Daemon in zIPL
oval:ssg-use-pam_wheel_for_su:def:1	false	compliance	[CCE-90085-2], [use_pam_wheel_for_su]	Enforce usage of pam_wheel for su authentication
oval:ssg-usbguard_rules_not_empty_not_missing:def:1	false	compliance	[usbguard_rules_not_empty_not_missing]	Check that file storing USBGuard rules exists and is not empty
oval:ssg-usbguard_generate_policy:def:1	false	compliance	[CCE-88882-6], [usbguard_generate_policy]	Generate USBGuard Policy

Figure 11.6 – OpenSCAP test scan results

With this test, we have run a vulnerability scan on our system, obtaining a set of results that—depending on the usage of the system—will need to be addressed. In many cases, the warnings received do not apply, so we need to review them carefully. This kind of exercise should be done carefully on production systems, taking care of having a proper backup and snapshot of the system before proceeding to apply changes. It is recommended to run the hardening in test environments while building the service before moving it to production when possible.

> **Important Note**
> *Security hardening* for RHEL 9 is a great piece of documentation to get started on system security. It is recommended to read through it in order to extend the knowledge acquired in this chapter. It's available at https://access.redhat.com/documentation/en-us/red_hat_enterprise_linux/9/html/security_hardening/index.

Let's learn more about the basics. For this scan, we have used the **Red Hat Security Advisories'** (RHSAs') **Open Vulnerability and Assessment Language** (OVAL) feed, as provided by system packages. To check, we have run the OpenSCAP tool to review different security advisories and vulnerabilities as written in OVAL.

OVAL requires that the analyzed resources are in a certain state to consider them correct. It does so in a declarative manner, which means the end state is described and reviewed, not the manner in how to get to it.

The Red Hat security team generates RHSAs to address the different vulnerabilities that the system may be incurring and releases an OVAL definition for each one of them. These are released openly and are available at `https://www.redhat.com/security/data/oval/v2/`.

Now, let's take a look at one example found in our report, as follows:

- **ID**: `oval:ssg-accounts_logon_fail_delay:def:1`
- **Result**: `false`
- **Class**: `compliance`
- **Reference ID**: `[CCE-83635-3]`, `[accounts_logon_fail_delay]`
- **Title**: `Ensure that FAIL_DELAY is Configured in login.defs`

We can check the manual page for it by running `man login.defs`. In it, we will find the following information:

```
FAIL_DELAY (number)
Delay in seconds before being allowed another attempt after
a  login failure.
```

This is the value to establish how long a user will have to wait after a failed login attempt. It is intended to avoid brute-force attacks on accounts in the system. We may take, for example, two approaches to fix it, as outlined here:

- Add the `FAIL_DELAY` variable and value to `login.defs`.
- Enforce access to the system by only allowing login access using **Secure Shell (SSH)** keys and not passwords.

Or, even better, we could do both (**security in depth**, or **SID**). We could continue to review each of the entries in the list and understand each of them to complete the hardening of the system to avoid as much exposure as possible. This is a task that is usually run in coordination with the security teams and is continuously being reviewed.

In a case where you need to scan remote servers, you can use the `oscap-ssh` tool that is included in the `openscap-utils` package. The `oscap-ssh` tool can be executed remotely in any Linux system that provides SSH access. The syntax is very similar to the `oscap` command. Let's try `oscap-ssh` with the latest RHSA OVAL definition file. First, we need to download the OVAL definition and decompress it. Here's how we can achieve this:

```
[root@rhel-instance ~]# wget -O - https://www.redhat.com/
security/data/oval/v2/RHEL9/rhel-9.oval.xml.bz2 |bzip2
--decompress > rhel-9.oval.xml
[skipped]
```

> **Tip**
>
> If the preceding command gives a wget: command not found error, we can install the
> wget package with the following command: dnf install wget -y.

And then, we can run the oscap-ssh command on a remote server, like so:

```
root@rhel-instance ~]# oscap-ssh root@remoteserver 22 oval eval
--report remote-report.html rhel-9.oval.xml
Connecting to 'root@remoteserver' on port '22'...
The authenticity of host 'remoteserver (::2)' can't be
established.
ED25519 key fingerprint is SHA256:nNhsxgLaScC3I5Is5SkKM091/
WmHQ7SboQvEiiqBfqo.
This key is not known by any other names
Are you sure you want to continue connecting (yes/no/
[fingerprint])? yes
Warning: Permanently added 'localhost' (ED25519) to the list of
known hosts.
root@remotehost's password:
Connected!
Copying input file 'rhel-9.oval.xml' to remote working
directory '/tmp/tmp.OXfIp4LxL6'...
rhel-9.oval.xml
100%  308KB  78.6MB/s   00:00
Starting the evaluation...
Definition oval:com.redhat.rhsa:def:20225267: false
Definition oval:com.redhat.rhsa:def:20225214: true
Definition oval:com.redhat.rhsa:def:20225099: true
Definition oval:com.redhat.rhsa:def:20225050: false
Definition oval:com.redhat.rhsa:def:20224990: false
Definition oval:com.redhat.rhsa:def:20224940: true
[skipped]
Definition oval:com.redhat.rhsa:def:20221728: false
Evaluation done.
oscap exit code: 0
```

```
Copying back requested files...
report.html
100%    33KB  36.0MB/s    00:00
Removing remote temporary directory...
Disconnecting ssh and removing master ssh socket directory...
Exit request sent.
[root@rhel-instance ~]# ls -la remote-report.html
-rw-r--r--. 1 root root 33558 Jun 29 23:12 remote-report.html
```

Now that we have run our first vulnerability scan, let's see how we can do it for compliance in the next section.

Using OpenSCAP with security profiles for OSPP and PCI DSS

There are several security profiles used for compliance in the industry. Two of the most common, which we will review here, are the OSPP and PCI DSS standards.

The OSPP standard is heavily used in the public sector, serving general-purpose systems and also as a baseline for other more restrictive environments (that is, defense-accredited systems).

PCI DSS is one of the most widely used standards in the finance sector, and also applies to other sectors that want to provide online payments using credit cards.

There are different types of descriptions that can be used with OpenSCAP. We already know OVAL. Let's check the most important ones here:

- **Extensible Configuration Checklist Description Format (XCCDF)**: XCCDF is used to build security checklists. It's very common for compliance testing and scoring.

- **Common Platform Enumeration (CPE)**: CPE helps identify systems by assigning **unique ID (UID)** names. This way, it can correlate tests and names.

- **Open Checklist Interactive Language (OCIL)**: OCIL is part of the SCAP standard. It is a way to aggregate other checks from different data stores.

- **DataStream (DS)**: DS is a format that puts together several components into a single file. It is used to distribute profiles easily.

- **Script Check Engine (SCE)**: This allows us to make scripts interoperable with a security policy SCAP extension to allow the script execution from a SCAP policy. For more information, please visit the SCE page (`https://www.open-scap.org/features/other-standards/sce/`).

- **Common Vulnerabilities and Exposures (CVE)**: CVE is a dictionary of reference for public security vulnerabilities and exposures. CVE allows interoperability between different security tools and provides standardized names of vulnerabilities (IDs).

- **Common Weakness Enumeration (CWE)**: CWE is a community-developed list of software weaknesses. CWE focuses its effort to describe in detail flaws and known security weaknesses. It provides information about the prevention, implementation, and mitigation of a weakness. The MITRE Corporation supports this project (`https://www.open-scap.org/features/scap-components/#`).

- **XCCDF 1.1 tailoring extension**: The SCAP standard offers tailoring for XCCDF 1.2. This extension was introduced by OpenSCAP to the XCCDF 1.1 specification, allowing tools and users to create tailoring files, even for XCCDF 1.1.

> **Tip**
>
> More information on the different security descriptions and components can be found on the OpenSCAP web page at the following link: `https://www.open-scap.org/features/scap-components/`.

In this case, we will use the `ssg-rhel9-ds.xml` file. Let's check the information related to it, as follows:

```
[root@rhel-instance ]# cd /usr/share/xml/scap/ssg/content/
[root@rhel-instance content]# oscap info ssg-rhel9-ds.xml
Document type: Source Data Stream
[omitted]
Profiles:
Title: ANSSI-BP-028 (enhanced)
Id: xccdf_org.ssgproject.content_profile_anssi_bp28_enhanced
Title: ANSSI-BP-028 (high)
Id: xccdf_org.ssgproject.content_profile_anssi_bp28_high
Title: ANSSI-BP-028 (intermediary)
Id: xccdf_org.ssgproject.content_profile_anssi_bp28_intermediar
Title: ANSSI-BP-028 (minimal)
Id: xccdf_org.ssgproject.content_profile_anssi_bp28_minimal
Title: [DRAFT] CIS Red Hat Enterprise Linux 9 Benchmark for
Level 2 - Server
Id: xccdf_org.ssgproject.content_profile_cis
Title: [DRAFT] CIS Red Hat Enterprise Linux 9 Benchmark for
Level 1 - Server
```

```
Id: xccdf_org.ssgproject.content_profile_cis_server_l1
[omitted]
Checks:
Ref-Id: scap_org.open-scap_cref_ssg-rhel9-oval.xml
Ref-Id: scap_org.open-scap_cref_ssg-rhel9-ocil.xml
Ref-Id: scap_org.open-scap_cref_--builddir--build--BUILD--scap-
security-guide-0.1.60--build--ssg-rhel9-cpe-oval.xml
Ref-Id: scap_org.open-scap_cref_security-data-oval-com.redhat.
rhsa-RHEL9.xml.bz2
Dictionaries:
Ref-Id: scap_org.open-scap_cref_--builddir--build--BUILD--scap-
security-guide-0.1.60--build--ssg-rhel9-cpe-dictionary.xml
```

As you can see, it includes profiles for both OSPP and PCI DSS for RHEL 9. Let's give them a try.

Scanning for OSPP compliance

We can use the --profile option for oscap to get information specific to the **OSPP** profile, as follows:

```
[root@rhel-instance content]# oscap info --profile \
ospp ssg-rhel9-ds.xml
Document type: Source Data Stream
Imported: 2022-03-28T09:05:12
Stream: scap_org.open-scap_datastream_from_xccdf_ssg-rhel9-
xccdf-1.2.xml
Generated: (null)
Version: 1.3
WARNING: Datastream component 'scap_org.open-scap_cref_
security-data-oval-com.redhat.rhsa-RHEL9.xml' points out to the
remote 'https://www.redhat.com/security/data/oval/com.redhat.
rhsa-RHEL9.xml'. Use '--fetch-remote-resources' option to
download it.
WARNING: Skipping 'https://www.redhat.com/security/data/
oval/com.redhat.rhsa-RHEL9.xml' file which is referenced from
datastream
Profile
Title: [DRAFT] Protection Profile for General Purpose Operating
Systems
Id: xccdf_org.ssgproject.content_profile_ospp
```

```
Description: This profile is part of Red Hat Enterprise Linux 9
Common Criteria Guidance documentation for Target of Evaluation
based on Protection Profile for General Purpose Operating
Systems (OSPP) version 4.2.1 and Functional Package for SSH
version 1.0.  Where appropriate, CNSSI 1253 or DoD-specific
values are used for configuration, based on Configuration Annex
to the OSPP.
```

We can see in the preceding information that the OSPP profile comes described as xccdf. We can now run oscap, indicating that we want to use that format with the xcddf option and that the action we want to take is to evaluate the system with eval. Here's the command we need to execute:

```
[root@rhel-instance content]# oscap xccdf eval \
--report /tmp/ospp-report.html --profile ospp ssg-rhel9-ds.xml
[omitted]
Title    Set Password Maximum Consecutive Repeating Characters
Rule     xccdf_org.ssgproject.content_rule_accounts_password_
pam_maxrepeat
Ident    CCE-82066-2
Result   fail
Title    Ensure PAM Enforces Password Requirements - Maximum
Consecutive Repeating Characters from Same Character Class
Rule     xccdf_org.ssgproject.content_rule_accounts_password_
pam_maxclassrepeat
Ident    CCE-81034-1
Result   fail
[omitted]
Title    Disable Kerberos by removing host keytab
Rule     xccdf_org.ssgproject.content_rule_kerberos_disable_no_
keytab
Ident    CCE-82175-1
Result   pass
```

We will obtain an `ospp-report.html` file with a full report on the OSPP rule results, as shown here:

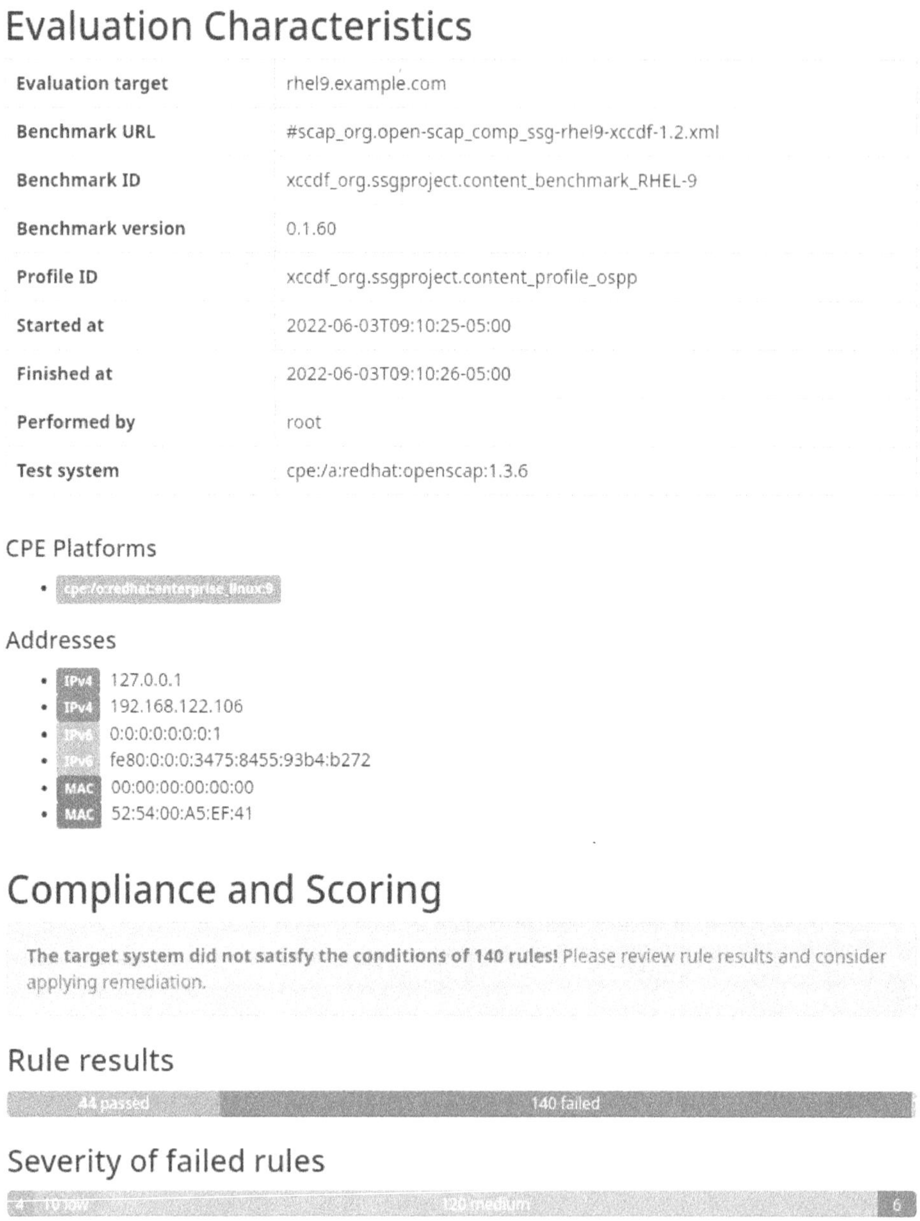

Evaluation Characteristics

Evaluation target	rhel9.example.com
Benchmark URL	#scap_org.open-scap_comp_ssg-rhel9-xccdf-1.2.xml
Benchmark ID	xccdf_org.ssgproject.content_benchmark_RHEL-9
Benchmark version	0.1.60
Profile ID	xccdf_org.ssgproject.content_profile_ospp
Started at	2022-06-03T09:10:25-05:00
Finished at	2022-06-03T09:10:26-05:00
Performed by	root
Test system	cpe:/a:redhat:openscap:1.3.6

CPE Platforms

- cpe:/o:redhat:enterprise_linux:9

Addresses

- **IPv4** 127.0.0.1
- **IPv4** 192.168.122.106
- **IPv6** 0:0:0:0:0:0:0:1
- **IPv6** fe80:0:0:0:3475:8455:93b4:b272
- **MAC** 00:00:00:00:00:00
- **MAC** 52:54:00:A5:EF:41

Compliance and Scoring

The target system did not satisfy the conditions of 140 rules! Please review rule results and consider applying remediation.

Rule results

44 passed   140 failed

Severity of failed rules

10 low   120 medium   6

Figure 11.7 – OpenSCAP OSPP scan results

It will show the points that require modification to be compliant with the profile, as indicated in the following screenshot:

Title	Severity	Result
▼ Guide to the Secure Configuration of Red Hat Enterprise Linux 9 (140x fail)		
▼ System Settings (126x fail)		
▼ Installing and Maintaining Software (17x fail)		
▼ System and Software Integrity (5x fail)		
▼ Software Integrity Checking (1x fail)		
▼ Verify Integrity with AIDE (1x fail)		
Install AIDE	medium	fail
▼ Federal Information Processing Standard (FIPS) (2x fail)		
Enable Dracut FIPS Module	medium	fail
Enable FIPS Mode	high	fail

Figure 11.8 – OpenSCAP OSPP scan results: details of rules that require action

We can now follow, step by step, the recommendations and fix them in order to be fully OSPP-compliant. Also, we can use this scan to harden systems that, even when they do not need to be OSPP-compliant, will be in an exposed network such as a **demilitarized zone (DMZ)**, and we want to have them hardened.

> **Important Note**
>
> Red Hat provides a way to automatically apply all those changes. It is based on the **Ansible** automation tool. It is provided in the form of a playbook—a set of descriptions for Ansible that will apply all the changes required to the system. The playbook for OSPP is located at `/usr/share/scap-security-guide/ansible/rhel9-playbook-ospp.yml`.

Now that we have reviewed the system for OSPP compliance, let's move to the next target—PCI DSS compliance.

Scanning for PCI DSS compliance

We can follow the same procedure as before, also using the `--profile` option for `oscap` to get information specific to the PCI DSS profile. Here's the code we need to execute:

```
[root@rhel-instance content]# oscap info --profile pci-dss \
ssg-rhel9-ds.xml
Document type: Source Data Stream
Imported: 2022-03-28T09:05:12
Stream: scap_org.open-scap_datastream_from_xccdf_ssg-rhel9-
xccdf-1.2.xml
Generated: (null)
Version: 1.3
WARNING: Datastream component 'scap_org.open-scap_cref_
security-data-oval-com.redhat.rhsa-RHEL9.xml' points out to the
remote 'https://www.redhat.com/security/data/oval/com.redhat.
rhsa-RHEL9.xml'. Use '--fetch-remote-resources' option to
download it.
WARNING: Skipping 'https://www.redhat.com/security/data/
oval/com.redhat.rhsa-RHEL9.xml' file which is referenced from
datastream
Profile
Title: PCI-DSS v3.2.1 Control Baseline for Red Hat Enterprise
Linux 9
Id: xccdf_org.ssgproject.content_profile_pci-dss
Description: Ensures PCI-DSS v3.2.1 security configuration
settings are applied.
```

We can run `oscap` with the same options as in the previous section but this time specify `pci-dss` as the profile, as illustrated in the following code snippet. This will generate a proper report:

```
[root@rhel-instance content]# oscap xccdf eval --report \
pci-dss-report.html --profile pci-dss ssg-rhel9-ds.xml
WARNING: Datastream component 'scap_org.open-scap_cref_
security-data-oval-com.redhat.rhsa-RHEL9.xml' points out to the
remote 'https://www.redhat.com/security/data/oval/com.redhat.
rhsa-RHEL9.xml'. Use '--fetch-remote-resources' option to
download it.
WARNING: Skipping 'https://www.redhat.com/security/data/
oval/com.redhat.rhsa-RHEL9.xml' file which is referenced from
datastream
```

```
WARNING: Skipping ./security-data-oval-com.redhat.rhsa-RHEL9.
xml file which is referenced from XCCDF content
Title    Verify File Hashes with RPM
Rule     xccdf_org.ssgproject.content_rule_rpm_verify_hashes
Ident    CCE-90841-8
Result   pass
[omitted]
Title    Enable Smartcards in SSSD
Rule     xccdf_org.ssgproject.content_rule_sssd_enable_
smartcards
Ident    CCE-89155-6
Result   pass
[root@rhel-instance content]# ls -l /tmp/pci-dss-report.html
-rw-r--r--. 1 root root 4005535 Jun  3 09:17 /tmp/pci-dss-
report.html
```

We can start reviewing the items in the report and begin fixing them.

> **Important Note**
>
> As in the previous section, Red Hat also provides a way to automatically apply all those changes with Ansible. The playbook for PCI DSS is located at /usr/share/scap-security-guide/ansible/rhel9-playbook-pci-dss.yml.

We have seen that changing from one profile to a different one with OpenSCAP is very easy, and we can scan for as many profiles as available.

Summary

By learning the basics of **OpenSCAP**, we are ready to review and harden systems to make them compliant with the regulations we need them to run under.

Now, if you are requested to comply with any regulatory requisitions, you can find the right SCAP profile for it (or build it if it doesn't exist) and ensure that your systems are fully compliant.

Also, even when no regulatory requirements apply, the use of OpenSCAP can help you find vulnerabilities in a system, or apply a more secure (and restrictive) configuration to your systems in order to reduce the risks.

There are ways to extend our knowledge and skills by learning Ansible and being able to automatically apply changes to our systems in a way that is easy to scale, as well as Red Hat Satellite, which can help run SCAP scans to the whole **information technology** (**IT**) base we are managing, even when we could be talking about thousands of systems.

Now that our security skills are improving and being consolidated, let's dive deeper into more low-level topics such as local storage and filesystems, as described in the next chapter.

Part 3 –
Resource Administration –
Storage, Boot Process, Tuning, and Containers

Managing the resources of machines running RHEL is fundamental to have a performant, efficient IT environment. Understanding the storage, tuning the performance (including the configuration required to make it permanent in the boot process), and then using containers to isolate processes and assign resources more efficiently are areas in which a system administrator will surely step in their daily work.

The following chapters are included in this part:

- *Chapter 12, Managing Local Storage and Filesystems*
- *Chapter 13, Flexible Storage Management with LVM*
- *Chapter 14, Advanced Storage Management with Stratis and VDO*
- *Chapter 15, Understanding the Boot Process*
- *Chapter 16, Kernel Tuning and Managing Performance Profiles with tuned*
- *Chapter 17, Managing Containers with Podman, Buildah, and Skopeo*

12

Managing Local Storage and Filesystems

In the previous chapters, we learned about security and system administration. In this chapter, we will focus on the administration of resources—specifically, storage administration.

Storage administration is an important part of keeping a system running: the system logs can eat available space, new applications might require additional storage to be set up for them (even on separate disks to improve performance), and such issues might require our action in order to solve them.

In this chapter, we will learn about the following topics:

- Partitioning disks (MBR and GPT disks)
- Formatting and mounting filesystems
- Setting default mounts and options in `fstab`
- Using network filesystems with **Network File System (NFS)**

This will provide us with basic knowledge to build upon our storage administration skills to keep our systems running.

Let's get hands-on!

Technical requirements

You can continue the practice using the **virtual machine (VM)** created at the beginning of this book in *Chapter 1, Getting RHEL Up and Running*. Any additional packages required for this chapter will be indicated alongside the text. You will also need additional disks for partitioning (MBR and GPT).

Let's start with a definition

A partition is a logical division of a storage device, and it's used to logically separate the available storage into smaller pieces.

Now, let's move on to learning a bit about the origins of storage to better understand it.

A bit of history

Storage is also related to a system's ability to use it, so let's explain a bit about the history of **personal computers** (**PCs**), the software that allows them to boot (**Basic Input/Output System** or **BIOS**), and how that influenced storage administration.

It might sound slightly strange, but initial storage needs were just a small amount of **kilobytes** (**KB**), and for the first hard drives in PCs, storage was just a few **megabytes** (**MB**).

Additionally, PCs come with one feature and limitation: PCs were compatible, which means that subsequent models had compatibility with the initial **International Business Machines** (**IBM**) PC design.

Traditional disk partitioning uses a space at the beginning of the disks after the MBR that allows four partition registers (start, end, size, partition type, and active flag), called **primary** partitions.

When the PC is booting, BIOS will check the partition table of the disk by running a small program in the MBR, and then it loads the boot area of the active partition and executes it to get the operating system booting.

The IBM PC that contained a **Disk Operating System** (**DOS**) and compatibles (MS-DOS, DR-DOS, FreeDOS, and more) also used a filesystem named a **File Allocation Table** (**FAT**). The FAT contained several structures based on its evolution, like the cluster addressing size (alongside some other features).

With a limit on the number of clusters, having bigger disks meant having bigger blocks, so if a file was only using a limited amount of space, the remaining ones couldn't be used by other files. Therefore, it became more or less normal to break bigger hard drives into smaller logical partitions so that small files would not eat up the available space because of any limits.

Let's think about this as an agenda with a maximum number of entries, similar to speed dial on your phone: if you only have nine slots for speed dial, a short number such as the number to call your voicemail will still count the same as having stored a long international number, as both use one slot, because entries are counted by slots, not just exact size.

Some of those limitations became reduced by subsequent versions of the FAT sizing, which, in parallel, increased the maximum supported disk size.

Of course, other operating systems introduced their own filesystems while using this same partitioning schema.

Later, a new partition type was created—the **extended partition**—which used one of the four available **primary partition** slots and allowed extra partitions to be defined inside it, enabling us to create logical disks to be assigned as needed.

Additionally, having several primary partitions allowed the installation of different operating systems on the same computer with their own dedicated space that was completely independent of other operating systems.

So... partitions allowed computers to have different operating systems, have better usage of the available storage, and even logically sort the data by keeping it in different areas, such as keeping operating system space separate from user data so that a user who was filling the available space would not affect the computer's operation.

As we said, many of those designs came with the compatibility restriction of the original IBM PC, so when new computers using the **Extensible Firmware Interface** (**EFI**) appeared to overcome the limitations of traditional BIOS, a new partition table format called **Globally Unique Identifier** (**GUID**) **Partition Table** (**GPT**) arrived.

Systems using GPT make use of 32-bit and 64-bit support versus the 16-bit support used by BIOS (inherited from IBM PC compatibility), so bigger addressing can be used for the disks, along with extra features such as extended controller loading.

In the next section, let's learn about disk partitioning.

Partitioning disks (MBR and GPT disks)

As mentioned earlier, using disk partitions allows us to more efficiently use the space available on our computers and servers.

Let's dig into disk partitioning by first identifying the disks to act on.

> **Important Note**
>
> Once we have learned what caused disks to be partitioned and the limitations of this, we should follow one schema or another based on our system specifications. However, bear in mind that EFI requires GPT and BIOS requires MBR, so a system supporting UEFI, but having a disk partitioned with MBR, will boot the system into BIOS-compatible mode. **Unified EFI** (**UEFI**) is the industry standard that was known previously as EFI.

Linux uses a different notation for the disks based on the way those are connected to the system, so—for example—you can see disks as hda, sda, or mmbclk0 depending on that connection being used. Traditionally, disks connected using the **Integrated Drive Electronics** (**IDE**) interface used to have disks named hda, hdb, and so on, while disks using the **Small Computer System Interface** (**SCSI**) used to have disks named sda, sdb, and so on.

We can list the available devices with `fdisk -l` or `lsblk -fp`, as you can see in the following screenshot:

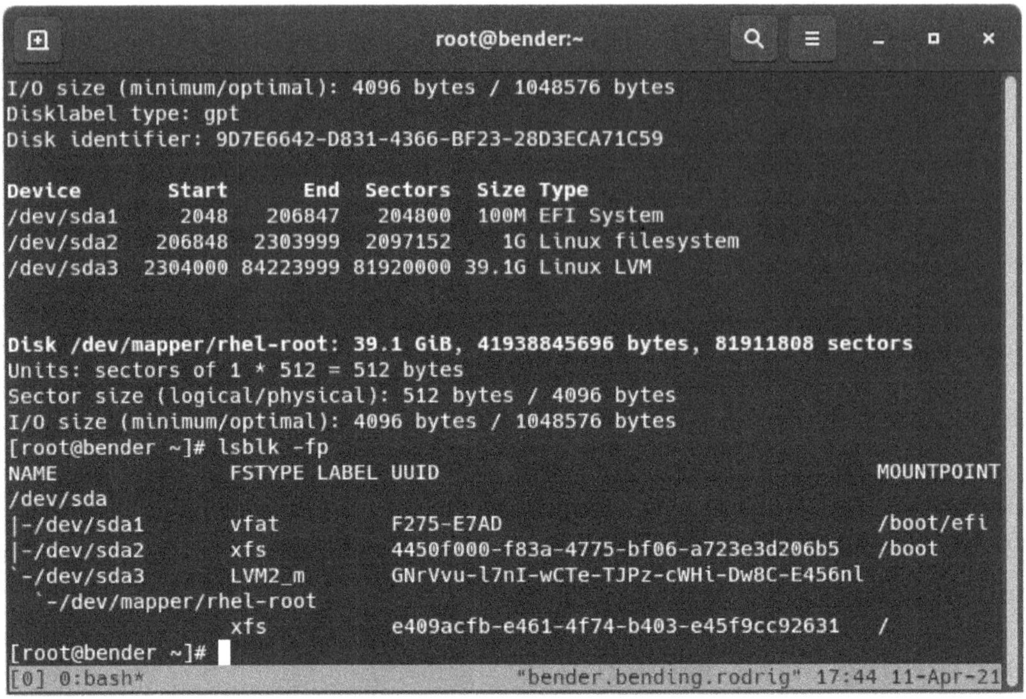

Figure 12.1 – The lsblk -fp and fdisk –l output

As you can see, the disk named `/dev/sda` has three partitions—`sda1`, `sda2`, and `sda3`—with `sda3` being an LVM volume group that has a volume named `/dev/mapper/rhel-root`.

To demonstrate disk partitioning in a safe way and to make it easier for readers using a VM for testing, we will create a fake **virtual hard drive** (**VHD**) for testing. In doing so, we will use the `truncate` utility that comes with the `coreutil` package and the `losetup` utility that comes with the `util-linux` package.

In order to create a VHD, we will execute the following sequence of commands as they appear in *Figure 12.2*:

1. `truncate -s 20G myharddrive.hdd`: This creates a 20-**gigabyte** (**GB**)-sized file, but this will be an empty file, which means that the file is not really using 20 GB on our disk but is just showing that size. Unless we use it, it will not consume more disk space (this is called a **sparse file**).

2. `losetup -f`: This will find the next available device.

3. `losetup /dev/loop0 myharddrive.hdd`: This will associate `loop0` with the file that has been created.
4. `lsblk -fp`: This will be used to validate the newly looped disk.
5. `fdisk -l /dev/loop0`: This will be used to list the available space in the new disk.

The following screenshot shows the output of the preceding sequential commands:

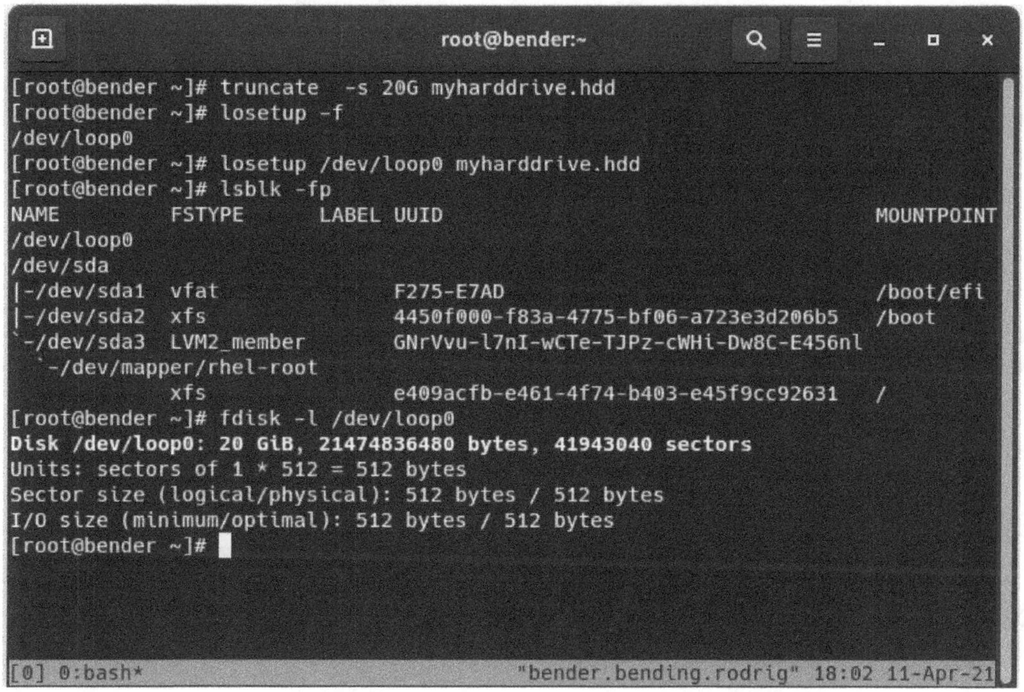
root@bender:~
[root@bender ~]# truncate -s 20G myharddrive.hdd
[root@bender ~]# losetup -f
/dev/loop0
[root@bender ~]# losetup /dev/loop0 myharddrive.hdd
[root@bender ~]# lsblk -fp
NAME FSTYPE LABEL UUID MOUNTPOINT
/dev/loop0
/dev/sda
|-/dev/sda1 vfat F275-E7AD /boot/efi
|-/dev/sda2 xfs 4450f000-f83a-4775-bf06-a723e3d206b5 /boot
`-/dev/sda3 LVM2_member GNrVvu-l7nI-wCTe-TJPz-cWHi-Dw8C-E456nl
`-/dev/mapper/rhel-root
xfs e409acfb-e461-4f74-b403-e45f9cc92631 /
[root@bender ~]# fdisk -l /dev/loop0
Disk /dev/loop0: 20 GiB, 21474836480 bytes, 41943040 sectors
Units: sectors of 1 * 512 = 512 bytes
Sector size (logical/physical): 512 bytes / 512 bytes
I/O size (minimum/optimal): 512 bytes / 512 bytes
[root@bender ~]#
[0] 0:bash*
"bender.bending.rodrig" 18:02 11-Apr-21

Figure 12.2 – The execution of the indicated commands for creating a fake hard drive

The `losetup -f` command finds the next available loopback device, which is a device used for looping back accesses to a backing file. This is often used for mounting ISO files locally, for example.

With the third command, we use the previously available loopback device to set up a loop connection between the `loop0` device and the file we created with the first command.

As you can see, in the remaining commands, the device now appears when running the same commands that we executed in *Figure 12.1*, showing that we have a disk of 20 GB available.

> **Important Note**
>
> Partitioning operations on disks can be dangerous and can render a system unusable and in need of restoration or reinstallation. To reduce that chance, the examples in this chapter will use the fake created disk, /dev/loop0, and only interact with that. Pay attention when performing this over real volumes, disks, and more.

Let's start creating partitions by executing fdisk /dev/loop0 on our newly created device, as shown in the following screenshot:

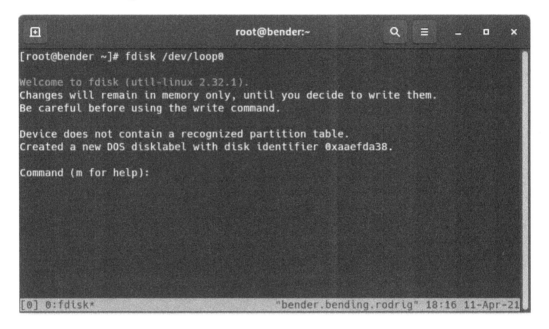

Figure 12.3 – The fdisk execution over /dev/loop0

As you can see in *Figure 12.3*, the disk doesn't contain a recognized partition table, so a new DOS partition disk label is created, but the changes only remain in memory until they are written back to the disk.

Inside the fdisk command, we can use several options to create a partition. The first one we should be aware of is m, as indicated in *Figure 12.3*, which shows the help functionality and available commands.

Additionally, the first thing to take into consideration is our previous explanation about UEFI, BIOS, and more. By default, fdisk is creating a DOS partition, but as we can see inside the manual (m), we can create a GPT one by running the g command inside fdisk.

Another important command to remember is p, which prints the current disk layout and partition, as defined in the next screenshot:

```
Device does not contain a recognized partition table.
Created a new DOS disklabel with disk identifier 0x7cde2e28.

Command (m for help): p
Disk /dev/loop0: 20 GiB, 21474836480 bytes, 41943040 sectors
Units: sectors of 1 * 512 = 512 bytes
Sector size (logical/physical): 512 bytes / 512 bytes
I/O size (minimum/optimal): 512 bytes / 512 bytes
Disklabel type: dos
Disk identifier: 0x7cde2e28

Command (m for help): g
Created a new GPT disklabel (GUID: 6EC8ADC1-B673-9B49-AFA7-65606D6122DE).

Command (m for help): p
Disk /dev/loop0: 20 GiB, 21474836480 bytes, 41943040 sectors
Units: sectors of 1 * 512 = 512 bytes
Sector size (logical/physical): 512 bytes / 512 bytes
I/O size (minimum/optimal): 512 bytes / 512 bytes
Disklabel type: gpt
Disk identifier: 6EC8ADC1-B673-9B49-AFA7-65606D6122DE

Command (m for help):
[0] 0:fdisk* 1:bash-                    "bender.bending.rodrig" 18:32 11-Apr-21
```

Figure 12.4 – fdisk creating a new partition table

As we can see, the initial disklabel type was dos and is now gpt, which is compatible with EFI/UEFI.

Let's review some of the basic commands we can use, as follows:

- n: This creates a new partition.
- d: This deletes a partition.
- m: This shows the manual page (help).
- p: This prints the current layout.
- x: This enters advanced mode (extra functionality intended for experts).
- q: This quits without saving.
- w: This writes changes to disk and exits.
- g: This creates a new GPT disk label.
- o: This creates a DOS disk label.
- a: In DOS mode, this sets the bootable flag to one of the primary partitions.

So, what will be the sequence for creating a new traditional disk partition layout with a bootable partition for the operating system and another one for the user data with half the disk size each?

The following list indicates the sequence of commands (these are also shown in *Figure 12.5*):

1. Type `fdisk /dev/loop0` and press *Enter* to continue.
2. Type o and press *Enter* to create a new DOS disk label.
3. Type n and press *Enter* to create a new partition.
4. Press *Enter* to accept a primary partition type.
5. Press *Enter* to confirm the use of the first partition (1).
6. Press *Enter* to accept the initial sector.
7. Type +10G and press *Enter* to indicate 10 GB in size from the first sector.
8. Type n and press *Enter* to create a second new partition.
9. Press *Enter* to accept it as a primary partition type.
10. Press *Enter* to accept the partition number (2).
11. Press *Enter* to accept the first sector as the default proposed by `fdisk`.
12. Press *Enter* to accept the end sector as the default proposed by `fdisk`.
13. Type a and press *Enter* to mark a partition as bootable.
14. Type 1 and press *Enter* to mark the first partition.

As you can see, most of the options accept the defaults; the only change was to specify a partition size of +10G, meaning it should be 10 GB (the disk was 20 GB), and then start with the second partition using the new n command and no longer specifying the size, as we want to use all the remaining ones. The last step is to mark the first partition as active for booting.

Of course, remember what we said earlier: unless we execute the w command, the changes are not written to disk, and we can use p to review them, as shown in the following screenshot:

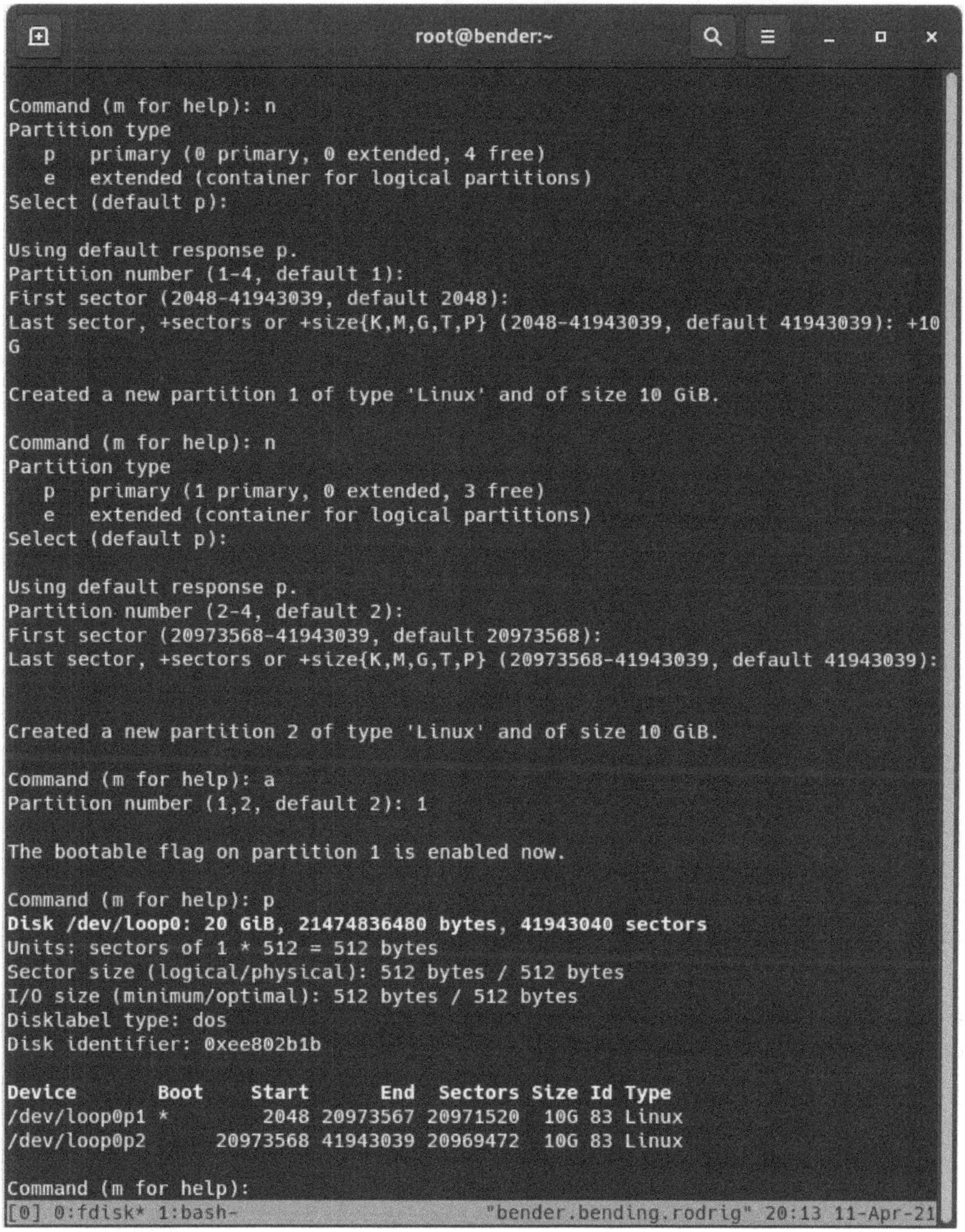

```
Command (m for help): n
Partition type
   p    primary (0 primary, 0 extended, 4 free)
   e    extended (container for logical partitions)
Select (default p):

Using default response p.
Partition number (1-4, default 1):
First sector (2048-41943039, default 2048):
Last sector, +sectors or +size{K,M,G,T,P} (2048-41943039, default 41943039): +10
G

Created a new partition 1 of type 'Linux' and of size 10 GiB.

Command (m for help): n
Partition type
   p    primary (1 primary, 0 extended, 3 free)
   e    extended (container for logical partitions)
Select (default p):

Using default response p.
Partition number (2-4, default 2):
First sector (20973568-41943039, default 20973568):
Last sector, +sectors or +size{K,M,G,T,P} (20973568-41943039, default 41943039):


Created a new partition 2 of type 'Linux' and of size 10 GiB.

Command (m for help): a
Partition number (1,2, default 2): 1

The bootable flag on partition 1 is enabled now.

Command (m for help): p
Disk /dev/loop0: 20 GiB, 21474836480 bytes, 41943040 sectors
Units: sectors of 1 * 512 = 512 bytes
Sector size (logical/physical): 512 bytes / 512 bytes
I/O size (minimum/optimal): 512 bytes / 512 bytes
Disklabel type: dos
Disk identifier: 0xee802b1b

Device        Boot    Start      End  Sectors Size Id Type
/dev/loop0p1  *        2048 20973567 20971520  10G 83 Linux
/dev/loop0p2        20973568 41943039 20969472  10G 83 Linux

Command (m for help):
[0] 0:fdisk* 1:bash-                    "bender.bending.rodrig" 20:13 11-Apr-21
```

Figure 12.5 – Disk partition layout creation and verification before writing it back to disk

To conclude this section, let's write the changes to disk with the w command, and let's move on to discuss filesystems in the next section. However, before that, let's execute `partprobe /dev/loop0` to make the kernel update its internal view on the disk and find the two new partitions. Without this, the `/dev/loop0p1` and `/dev/loop0p2` special files might not be created and will not be usable.

Note that some partition modifications will not be updated even after a `partprobe` execution and might require the system to be rebooted. For example, this is happening in disks that have partitions in use, such as the one holding the root filesystem in our computer.

Formatting and mounting filesystems

In the previous section, we learned how to logically divide our disk, but that disk is still not usable for storing data. To enable this to do so, we need to define a **filesystem** on it as the first step to make it available to our system.

A filesystem is a logical structure that defines how files, folders, and more are stored and provides, based on each type, a different set of features.

The number and types of filesystems supported depend on the operating system version, as during its evolution, new filesystems might be added, removed, and more.

> **Tip**
> Remember that **Red Hat Enterprise Linux** (**RHEL**) focuses on stability. So, there are strict controls about which features are added or phased out for newer releases, but not within the current release. You can read more about this at `https://access.redhat.com/articles/rhel9-abi-compatibility`.

In RHEL 9, the default filesystem is the **eXtended File System** (**XFS**), but you can see a list of available ones it the RHEL documentation, which can be found at `https://access.redhat.com/documentation/en-us/red_hat_enterprise_linux/8/html/system_design_guide/index`, and of course, others such as **Fourth Extended Filesystem** (**EXT4**) can be used.

The choice of the filesystem depends on several factors such as usage intention, the type of files that are going to be used, and more, as different filesystems might have performance implications.

For example, both EXT4 and XFS are journaled filesystems that provide more protection against power failures, but the maximum filesystem size differs in terms of other aspects such as the likeliness of becoming fragmented and more.

Before choosing a filesystem, it is a good practice to get an idea of the kind of files being deployed and their usage patterns, as choosing the wrong one might affect system performance.

As we defined our VHD in the previous section with two partitions, we can try creating both XFS and EXT4 filesystems. However, again, be very careful when performing operations, as filesystem creation is a destructive operation that writes new structures back to the disk, and when operating as a root user of the system, which is required, selecting the wrong one can destroy the available data we had on our system within seconds.

> **Important Note**
>
> Remember to check the man page for the commands being used in order to get familiar with the different recommendations and options available for each one.

Then, let's use the two partitions we created to test with two filesystems, XFS and EXT4, by using the mkfs.xfs and mkfs.ext4 commands against each one of the devices respectively, as follows:

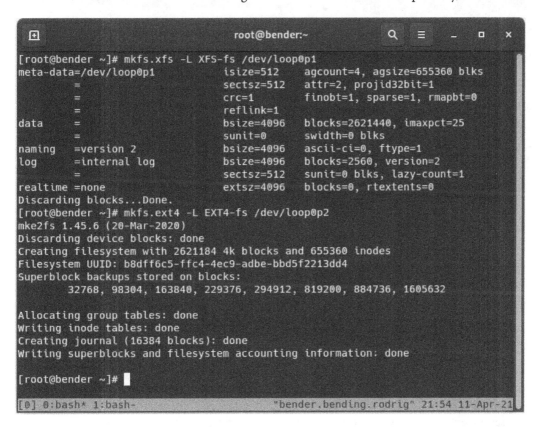

Figure 12.6 – Filesystem creation on the VHD created

Note that we have specified the different loop device partitions, and we also specified one -L parameter for each command. We will look at this again later.

Now that the filesystem has been created, we can run `lsblk -fp` to verify this, and we can see both devices, now indicating the filesystem is in use along with the LABEL and UUID values (the ones shown when we created the filesystem with `mkfs`), as you can see in the following screenshot:

```
[root@bender ~]# lsblk -fp
NAME                  FSTYPE LABEL UUID                                 MOUNTPOINT
/dev/loop0
|-/dev/loop0p1        xfs    XFS-fs
|                                   08e6169b-592e-432a-afd0-179eb1866555
`-/dev/loop0p2        ext4   EXT4-fs
                                    b8dff6c5-ffc4-4ec9-adbe-bbd5f2213dd4
/dev/sda
|-/dev/sda1           vfat         F275-E7AD                            /boot/efi
|-/dev/sda2           xfs          4450f000-f83a-4775-bf06-a723e3d206b5 /boot
`-/dev/sda3           LVM2_m       GNrVvu-l7nI-wCTe-TJPz-cWHi-Dw8C-E456nl
  `-/dev/mapper/rhel-root
                      xfs          e409acfb-e461-4f74-b403-e45f9cc92631 /
[root@bender ~]#
```
```
[0] 0:lsblk* 1:bash-                          "bender.bending.rodrig" 16:48 17-Apr-21
```

Figure 12.7 – Output of lsblk -fp after creating the filesystems

From the preceding output, it's important to pay attention to the UUID and LABEL values (if you remember, the value listed is the one we specified in the `mkfs` command with the `-L` option), as we will be using them later in this chapter.

Now that the filesystems have been created, in order to use them, we need to mount them. This means making the filesystem available at a path in our system so that every time we store inside that path, we will be using that device.

Mounting a filesystem can be done in several ways, but the simplest way is to use autodetection and just specify the device to mount and the local path to mount it at. However, more complex ones that allow several options to be defined can be found when checking the `man mount` help page.

To mount our two created filesystems, we will create two folders and then proceed to mount each device by executing the following commands:

```
cd
mkdir first second
mount /dev/loop0p1 first/
mount /dev/loop0p2 second/
```

At this point, the two filesystems will be available in our home folder (the root user) in the subfolders named first and second.

The kernel automatically found which filesystem is in use for each device and loaded it via the appropriate controller, and this worked. However, sometimes, we might want to define specific options—for example, forcing the filesystem type, which was used in the past when ext2 and ext3 were the common filesystems to enable or disable journaling, or, for example, to disable the built-in features that update the file or directory access time to reduce the disk I/O and increase performance.

All options specified on the command line, or the filesystems that are mounted, will not be available once the system has been rebooted, as those are only runtime changes. Let's move on to the next section to learn how to define default options and filesystem mounts when the system is being started.

Setting default mounts and options in fstab

In the previous section, we introduced how disks and partitions can be mounted so that our services and users can make use of them. In this section, we will learn how to make those filesystems available in a persistent way.

The /etc/fstab file contains the filesystem definitions for our system. Of course, it also has a dedicated manual page that can be checked with man fstab. It contains useful information about the formatting, fields, ordering, and more that must be taken into consideration, as this file is critical for the smooth operation of the system.

The file format is defined by several fields separated by tabs or spaces, with lines starting with a # character considered as comments.

For example, we will use this line to look at each field description:

```
LABEL=/ / xfs defaults 0 0
[.1.] [2] [3] [..4..] [5] [6]
```

The first field is the device definition, which can be a special block device, a remote filesystem, or—as we can see—a selector made by LABEL, UUID, or, for **GPT** systems, PARTUUID or PARTLABEL. The man page for mount, blkid, and lsblk provides more information about device identifiers.

The second field is the mount point for the filesystem, which is where to make the contents of that filesystem available based on our system directory hierarchy. Some special devices/partitions such as swap areas have this defined as none, as effectively, the contents are not made available via the filesystem.

The third field is the filesystem type, which is supported by the mount command or swap for swap partitions.

The fourth field is the mount option, which is supported by the mount or swapon commands (check their man pages for more details), being at its defaults setting an alias for most common options (such as read/write, allow devices, allow execution, automount on boot, async access, and more). Other common options might be noauto, which defines the filesystem but doesn't mount at boot (it is often used with removable devices); user, which allows users to mount and unmount it; and _netdev, which defines remote paths that require networking to be up before attempting the mount.

The fifth field is used by dump to determine which filesystems should be used—its value defaults to 0.

The sixth field is used by fsck to determine the order for filesystems to be checked on boot. The root filesystem should have a value of 1, and the others should have a value of 2 (the default is 0, not fsck). Checks are performed in parallel to speed up the booting process. Note that with filesystems that have a journal, the filesystem itself can perform a fast validation instead of a full one.

In the following screenshot, let's see what it looks like on our system with the output of cat /etc/ fstab:

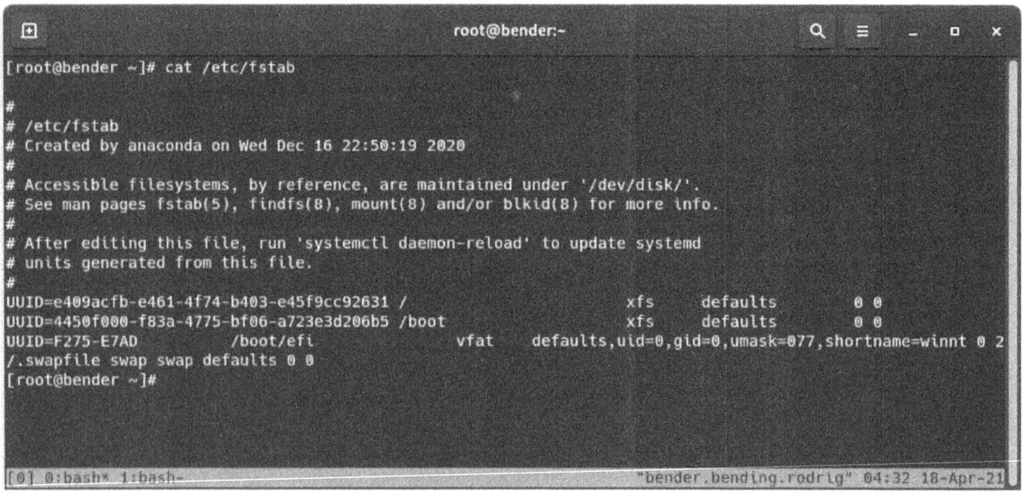

Figure 12.8 – The fstab example from our system

So, why should we use UUID or LABEL instead of devices such as /dev/sda1?

Disk ordering might change when a system is booting, as some kernels might introduce differences in the devices in terms of how they are accessed and more, causing changes in the enumeration of the devices. This happens not only for removable devices, such as **Universal Serial Bus** (**USB**), but also to internal devices such as network interfaces or hard drives.

When, instead of specifying the devices, we use UUID or LABEL, even in the event of a device reordering, the system will still be able to find the right device to use and boot from it. This was especially important when systems used to have **IDE** and **Serial Advanced Technology Attachment** (**SATA**) drives and **SCSI** drives, or even today when **internet SCSI** (**iSCSI**) devices might be connected in a different order than expected, resulting in device name changes and failures when reaching them.

Remember to use the blkid or lsblk -fp commands to check the filesystems' labels and **universally unique identifiers** (**UUIDs**) that could be used when referring to them.

> **Important Note**
> When editing the /etc/fstab file, be extremely careful: altering the mount points used by the system might render your system unusable. If in doubt, double-check for any changes and ensure you are familiar with the system recovery methods and have rescue media available if this is needed.

In the next section, let's learn about mounting a remote NFS.

Using network filesystems with NFS

Mounting a remote NFS is not very different from mounting local devices, but instead of specifying a local device as we did in the previous section with our /dev/loop0p1 file, we provide server:export as a device.

We can find a range of available options by checking the manual page via man mount, and this will show us several of the options and the way the device looks.

When an NFS mount is going to be used, the administrator will need to use the host and the export name to mount that device. Let's look at an example based on the following data about the NFS export:

- **Server**: server.example.com
- **Export**: /isos
- **Mount point**: /mnt/nfs

With the preceding data, it's easy to construct the mount command, which will look like this:

```
mount -t nfs sever.example.com:/isos /mnt/nfs
```

If we analyze the preceding command, it will define the type of filesystem to mount as `nfs`, provided by a `server.example.com` hostname, and using the `/isos` NFS export. It will be made available locally under the `/mnt/nfs` folder.

If we want to define this filesystem as available at boot, we should add an entry in `/etc/fstab`, but how should we indicate this?

Based on the settings explained during this chapter, the constructed entry would look something like this:

```
server.example.com:/isos /mnt/nfs nfs defaults,_netdev 0 0
```

The preceding line of code contains the parameters we indicated on the command line. But it also adds that it is a resource that requires network access before attempting to mount it. This is because networking is required to be able to reach the NFS server, similar to what will be required for other network-based storage, such as Samba mounts or iSCSI.

> **Important Note**
> Reinstating the idea of keeping our system bootable, once we make modifications to the `/etc/fstab` configuration file, it is recommended that you execute `mount -a` so that the validation is performed from a running system. If after the execution the new filesystems are available and shown when executing—for example, `df`—and no error has appeared, it should be safe.

Summary

In this chapter, we learned how a disk can be divided logically for optimal use of storage and how to later create a filesystem in that disk division so that it can be used to actually store data.

Once the actual filesystem had been created, we learned how to make it accessible in our system and how to ensure that it would be available after the next system restart via modification of the `/etc/fstab` configuration file.

Finally, we also learned how to use a remote filesystem with NFS based on the data that was provided for us and how to add it to our `fstab` file to make it persistent.

In the next chapter, we will learn how to make storage even more useful via **Logical Volume Management** (**LVM**), which empowers the definition of different logical units that can be resized, combined to provide data redundancy, and more.

13

Flexible Storage Management with LVM

Managing local storage can be done more flexibly than in *Chapter 12, Managing Local Storage and Filesystems*, by using **Logical Volume Manager** (**LVM**). LVM allows you to assign more than one disk to the same logical volume (the equivalent in LVM to a partition), replicate data across different disks, and make snapshots of a volume.

In this chapter, we will review the basic usage of LVM and the main objects that are used to manage storage. We will learn how to prepare disks to be used with LVM and then aggregate them into a pool, thereby not only increasing the available space but also enabling you to use it consistently. We will also learn how to distribute that aggregated disk space into partition-like chunks that can easily be extended if necessary. To do so, we will go through the following topics:

- Understanding LVM
- Creating, moving, and removing physical volumes
- Combining physical volumes into volume groups
- Creating and extending logical volumes
- Adding new disks to a volume group and extending a logical volume
- Removing logical volumes, volume groups, and physical volumes
- Reviewing LVM commands

Technical requirements

For this chapter, we will add two more disks to the machine we are currently working with to be able to follow the examples mentioned. These are your options:

- If you are using a physical machine, you may add a couple of USB drives.
- If you are using a local virtual machine, you will need to add two new virtual drives.
- If you are using a cloud instance, you can add two new block devices to it.

As an example, let's see how these disks are added to our virtual machine in Linux. First, we power off the virtual machine we installed in *Chapter 1*, *Getting RHEL Up and Running*, called `rhel-instance`. Then, we open the characteristics page of the virtual machine. There we find the **Add Hardware** button:

Figure 13.1 – Editing the virtual machine properties

> **Tip**
>
> Depending on the virtualization platform you are using, there are different paths to reach the virtual machine characteristics. However, it's very common that there is an option directly accessible from the virtual machine menu.

Clicking on **Add Hardware** will open the dialog in the following screenshot. In it, we will select the **Storage** option and specify the size of the virtual disk to be created and attached to the virtual machine, in this case, 1 GiB, and then click **Finish**:

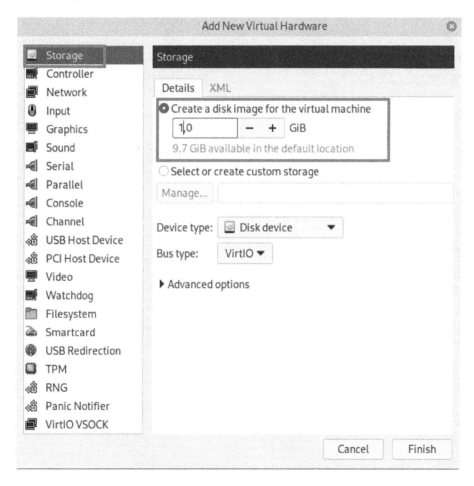

Figure 13.2 – Adding a disk to a virtual machine

We will repeat this procedure twice to add two disks. The end result will look as follows:

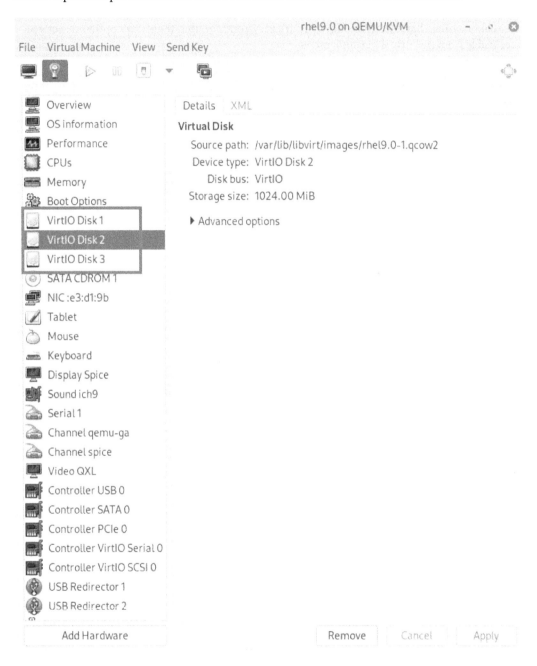

Figure 13.3 – Two new disks added to a virtual machine, making a total of three

We will now power on the virtual machine and log in to it to check the availability of new devices:

```
[root@rhel-instance ~]# lsblk
NAME              MAJ:MIN RM  SIZE RO TYPE MOUNTPOINTS
sr0                 11:0   1 1024M  0 rom
vda                252:0   0   10G  0 disk
├─vda1             252:1   0    1G  0 part /boot
└─vda2             252:2   0    9G  0 part
  ├─rhel-root 253:0   0    8G  0 lvm  /
  └─rhel-swap 253:1   0    1G  0 lvm  [SWAP]
vdb                252:16  0    1G  0 disk
vdc                252:32  0    1G  0 disk
```

We can see that the new 1 GiB disks, vdb and vdc, are available. Now that we have a system disk, where we installed the RHEL 9 operating system and two more disks to work with, we are ready to continue with this chapter.

> **Tip**
>
> The naming of the disk devices in Linux depends on the driver they use. Devices attached as SATA or SCSI show as sd and a letter, such as sda or sdb. Devices connected with IDE buses use hd and a letter, such as hda or hdb. Devices, as in the example, that use the VirtIO paravirtualized drivers, use vd and a letter – for example, vda or vdb.

Understanding LVM

LVM uses three layers to manage the storage devices in our systems. These layers are as follows:

- **Physical Volumes** (**PVs**): The first layer of LVM. Assigned to the block devices directly. A physical volume can be either a partition on a disk or a full raw disk itself.

- **Volume Groups** (**VGs**): The second layer of LVM. It groups the physical volumes to aggregate space. This is an intermediate layer and not very visible, but its role is very important.

- **Logical Volumes** (**LVs**): The third layer of LVM. It distributes the space that the volume groups aggregate.

Let's see the example that we want to implement using the two newly added disks:

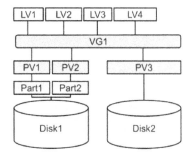

Figure 13.4 – An example of LVM using two disks

Let's explain this example diagram to understand all the layers:

- We have two disks, which are **Disk1** and **Disk2** in the diagram.
- **Disk1** is partitioned into two partitions, **Part1** and **Part2**.
- **Disk2** is not partitioned.
- There are three PVs. The purpose of these is to prepare the disk space to be used in LVM. The PVs are as follows:

  - **PV1**, created on the **Part1** partition of **Disk1**
  - **PV2**, created on the **Part2** partition of **Disk1**
  - **PV3**, created directly on **Disk2**

- One single VG, **VG1**, aggregates all three PVs, **PV1**, **PV2**, and **PV3**. Now, all that disk space is consolidated and can be easily redistributed.
- To distribute the space, there are four LVs – **LV1**, **LV2**, **LV3**, and **LV4**. Please note that the LVs do not use the whole disk. This way, if we need to extend a volume or create a snapshot, it will be possible.

This is a basic description of how the layers are distributed without diving into complex cases such as mirroring, thin provisioning, or snapshotting.

As a rule of thumb, we need to understand that PVs are designed to prepare devices to be used by LVM, VGs to aggregate PVs, and LVs to distribute the aggregated space.

It is interesting to see that if we create a VG, we can add an extra disk to it, thereby increasing its size without having to stop or reboot the machine. Equally, we can distribute the added space along the LVs that require it without having to stop or reboot the machine. This is one of the main reasons why LVM is so powerful and recommended for every server, with very few exceptions.

Now that we know the layers into which LVM is divided, let's begin using them to start understanding how they work.

Creating, moving, and removing physical volumes

Having our machine ready with the two new disks, vdb and vdc, as the *Technical requirements* section explains, we can get started on implementing the example diagram, as shown in *Figure 13.4*, on our machine.

The first step is not directly related to LVM, but it is still important in order to continue with the example. This first step involves partitioning the vdb disk. Let's take a look at this with the tool to manage partitions, parted:

```
[root@rhel-instance ~]# parted /dev/vdb print
Error: /dev/vdb: unrecognised disk label
Model: Virtio Block Device (virtblk)
Disk /dev/vdb: 1074MB
Sector size (logical/physical): 512B/512B
Partition Table: unknown
Disk Flags:
```

> **Important Note**
> Your disk device, if you are using a physical machine or a different disk driver, may be different. For example, if we were using SATA disks, it would be /dev/sdb instead of /dev/vdb.

The disk is completely unpartitioned, as we can see in the unrecognised disk label message. As is explained in *Chapter 12, Managing Local Storage and Filesystems*, there are two types of disk labels that we can use; msdos (also referred to as **MBR**), the old type that machines with a **Basic Input Output System** (**BIOS**) can use to boot, and gpt, the new type that machines with a **Unified Extensible Firmware Interface** (**UEFI**) can use to boot. In case of doubt, use gpt, as we will do in this example. The option to be used with parted to create a new label is mklabel:

```
[root@rhel-instance ~]# parted /dev/vdb mklabel gpt
Information: You may need to update /etc/fstab.
[root@rhel-instance ~]# parted /dev/vdb print
Model: Virtio Block Device (virtblk)
```

```
Disk /dev/vdb: 1074MB
Sector size (logical/physical): 512B/512B
Partition Table: gpt
Disk Flags:
Number  Start  End  Size  File system  Name  Flags
```

> **Tip**
>
> To create an msdos label, the command would be parted /dev/vdb mklabel msdos.

Now, we have a disk with a gpt label but without partitions. Let's create a partition using the mkpart option in the interactive mode:

```
[root@rhel-instance ~]# parted /dev/vdb mkpart
```

Now, we can enter the partition name, mypart0:

```
Partition name?   []? mypart0
```

For the next step, specifying the filesystem, we will use ext2:

```
File system type?   [ext2]? ext2
```

Now, it is time to set the start point. We will use the first sector available, which is 2048s:

```
Start? 2048s
```

> **Tip**
>
> The first sector in modern disks is 2048s by definition. This is not provided by the tool. We can review other existing disks when in doubt by running parted /dev/vda unit s print.

And then we come to the final step, setting the endpoint, which can be described as the size of the partition we want to make:

```
End? 200MB
```

The command is complete with the following warning:

```
Information: You may need to update /etc/fstab.
```

To ensure that the partition table is refreshed in the system, and to allow the devices to be generated under /dev, we can run the following command:

```
[root@rhel-instance ~]# udevadm settle
```

> **Tip**
> The full command to run in non-interactive mode is parted /dev/vdb mkpart mypart0 ext2 2048s 200MB.

We can see the new partition available:

```
[root@rhel-instance ~]# parted /dev/vdb print
Model: Virtio Block Device (virtblk)
Disk /dev/vdb: 1074MB
Sector size (logical/physical): 512B/512B
Partition Table: gpt
Disk Flags:
Number  Start   End    Size   File system  Name     Flags
1       1049kB  200MB  199MB               mypart0
```

We need to change the partition to be able to host LVM physical volumes. The parted command uses the set option to change the partition type. We need to specify the number of the partition, which is 1, and then type lvm and on to activate it:

```
[root@rhel-instance ~]# parted /dev/vdb set 1 lvm on
Information: You may need to update /etc/fstab.
[root@rhel-instance ~]# udevadm settle
[root@rhel-instance ~]# parted /dev/vdb print
Model: Virtio Block Device (virtblk)
Disk /dev/vdb: 1074MB
Sector size (logical/physical): 512B/512B
Partition Table: gpt
Disk Flags:
Number  Start   End    Size   File system  Name     Flags
1       1049kB  200MB  199MB               mypart0  lvm
```

We see the flag of the partition is now set as lvm.

Let's add the second partition, `mypart1`:

```
[root@rhel-instance ~]# parted /dev/vdb mkpart mypart1 \
xfs 200MB 100%
Information: You may need to update /etc/fstab.
[root@rhel-instance ~]# parted /dev/vdb set 2 lvm on
Information: You may need to update /etc/fstab.
[root@rhel-instance ~]# parted /dev/vdb print
Model: Virtio Block Device (virtblk)
Disk /dev/vdb: 1074MB
Sector size (logical/physical): 512B/512B
Partition Table: gpt
Disk Flags:
Number  Start    End      Size    File system  Name      Flags
1       1049kB   200MB    199MB   mypart0                lvm
2       200MB    1073MB   872MB   mypart1                lvm
```

Now that we have created two partitions, /dev/vdb1 (named `mypart0`) and /dev/vdb2 (named `mypart1`), this is how our storage looks:

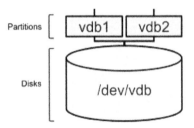

Figure 13.5 – The partitions created in our two new disks

Tip

There is another tool that comes by default in RHEL 9 to manage partitions, which is `fdisk`. You may want to give it a try to see whether you find it easier to use. If you want to learn more about partitions, you can check the RedHat official documentation here: `https://access. redhat.com/documentation/en-us/red_hat_enterprise_linux/9/html/ managing_storage_devices/disk-partitions_managing-storage- devices#overview-of-partitions_disk-partitions`.

Now, it is time to create PVs. We will do it only on the newly created partitions. First, we check the available PVs with the pvs command:

```
[root@rhel-instance ~]# pvs
  PV           VG    Fmt  Attr PSize  PFree
  /dev/vda2    rhel  lvm2 a--  <9,00g     0
```

Now, we proceed to create the PVs with pvcreate:

```
[root@rhel-instance ~]# pvcreate /dev/vdb1
  Physical volume "/dev/vdb1" successfully created.
[root@rhel-instance ~]# pvcreate /dev/vdb2
  Physical volume "/dev/vdb2" successfully created.
```

We check they have been created correctly with pvs again:

```
[root@rhel-instance ~]# pvs
  PV           VG    Fmt  Attr PSize    PFree
  /dev/vda2    rhel  lvm2 a--  <9,00g        0
  /dev/vdb1          lvm2 ---  190,00m 190,00m
  /dev/vdb2          lvm2 ---  832,00m 832,00m
```

Notice that PVs do not have their own name, but the name of the partition (or device) they are created on. We can name them as PV1 and PV2 to refer to the diagram.

This is now the status:

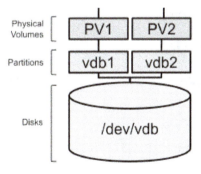

Figure 13.6 – The PVs created in the two new partitions

We can also create a PV directly on the disk device, vdc. Let's do it:

```
[root@rhel-instance ~]# pvcreate /dev/vdc
  Physical volume "/dev/vdc" successfully created.
[root@rhel-instance ~]# pvs
  PV          VG   Fmt  Attr PSize   PFree
  /dev/vda2   rhel lvm2 a--   <9,00g        0
  /dev/vdb1        lvm2 ---   190,00m 190,00m
  /dev/vdb2        lvm2 ---   832,00m 832,00m
  /dev/vdc         lvm2 ---     1,00g   1,00g
```

As in the previous example, there is no name for the PV, which we will refer to as PV3. The result looks as follows:

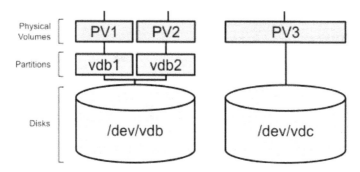

Figure 13.7 – PVs created in the two new partitions and the new disk device

Now that we have the PVs, let's group them using VGs in the next section.

Combining physical volumes into volume groups

It's time to create a new VG with the PVs added previously. Before doing that, we can check the VGs available with the vgs command:

```
[root@rhel-instance ~]# vgs
  VG   #PV #LV #SN Attr   VSize  VFree
  rhel   1   2   0 wz--n- <9,00g     0
```

We can see that only the VG created during the installation of the operating system is available. Let's create our `storage` VG with the `/dev/vdb1` and `/dev/vdb2` PVs or partitions using the `vgcreate` command:

```
[root@rhel-instance ~]# vgcreate storage /dev/vdb1 /dev/vdb2
  Volume group "storage" successfully created
[root@rhel-instance ~]# vgs
  VG           #PV #LV #SN Attr   VSize     VFree
  rhel           1   2   0 wz--n-  <9,00g         0
  storage        2   0   0 wz--n- 1016,00m 1016,00m
```

As you can see, the new `storage` VG has been created. The diagram of the current status would now look as follows:

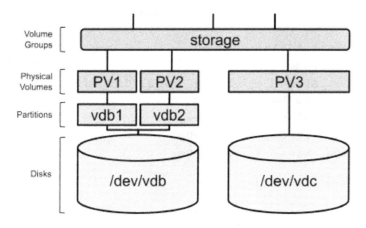

Figure 13.8 – The first VG created with two PVs

> **Important Note**
>
> VGs are a very thin layer in LVM whose only goal is to aggregate disks or partitions into a pool of storage. The advanced management of that storage, such as, for example, having data mirrored in two different disks, is done with LVs.

We have already prepared the partitions and disks as PVs and aggregated them into VGs, so we have a pool of disk space. Let's move on to the next section to learn how the distribution of that disk space can be done using LVs.

Creating and extending logical volumes

We currently have several PVs created and two of them grouped into a VG. Let's move to the next layer and check the LVs with the `lvs` command:

```
[root@rhel-instance ~]# lvs
  LV   VG   Attr       LSize   Pool Origin Data%  Meta%  Move
Log Cpy%Sync Convert
  root rhel -wi-ao---- <8,00g
  swap rhel -wi-ao----  1,00g
```

We see the `root` and `swap` volumes on the `rhel` VG, which hosts the operating system.

Now, we can create a simple LV called `data`, 200 MB in size, on the `storage` VG:

```
[root@rhel-instance ~]# lvcreate --name data --size 200MB
storage
  Logical volume "data" created.
[root@rhel-instance ~]# lvs
  LV   VG       Attr       LSize    Pool Origin
Data%  Meta%  Move Log Cpy%Sync Convert
  root rhel     -wi-ao----  <8,00g
  swap rhel     -wi-ao----   1,00g
  data storage -wi-a-----  200,00m
```

Our configuration now looks as follows:

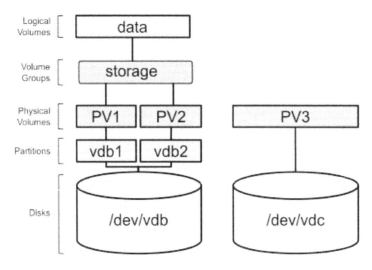

Figure 13.9 – The first LV created using space from a VG

The LV created is a block device and behaves similarly to a disk partition. Therefore, in order to use it, we need to format it with a filesystem. Let's do this by formatting it with the xfs format:

```
[root@rhel-instance ~]# mkfs.xfs /dev/storage/data
meta-data=/dev/storage/data        isize=512 agcount=4,
agsize=12800 blks
         =                         sectsz=512    attr=2,
projid32bit=1
         =                         crc=1 finobt=1, sparse=1,
rmapbt=0
         =                         reflink=1
data     =                         bsize=4096
blocks=51200,imaxpct=25
         =                         sunit=0       swidth=0 blks
naming   =version 2                bsize=4096    ascii-ci=0,
ftype=1
log      =internal log             bsize=4096    blocks=1368,
version=2
         =                         sectsz=512 sunit=0 blks,
lazy-count=1
realtime =none                     extsz=4096    blocks=0,
rtextents=0
Discarding blocks...Done.
```

Now, it's ready to be mounted. We can create the /srv/data directory and mount it there:

```
[root@rhel-instance ~]# mkdir /srv/data
[root@rhel-instance ~]# mount -t xfs /dev/storage/data /srv/
data
[root@rhel-instance ~]# df -h /srv/data/
Filesystem              Size  Used Avail Use% Mounted on
/dev/mapper/storage-data  195M   12M  184M   6% /srv/data
```

We have set up our LVM-enabled space available in our system. Mounting a filesystem manually, as in the previous example, works while the system is not powered down or rebooted. To make it persistent, we need to add the following line to /etc/fstab:

```
/dev/storage/data    /srv/data    xfs    defaults    0 0
```

To test that the line is correctly written, we can run the following commands. First, dismount the filesystem:

```
[root@rhel-instance ~]# umount /srv/data
```

Check the available space in the mount point:

```
[root@rhel-instance ~]# df -h /srv/data/
Filesystem              Size  Used Avail Use% Mounted on
/dev/mapper/rhel-root   8,0G  2,8G  5,3G  35% /
```

The output of the df (which stands for *disk-free*) command shows that the space in the /srv/data/ directory is related to the root partition, meaning that the folder does not have any filesystem associated with it. Let's now run the mount command when the system is starting:

```
[root@rhel-instance ~]# mount -a
```

All filesystems in /etc/fstab that are not mounted will be mounted, or an error will be shown if there is any issue with them (such as a typo in /etc/fstab). Let's check that it is mounted:

```
[root@rhel-instance ~]# df -h /srv/data/
Filesystem                Size  Used Avail Use% Mounted on
/dev/mapper/storage-data  195M   12M  184M   6% /srv/data
```

> **Important Note**
> The /dev/storage/data and /dev/mapper/storage-data devices are aliases (or symbolic links, to be more precise) of the same device generated by a component called a **device mapper**. They are fully interchangeable.

As we can see, the filesystem is properly mounted. Now that we know how to create a LV and assign a filesystem and mount point to it, we can move on to more advanced tasks, such as extending disk space in our LVM layers and beyond.

Adding new disks to a volume group and extending an logical volume

One of the great things about LVM, and more specifically, VGs, is that we can add a new disk to it and start using that newly extended space. Let's try it by adding the PV in /dev/vdc to the storage volume group:

```
[root@rhel-instance ~]# vgs
  VG       #PV #LV #SN Attr   VSize    VFree
```

```
  rhel      1  2   0 wz--n-   <9,00g      0
  storage   2  1   0 wz--n- 1016,00m 816,00m
[root@rhel-instance ~]# vgextend storage /dev/vdc
  Volume group "storage" successfully extended
[root@rhel-instance ~]# vgs
  VG        #PV #LV #SN Attr   VSize  VFree
  rhel       1   2  0 wz--n- <9,00g   0
  storage    3   1  0 wz--n- <1,99g 1,79g
```

Now, our disk distribution looks as follows:

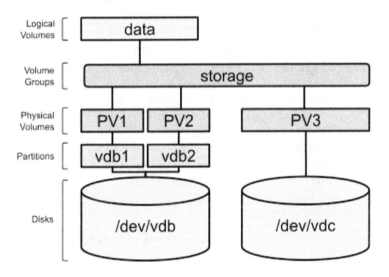

Figure 13.10 – An extended VG with three PVs

Let's now extend the data LV by adding 200 MB to it:

```
[root@rhel-instance ~]# lvs
  LV    VG       Attr       LSize    Pool Origin
Data%  Meta%  Move Log Cpy%Sync Convert
  root rhel     -wi-ao----   <8,00g
  swap rhel     -wi-ao----   1,00g
  data storage -wi-ao---- 200,00m
[root@rhel-instance ~]# lvextend --size +200MB /dev/storage/
data
```

```
   Size of logical volume storage/data changed from 200,00 MiB
(50 extents) to 400,00 MiB (100 extents).
   Logical volume storage/data successfully resized.
[root@rhel-instance ~]# lvs
  LV   VG      Attr       LSize    Pool Origin
Data%  Meta%  Move Log Cpy%Sync Convert
  root rhel    -wi-ao---- <8,00g
  swap rhel    -wi-ao----  1,00g
  data storage -wi-ao---- 400,00m
```

The LV has been extended. However, the filesystem on top of it hasn't:

```
[root@rhel-instance ~]# df -h /srv/data/
Filesystem                  Size  Used Avail Use% Mounted on
/dev/mapper/storage-data    195M   12M  184M   6% /srv/data
```

We need to extend the filesystem. The tool to do so depends on the type of filesystem. In our case, as it is xfs, the tool to extend it is xfs_growfs. Let's do it:

```
[root@rhel-instance ~]# xfs_growfs /dev/storage/data
meta-data=/dev/mapper/storage-data isize=512     agcount=4,
agsize=12800 blks
         =                          sectsz=512    attr=2,
projid32bit=1
         =                          crc=1 finobt=1, sparse=1,
rmapbt=0
         =                          reflink=1
data     =                         bsize=4096 blocks=51200
imaxpct=25
         =                          sunit=0       swidth=0 blks
naming   =version 2                bsize=4096     ascii-ci=0,
ftype=1
log      =internal log             bsize=4096     blocks=1368
version=2
         =                          sectsz=512    sunit=0 blks,
lazy-count=1
realtime =none                     extsz=4096     blocks=0,
rtextents=0
```

```
data blocks changed from 51200 to 102400
[root@rhel-instance ~]# df -h /srv/data/
Filesystem                   Size  Used Avail Use% Mounted on
/dev/mapper/storage-data     395M   14M  382M   4% /srv/data
```

And now, the filesystem has some extra space added and available.

> **Important Note**
>
> When doing this task, the LV can be mounted and used by the system. LVM is ready to do volume extensions on production systems while running.

It's very easy to redistribute the space and add another LV:

```
[root@rhel-instance ~]# lvcreate --size 100MB --name img
storage
  Logical volume "img" created.
[root@rhel-instance ~]# lvs
  LV   VG      Attr       LSize   Pool Origin
Data%  Meta%  Move Log Cpy%Sync Convert
  root rhel    -wi-ao---- <8,00g
  swap rhel    -wi-ao----  1,00g
  data storage -wi-ao---- 400,00m
  img  storage -wi-a----- 100,00m
[root@rhel-instance ~]# mkfs.xfs /dev/storage/img
meta-data=/dev/storage/img       isize=512     agcount=4,
agsize=6400 blks
         =                       sectsz=512    attr=2,
projid32bit=1
         =                       crc=1 finobt=1, sparse=1,
rmapbt=0
         =                       reflink=1
data     =                       bsize=4096 blocks=25600
imaxpct=25
         =                       sunit=0       swidth=0 blks
naming   =version 2              bsize=4096    ascii-ci=0,
ftype=1
log      =internal log           bsize=4096    blocks=1368,
version=2
```

```
                =                            sectsz=512    sunit=0 blks,
lazy-count=1
realtime =none                              extsz=4096    blocks=0,
rtextents=0
Discarding blocks...Done.
[root@rhel-instance ~]# mkdir /srv/img
[root@rhel-instance ~]# mount -t xfs /dev/storage/img /srv/img
[root@rhel-instance ~]# df /srv/img/
Filesystem                 1K-blocks  Used Available Use% Mounted
on
/dev/mapper/storage-img      96928   6068     90860   7% /srv/img
[root@rhel-instance ~]# df -h /srv/img/
Filesystem                  Size  Used Avail Use% Mounted on
/dev/mapper/storage-img      95M  6,0M   89M   7% /srv/img
```

The --size and --extents options for the lvcreate command have several options that can be used to define the space to be consumed:

- **Human-readable**: We can define the size in human-readable blocks, such as gigabytes, using GB, or megabytes, using MB (in other words, --size 3GB).

- **Extents**: If we just provide a number after --extents, the command will use its internal measure, extents, which is similar to the block size for disk partitions (that is, --extents 125).

The --size and --extents options also apply to the lvextend command. In this case, we can use the options shown previously for lvcreate to define the new size of a LV. We also have other options to define increments of space to be assigned to them:

- **Adding space**: If we provide the + symbol before the number in lvextend, this will increase the size in the measurement provided (that is, --size +1GB adds one extra GB to the current logical volume).

- **Percentage of free space**: We can provide the percentage of free space to be created or extended by using --extents and the percentage of free space to be used followed by %FREE (that is, --extents 10%FREE).

> Tip
>
> As we have seen previously with other tools, we can use the manual pages to remind ourselves of the options available. Please run man lvcreate and man lvextend to get familiar with the pages for these tools.

We are going to create a LV to be used as **swap**, which is a part of the disk that the system uses as a parking space for memory. The system puts processes that consume memory and are not active there so that the physical memory (which is much faster than disk) is freed. It is also used when there is no more free physical memory in the system.

Let's create a `swap` device on LVM:

```
[root@rhel-instance ~]# lvcreate --size 100MB --name swap
storage
  Logical volume "swap" created.
[root@rhel-instance ~]# mkswap /dev/storage/swap
Setting up swapspace version 1, size = 100 MiB (104853504
bytes)
no label, UUID=70d07e58-7e8d-4802-8d20-38d774ae6c22
```

We can check the memory and `swap` status with the `free` command:

```
[root@rhel-instance ~]# free
               total        used        free      shared     buff/
cache     available
Mem:         1346424      218816      811372        9140
316236      974844
Swap:        1048572           0     1048572
[root@rhel-instance ~]# swapon /dev/storage/swap
[root@rhel-instance ~]# free
               total        used        free      shared     buff/
cache     available
Mem:         1346424      219056      811040        9140
316328      974572
Swap:        1150968           0     1150968
```

> **Important Note**
>
> The two new changes would require adding a line for each to `/etc/fstab` to make use of them persistently across reboots. An example `swap` line would be `/dev/storage/swap swap swap defaults 0 0`.

Our disk space distribution would now look as follows:

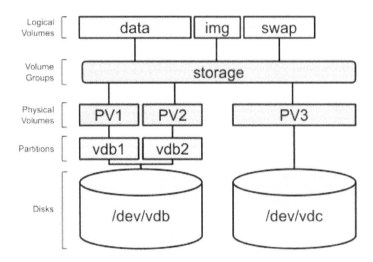

Figure 13.11 – An extended VG with three PVs

This distribution looks very similar to the initial example we used to describe the layers of LVM. We have now practiced with all the layers to create the pieces required in each one of them. We know how to create, so now it's time to learn how to remove them in the next section.

Removing logical volumes, volume groups, and physical volumes

To start with the commands used to remove, let's do the simple step of removing the img LV. First, we need to check whether it's mounted:

```
[root@rhel-instance ~]# mount | grep img
/dev/mapper/storage-img on /srv/img type xfs
(rw,relatime,seclabel,attr2,inode64,logbufs=8,logbsize=32k,
noquota)
```

As it is mounted, we need to dismount it:

```
[root@rhel-instance ~]# umount /srv/img
[root@rhel-instance ~]# mount | grep img
```

The last command shows an empty output, which means that it isn't mounted. Let's proceed to remove it:

```
[root@rhel-instance ~]# lvremove /dev/storage/img
Do you really want to remove active logical volume storage/img?
[y/n]: y
  Logical volume "img" successfully removed
```

Now, we can also remove the mount point:

```
[root@rhel-instance ~]# rmdir /srv/img
```

And with that, the removal of the LV is complete. This process is not reversible, so run it with care. Our disk distributions now look as follows:

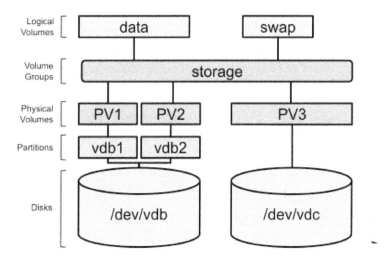

Figure 13.12 – A VG with the LV removed

Now, it's time for a more complex task, removing a PV from a volume group. The reason for doing so is that sometimes you want to transfer the data stored on a physical disk to a different disk, and then detach it and remove it from the system. This can be done, but first, let's add some files to the data LV:

```
[root@rhel-instance ~]# dnf install scap-security-guide -y
[omitted]
[root@rhel-instance ~]# cp -ar /usr/share/scap-security-guide \
/srv/data/
[root@rhel-instance ~]# ls /srv/data/
scap-security-guide
```

```
[root@rhel-instance ~]# du -sh /srv/data/
8.2M   /srv/data/
```

Now, let's evacuate the data from /dev/vdb1 using the pvmove command:

```
[root@rhel-instance ~]# pvmove /dev/vdb1
  /dev/vdb1: Moved: 7,75%
  /dev/vdb1: Moved: 77,52%
  /dev/vdb1: Moved: 100,00%
```

> **Important Note**
>
> Depending on the allocation of the extents, you may receive a message stating no data to move for storage. This means that the saved data was already allocated to the other disk. You can use pvmove with other devices to try it.

Now, there is no data stored in /dev/vdb1 and it can be removed from the VG. We can do so by using the vgreduce command:

```
[root@rhel-instance ~]# vgreduce storage /dev/vdb1
  Removed "/dev/vdb1" from volume group "storage"
```

We can see that there is now less space in the storage VG:

```
[root@rhel-instance ~]# vgs
  VG        #PV #LV #SN Attr   VSize  VFree
  rhel        1   2   0 wz--n- <9,00g     0
  storage     2   2   0 wz--n-  1,80g 1,32g
[root@rhel-instance ~]# vgdisplay storage
  --- Volume group ---
  VG Name               storage
  System ID
  Format                lvm2
  Metadata Areas        2
  Metadata Sequence No  8
  VG Access             read/write
  VG Status             resizable
  MAX LV                0
  Cur LV                2
  Open LV               2
```

```
    Max PV                  0
    Cur PV                  2
    Act PV                  2
    VG Size                 1,80 GiB
    PE Size                 4,00 MiB
    Total PE                462
    Alloc PE / Size         125 / 500,00 MiB
    Free  PE / Size         337 / 1,32 GiB
    VG UUID                 1B6Nil-rvcM-emsU-mBLu-wdjL-mDlw-66dCQU
```

We can also see that the PV, /dev/vdb1, is not attached to any VG:

```
[root@rhel-instance ~]# pvs
  PV             VG       Fmt  Attr PSize     PFree
  /dev/vda2      rhel     lvm2 a--   <9,00g         0
  /dev/vdb1               lvm2 ---   190,00m   190,00m
  /dev/vdb2      storage  lvm2 a--   828,00m   318,00m
  /dev/vdc       storage  lvm2 a--  1020,00m  1020,00m
[root@rhel-instance ~]# pvdisplay /dev/vdb1
  "/dev/vdb1" is a new physical volume of "190,00 MiB"
  --- NEW Physical volume ---
  PV Name                 /dev/vdb1
  VG Name
  PV Size                 190,00 MiB
  Allocatable             NO
  PE Size                 0
  Total PE                0
  Free PE                 0
  Allocated PE            0
  PV UUID                 veOsec-WV0n-JP9D-WMz8-UYeZ-Zjs6-sJSJst
```

> **Tip**
>
> The vgdisplay, pvdisplay, and lvdisplay commands show detailed information on any of the parts of LVM.

The most important part is that we can do these operations while the system is running production workloads with confidence. Our disk distribution now looks as follows:

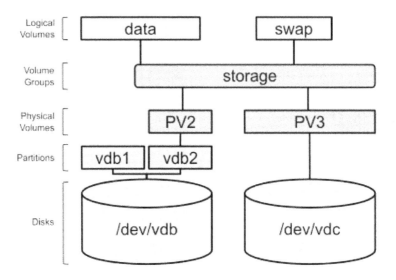

Figure 13.13 – A VG with the PVs removed

Now, it's time to remove the VG, but we need to remove the LVs first, just as we did before (feel free to run `lvs` and `vgs` before and after each command to check the progress):

```
[root@rhel-instance ~]# swapoff /dev/storage/swap
[root@rhel-instance ~]# lvremove /dev/storage/swap
Do you really want to remove active logical volume storage/
swap? [y/n]: y
  Logical volume "swap" successfully removed
```

With this, we have removed /dev/storage/swap. Let's now remove /dev/storage/data, using the --yes option so that we do not get asked for confirmation (important when using this command in a script):

```
[root@rhel-instance ~]# umount /dev/storage/data
[root@rhel-instance ~]# lvremove --yes /dev/storage/data
  Logical volume "data" successfully removed
```

Now, it's time to remove the `storage` VG:

```
[root@rhel-instance ~]# vgremove storage
```

The `storage` VG has been removed successfully.

And finally, clean up the PVs:

```
[root@rhel-instance ~]# pvremove /dev/vdb1 /dev/vdb2
  Labels on physical volume "/dev/vdb1" successfully wiped.
  Labels on physical volume "/dev/vdb2" successfully wiped.
```

And with this, we know how to work with each part of the LVM in our RHEL 9 systems. Let's review the commands used in the next section.

Reviewing LVM commands

As a summary of the commands used to manage PVs, let's take a look at the following table:

Command	Usage
pvcreate	Creates a physical volume on a partition or disk (also referred to as a block device)
pvs	Shows basic information of the physical volumes in the system
pvdisplay	Shows extended information of the physical volumes in the system
pvmove	Evacuates data from a physical volume, moving it to other available physical volumes
pvremove	Removes physical volumes

Table 13.1 – PV commands

Now, let's review the commands used to manage VGs:

Command	Usage
vgcreate	Creates a volume group aggregating different physical volumes
vgs	Shows basic information of the volume groups in the system
vgdisplay	Shows extended information of the volume groups in the system
vgextend	Extends an existing volume group by adding new physical volumes to it
vgreduce	Removes evacuated physical volumes from a volume group
vgremove	Removes a volume group

Table 13.2 – VG commands

And finally, let's review the commands used to manage LVs:

Command	Usage
lvcreate	Creates a logical volume on a volume group, preparing it to be formatted with a filesystem
lvs	Shows basic information of the logical volumes in the system
lvdisplay	Shows extended information of the logical volumes in the system
lvextend	Extends an existing logical volume by adding available space from the volume group to it
lvremove	Removes a logical volume

Table 13.3 – LV commands

Remember that you can always use the manual pages available for each command to get more information on the options you want to use and learn new ones, just by running man <command>.

> **Important Note**
>
> The web administration interface Cockpit has an extension for managing storage components. It can be installed as root (or with sudo) with the following command: dnf install cockpit-storaged. A good exercise for you would be to repeat the process done in this chapter using the storage interface in Cockpit.

Summary

The LVM is an incredibly useful part of RHEL and provides the capabilities to manage, reallocate, distribute, and assign disk space without having to stop anything in the system. Battle-tested over the years, it is a key component for system administrators, as well as facilitating the incorporation of other extended capabilities in our systems (a flexible way to provide storage to be shared via iSCSI, also know as Internet SCSI).

Practicing LVM on test machines is extremely important so that we can be sure that the command we will run on a production system will not mean the service being stopped or data being lost.

In this chapter, we have seen the most basic, yet important, tasks that can be done with LVM. We have learned how the different layers of LVM work: PVs, VGs, and LVs. We've also seen how they interact with each other and how they can be managed. We have practiced creating, extending, and removing LVs, VGs, and PVs. It will be important to practice them to consolidate the knowledge acquired and be able to use them in production systems. However, the basis for doing so is now already in place.

Now, let's move on to the next chapter to discover a new feature in RHEL 9 to improve the storage layer further by adding deduplication capabilities to it – **Virtual Data Optimizer** (**VDO**).

14

Advanced Storage Management with Stratis and VDO

In this chapter, we will learn about **Stratis** and **Virtual Data Optimizer** (VDO).

Stratis is a storage management tool to simplify running the most typical daily tasks. It uses the underlying technologies explained in the previous chapters, such as LVM, partition schemas, and filesystems.

VDO is a storage layer that includes a driver that sits between our applications and the storage devices to provide deduplication and compression of the data stored, as well as tools to manage this functionality. This will allow us, for example, to maximize the ability of our system to hold **virtual machine** (**VM**) instances that will only consume disk space for the bits that are different between them while only having one copy of the common ones.

We can also use VDO for storing different copies of our backups, knowing that disk usage will still be optimized. By the end of this chapter, we will know how VDO works and what is required to set it up for our system.

We will explore how to prepare, configure, and use our systems in the following sections:

- Understanding Stratis
- Installing and enabling Stratis
- Managing storage pools and filesystems with Stratis
- Preparing systems to use VDO
- Creating and using a VDO volume
- Testing a VDO volume and reviewing stats

Let's jump into preparing our systems to use Stratis.

Technical requirements

It is possible to continue the practice of using the VM created at the beginning of this book in *Chapter 1, Installing RHEL9*. Any additional packages required for this chapter will be indicated and can be downloaded from https://github.com/PacktPublishing/Red-Hat-Enterprise-Linux-RHEL-9-Administration.

For the *Understanding Stratis* section, we will need the same two disks that were added in *Chapter 13, Flexible Storage Management with LVM*, after all the LVM components have been cleaned up from them.

Understanding Stratis

As a new feature, to manage storage, **Stratis** was included in RHEL 8 as a technology preview (and is still a preview in RHEL 9). Stratis was created to manage local storage by combining a system service, **stratisd**, with the well-known tools in LVM (explained in *Chapter 13, Flexible Storage Management with LVM*) and the XFS filesystem (explained in *Chapter 12, Managing Local Storage and Filesystems*), which makes it very solid and reliable.

> **Important Note**
> The filesystems/pools created with Stratis should always be managed with it, and not with the LVM/XFS tools. In the same way, already-created LVM volumes should not be managed with Stratis.

Stratis combines local disks into **pools** and then distributes the storage in **filesystems**, as shown in the following diagram:

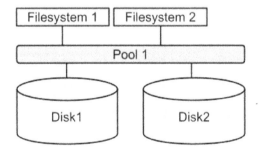

Figure 14.1 – Stratis simplified architecture diagram

As can be seen, when compared to LVM, Stratis provides a much simpler and easy-to-understand interface to storage management. In the following sections, we will install and enable Stratis and then use the same disks created in *Chapter 13, Flexible Storage Management with LVM*, to create a pool and a couple of filesystems.

Installing and enabling Stratis

To be able to work with Stratis, we will start by installing it. The two packages required to work with it are these:

- `stratis-cli`: A command-line tool to execute storage management tasks
- `stratisd`: A system service (also known as a daemon) that takes commands and executes the low-level tasks

To install them, we will use the `dnf` command:

```
[root@rhel-instance ~]# dnf install stratis-cli stratisd
Updating Subscription Management repositories.
Red Hat Enterprise Linux 9 for x86_64 - BaseOS
(RPMs)                 17 MB/s |   32 MB      00:01
Red Hat Enterprise Linux 9 for x86_64 - AppStream
(RPMs)                 12 MB/s |   30 MB      00:02
Dependencies resolved.
================================================================
===================================
Package                        Arch      Version
Repository                     Size
================================================================
===================================
Installing:
stratis-cli
noarch                               2.4.3-2.el9
                rhel-9-for-x86_64-appstream-rpms
                108 k
stratisd
x86_64                               2.4.2-3.el9
                rhel-9-for-x86_64-appstream-
rpms                                 2.8 M
[omitted]
Complete!
```

Now we can start the `stratisd` service with `systemctl`:

```
[root@rhel-instance ~]# systemctl start stratisd
[root@rhel-instance ~]# systemctl status stratisd
● stratisd.service - Stratis daemon
```

```
    Loaded: loaded (/usr/lib/systemd/system/stratisd.service;
enabled; vendor preset: enabled)
    Active: active (running) since Sat 2022-05-22 17:31:35 CEST;
53s ago
      Docs: man:stratisd(8)
Main PID: 17797 (stratisd)
     Tasks: 1 (limit: 8177)
    Memory: 1.2M
    CGroup: /system.slice/stratisd.service
            └─17797 /usr/libexec/stratisd --log-level debug
[omitted]
```

Now we shall enable it to start at boot:

```
[root@rhel-instance ~]# systemctl enable stratisd
[root@rhel-instance ~]# systemctl status stratisd
● stratisd.service - Stratis daemon
    Loaded: loaded (/usr/lib/systemd/system/stratisd.service;
enabled; vendor preset: enabled)
[omitted]
```

> **Tip**
> We can do both tasks with one command, which is `systemctl enable --now stratisd`.

Let's check with `stratis-cli` that the daemon (also known as system service) is running:

```
[root@rhel-instance ~]# stratis daemon version
2.4.2
```

We have Stratis ready, so it's time to start working on disks. Let's move on to the next subsection.

Managing storage pools and filesystems with Stratis

In order to have some storage available for Stratis, we will use the /dev/vdb and /dev/vdc disks. We need to be sure that they do not have any logical volumes or partitions on them. Let's review them:

```
[root@rhel-instance ~]# lvs
  LV   VG   Attr       LSize  Pool Origin Data%  Meta%  Move
Log Cpy%Sync Convert
```

```
   root rhel -wi-ao---- <8,00g
   swap rhel -wi-ao----  1,00g
[root@rhel-instance ~]# vgs
   VG   #PV #LV #SN Attr   VSize  VFree
   rhel  1   2   0 wz--n- <9,00g    0
[root@rhel-instance ~]# pvs
   PV         VG   Fmt  Attr PSize  PFree
   /dev/vda2  rhel lvm2 a--  <9,00g    0
```

We are good: all the LVM-created objects are on the /dev/vda disk. Let's check the other two disks, /dev/vdb and /dev/vdc:

```
[root@rhel-instance ~]# parted /dev/vdb print
Model: Virtio Block Device (virtblk)
Disk /dev/vdb: 1074MB
Sector size (logical/physical): 512B/512B
Partition Table: gpt
Disk Flags:
Number  Start  End  Size  File system  Name  Flags
<snip>
[root@rhel-instance ~]# parted /dev/vdc print
Error: /dev/vdc: unrecognised disk label
Model: Virtio Block Device (virtblk)
Disk /dev/vdc: 1074MB
Sector size (logical/physical): 512B/512B
Partition Table: unknown
Disk Flags:
```

The /dev/vdc disk has no partition table label. We are good with this one. However, the /dev/vdb disk has a partition table. Let's remove it:

```
[root@rhel-instance ~]# dd if=/dev/zero of=/dev/vdb count=2048
bs=1024
2048+0 records in
2048+0 records out
2097152 bytes (2,1 MB, 2,0 MiB) copied, 0,0853277 s, 24,6 MB/s
```

> **Tip**
>
> The dd command, which stands for disk dump, is used to dump data from devices and to devices. The special /dev/zero device simply generates zeroes, which we use to overwrite the initial sectors of the disk, where the label lives. Please use dd with care; it may overwrite anything without warning.

Now we are ready to create the first pool with the stratis command:

```
[root@rhel-instance ~]# stratis pool create mypool /dev/vdb
[root@rhel-instance ~]# stratis pool list
Name                     Total Physical      Properties
mypool    1 GiB / 37.63 MiB / 986.37 MiB         ~Ca,~Cr
```

We currently have the pool created, as shown in the following diagram:

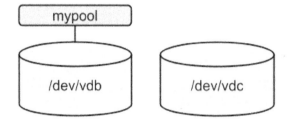

Figure 14.2 – Stratis pool created

We have the pool created; we can now create a filesystem on top of it:

```
[root@rhel-instance ~]# stratis filesystem create mypool data
[root@rhel-instance ~]# stratis filesystem list
Pool Name    Name    Used       Created              Device
  UUID
mypool       data    546 MiB    May 23 2022 19:16    /dev/stratis/
mypool/data    b073b6f1d56843b888cb83f6a7d80a43
```

The status of the storage is as follows:

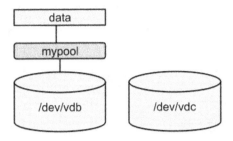

Figure 14.3 – Stratis filesystem created

Let's prepare to mount the filesystem. We need to add the following line in `/etc/fstab`:

```
/dev/stratis/mypool/data /srv/stratis-
data       xfs      defaults,x-systemd.requires=stratisd.
service          0 0
```

> **Important Note**
> In order for a Stratis filesystem to be mounted correctly during boot, we shall add the `x-systemd.requires=stratisd.service` option so it is mounted after the `stratisd` service is started.

Now we can mount it:

```
[root@rhel-instance ~]# mkdir /srv/stratis-data
[root@rhel-instance ~]# mount /srv/stratis-data/
```

Let's now extend the pool:

```
[root@rhel-instance ~]# stratis blockdev list mypool
Pool Name    Device Node    Physical Size    Tier
mypool       /dev/vdb            1 GiB    Data
[root@rhel-instance ~]# stratis pool add-data mypool /dev/vdc
[root@rhel-instance ~]# stratis blockdev list mypool
Pool Name    Device Node    Physical Size    Tier
mypool       /dev/vdb            1 GiB    Data
mypool       /dev/vdc            1 GiB    Data
```

As the underlying layer uses thin-pooling, we do not need to extend the filesystem. The storage is as follows:

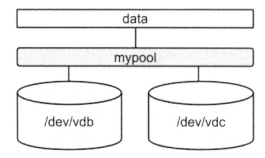

Figure 14.4 – Stratis pool extended

Time to use the `stratis snapshot` command to create a snapshot. Let's create some data and then snapshot it:

```
[root@rhel-instance ~]# stratis filesystem
Pool Name    Name    Used      Created            Device
UUID
mypool       data    546 MiB   May 23 2022 19:54    /dev/stratis/
mypool/data    08af5d5782c54087a1fd4e9531ce4943
[root@rhel-instance ~]# dd if=/dev/urandom of=/srv/stratis-
data/file bs=1M count=512
512+0 records in
512+0 records out
536870912 bytes (537 MB, 512 MiB) copied, 2,33188 s, 230 MB/s
[root@rhel-instance ~]# stratis filesystem
Pool Name    Name    Used      Created            Device
UUID
mypool       data    966 MiB   May 23 2022 19:54    /dev/stratis/
mypool/data    08af5d5782c54087a1fd4e9531ce4943
[root@rhel-instance ~]# stratis filesystem snapshot mypool data
data-snapshot1
[root@rhel-instance ~]# stratis filesystem
Pool Name    Name               Used        Created
Device                                  UUID
mypool       data               1.03 GiB    May
23 2022 19:54    /dev/stratis/mypool/
data               08af5d5782c54087a1fd4e9531ce4943
```

```
mypool      data-snapshot1   1.03 GiB    May
23 2022 19:56    /dev/stratis/mypool/data-
snapshot1    a2ae4aab56c64f728b59d710b82fb682
```

> **Tip**
>
> To see the internal pieces of Stratis, you can run the `lsblk` command. With it, you will see the components used by Stratis in a tree: physical devices, allocations for metadata and data, pools, and filesystems. All of that is abstracted by Stratis.

With this, we have seen an overview of Stratis in order to cover the basics of its management. Remember that Stratis is in preview and therefore it should not be used in production systems.

Let's move on now to other advanced topics in storage management by reviewing data deduplication with VDO.

Preparing systems to use VDO

As mentioned earlier, VDO is a driver, specifically a Linux device-mapper driver, that uses two kernel modules:

- `kvdo`: This does data compression.
- `uds`: This is in charge of deduplication.

Regular storage devices such as local disks, **Redundant Array of Inexpensive Disks** (**RAIDs**), and so on are the final backend where data is stored; the VDO layer on top reduces disk usage via the following:

- The removal of zeroed blocks, only storing them in the metadata.
- Deduplication: Duplicate data blocks are referenced in the metadata but stored only once.
- Compression, using 4 KB data blocks with a lossless compression algorithm (LZ4: `https://lz4.github.io/lz4/`).

These techniques have been used in the past in other solutions, such as in thin-provisioned VMs that only kept the differences between VMs, but VDO makes this happen transparently.

Similar to thin-provisioning, VDO can mean faster data throughput, as data can be cached by the system controller, and several services or even VMs can use that data without there being a need for additional disk reads to access it.

Let's install the required packages on our system in order to create VDO volumes by installing the vdo and `kmod-kvdo` packages:

```
dnf install vdo kmod-kvdo
```

Now, with the packages installed, we're ready to create our first volume in the next section.

Creating and using a VDO volume

To create a VDO device, we will make use of the loopback device we created in *Chapter 12, Managing Local Storage and Filesystems*, so we will check first whether it's mounted or not by executing this:

```
mount|grep loop
```

If there's output, we might need to run umount /dev/loop0p1 (or relevant mounted filesystems); if no output is shown, we're set for creating our vdo volume on top of it. Note that we should have a volume with enough space for this test, in this case, we're using a 10 GB file to simulate a disk.

Since RHEL8, VDO has been integrated into the LVM, so for creating a VDO volume, we'll need to prepare our loop device to be a physical volume with a volume group, which can be done with the following commands:

```
pvcreate /dev/loop0
vgcreate vdo /dev/loop0
lvcreate --type vdo --name myvdo vdo  -L 20G
```

The output is shown in the following screenshot:

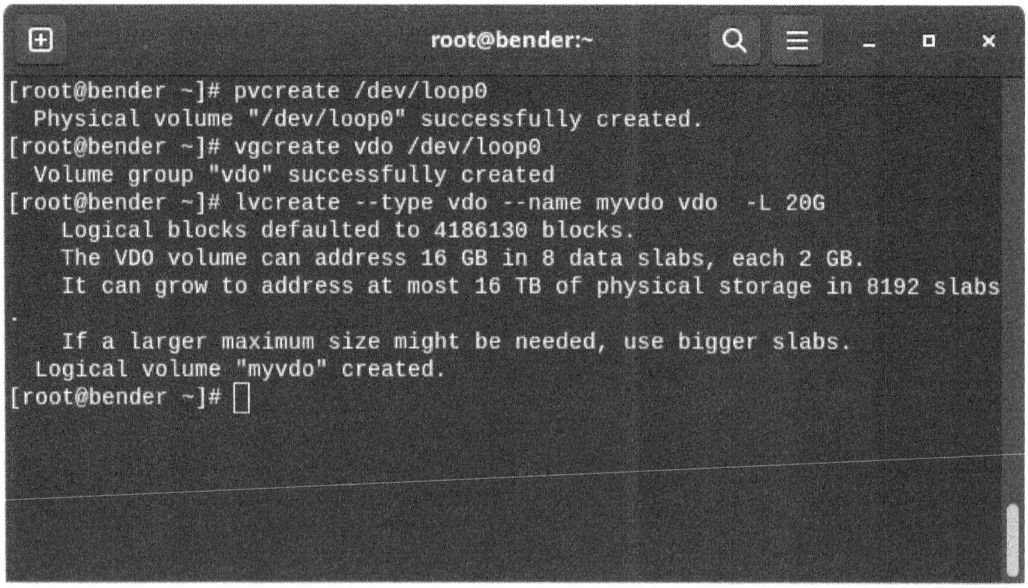

```
[root@bender ~]# pvcreate /dev/loop0
  Physical volume "/dev/loop0" successfully created.
[root@bender ~]# vgcreate vdo /dev/loop0
  Volume group "vdo" successfully created
[root@bender ~]# lvcreate --type vdo --name myvdo vdo  -L 20G
    Logical blocks defaulted to 4186130 blocks.
    The VDO volume can address 16 GB in 8 data slabs, each 2 GB.
    It can grow to address at most 16 TB of physical storage in 8192 slabs

    If a larger maximum size might be needed, use bigger slabs.
  Logical volume "myvdo" created.
[root@bender ~]# 
```

Figure 14.5 – vdo volume creation

Once the volume has been created, we can execute `lvs -o+vdo_compression,vdo_deduplication` to get detailed information about the volume created, as seen in the following screenshot:

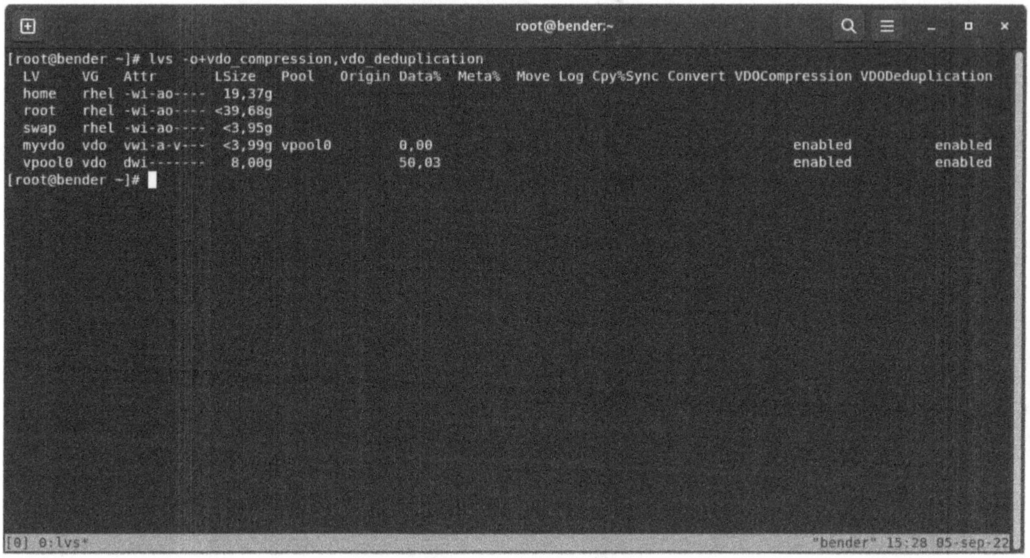

```
[root@bender ~]# lvs -o+vdo_compression,vdo_deduplication
  LV     VG    Attr       LSize   Pool    Origin Data%  Meta%  Move Log Cpy%Sync Convert VDOCompression VDODeduplication
  home   rhel  -wi-ao----  19,37g
  root   rhel  -wi-ao---- <39,68g
  swap   rhel  -wi-ao----  <3,95g
  myvdo  vdo   vwi-a-v---  <3,99g vpool0         0,00                                     enabled        enabled
  vpool0 vdo   dwi-------   8,00g                50,03                                    enabled        enabled
[root@bender ~]#
```
```
[0] 0:lvs*                                                                          "bender" 15:28 05-sep-22
```

Figure 14.6 – Output of vdo status

As we can see, there's information about the volume (size, compression status, and so on).

The new volume can now be seen via /dev/vdo/myvdo (the name we assigned) and it's ready to be used.

As we saw in the prior screenshot, both compression and deduplication are enabled on our volume, so we're ready to test the functionality. At this point, our /dev/vdo/myvdo volume is ready to be formatted. Let's use the XFS filesystem:

```
mkfs.xfs /dev/vdo/myvdo
```

Once the filesystem has been created, let's put some data on it by mounting, as follows:

```
mount /dev/vdo/myvdo /mnt
```

Now, let's test the VDO volume in the next section.

Testing a VDO volume and reviewing the stats

In order to test deduplication and compression, we will test with a big file, such as the RHEL 9 KVM guest image or the installation ISO available at https://access.redhat.com/downloads/ content/479/ver=/rhel---9/9.0/x86_64/product-software.

Once downloaded, save it as rhel-9.0-x86_64.iso and copy it four times to our VDO volume:

```
cp rhel-9.0-x86_64.iso /mnt/vm1.iso
cp rhel-9.0-x86_64.iso /mnt/vm2.iso
cp rhel-9.0-x86_64.iso /mnt/vm3.iso
cp rhel-9.0-x86_64.iso /mnt/vm4.iso
```

This would be the typical case for a server holding VMs that start with the same base disk image, but do we see any improvement?

Let's execute vdostats --si to verify the data. Note that the image downloaded is 8.5 GB, as reported by ls -si. The output obtained from vdostats --human-readable is as follows:

```
Device                Size      Used Available Use% Space
saving%
vdo-vpool0-
vpool      20.0G     11.8G      8.2G  59%          2%
```

The original volume (LV) was 20 GB, so that's the size we can see from the output, and we see that approximately only 11.8 GB has been consumed, even if we've got four files of 8.5 GB each.

The percentage is also very clear. We have a reported saved 2% of the space (three files out of four are exact copies, and the used disk matches a bit over the usage of an image). If we make an additional copy, we will see that the values increase to keep 11.8 GB used and 8.2 GB available, so no changes here.

Let's check out one of the other approaches, by creating an empty file (filled with zeros):

```
[root@bender mnt]# dd if=/dev/zero of=emptyfile bs=1G
count=16384
dd: error writing 'emptyfile': No space left on device
8+0 records in
7+0 records out
8489861120 bytes (8,5 GB, 7,9 GiB) copied, 89,5963 s, 94,8 MB/s
```

As we can see, we were able to write 8.5 GB before the disk completely filled, but let's check the vdo stats again with `vdostats --human-readable`, as seen in the following screenshot:

```
[root@bender mnt]# ls -l
total 49721920
-rw-r--r--. 1 root root 8489861120 sep  5 21:58 emptyfile
-rw-r--r--. 1 root root 8485076992 sep  5 16:38 vm1.iso
-rw-r--r--. 1 root root 8485076992 sep  5 16:34 vm2.iso
-rw-r--r--. 1 root root 8485076992 sep  5 16:42 vm3.iso
-rw-r--r--. 1 root root 8485076992 sep  5 16:50 vm4.iso
-rw-r--r--. 1 root root 8485076992 sep  5 21:50 vm5.iso
[root@bender mnt]# vdostats --human-readable
Device             Size    Used Available Use% Space saving%
vdo-vpool0-vpool   20.0G   11.8G     8.2G  59%          51%
[root@bender mnt]#



[0] 0:journalctl  1:journalctl  2:bash*                    "bender" 22:01 05-sep-22
```

Figure 14.7 – Checking the vdostats output

As we can see, we still have 8.2 GB available and we've increased the disk space saved from 2% to 51% because this big file is empty.

Wait – how, if we're using deduplication and compression, have we filled the volume if 51% of it has been saved?

As we did not indicate the logical size of the VDO volume, it set by default a 1:1 ratio with the underlying device. This is the safest approach, but we're not taking real advantage of the compression and deduplication beyond performance.

We can tune this with the `--virtualsize` parameter of `lvcreate` on LV creation.

> **Important Note**
> It's tempting to set a really big logical volume out of our real physical disk space, but we should plan ahead and think about avoiding future problems, such as the likelihood of compression ratios not being as high as our optimism. Adequately profiling the actual data being stored and the typical compression ratios for it can give us a better idea of a safe approach to be used while we continue to actively monitor disk usage evolution, both for the logical volume and for the physical one.

Long ago, when disk space was really expensive (and hard drives were 80 MB in total), it became very popular to use tools that allowed an *increase* in disk space by using a transparent layer of compression that could make some estimations and report bigger space; in reality, we know that content such as images and movies don't compress as well as other document formats such as text files. Some document formats, such as the ones used by LibreOffice, are already compressed files, so no extra compression benefits are gained.

But this changes when we speak about VMs, where the base for each one is more or less equal (based on company policies and standards) and are deployed via cloning disk images and later performing small customization, but in essence, sharing most of the disk contents.

> **Tip**
> In general, bear in mind that optimizations really just mean trade-offs. In the case of tuned profiles, you're adjusting throughput for latency, and in our case, you're trading CPU and memory resources for disk availability. The only way to tell whether something's a worthwhile trade-off is to implement it and see how it performs, look at the benefits gained, and then continue to monitor performance over time.

Summary

In this chapter, we have learned about VDO and Stratis. We've looked at simple ways to manage storage, how to save disk space transparently, and how to gain some throughput in the process.

With Stratis, we have created a pool with two disks and assigned it to a mount point. It takes fewer steps than doing so with LVM, but on the other hand, we have less control over what we are doing. In any case, we learned how to use this preview technology in RHEL 9.

With VDO, we used the volume we created to define an LVM PV and, on top of it, a volume group and a logical volume that we've formatted using the knowledge gained in previous chapters to store a VM disk image several times, to simulate a scenario where several VMs are started from the same base.

We also learned how to check the VDO optimizations and the amount of disk saved.

Now, we're ready to use Stratis instead of LVM to group and distribute storage (though not for production). We can also implement VDO for our servers to start optimizing disk usage via deduplication and compression.

In the next chapter, we will learn about the boot process.

15

Understanding the Boot Process

The boot process is what happens between the moment you power on a machine (physical or virtual) and when the operating system has been completely loaded.

Like in many good video games, there are three stages for it: the initial startup performed by the hardware (again physical or virtual), the loading of the initial stages of the operating system, and then the mechanism that helps run the required services in the system. We will review the three stages in this chapter, and we will also add tips and tricks in order to intervene in a system and perform rescue operations.

In this chapter, we will cover the following topics:

- Understanding the boot process – BIOS and **Unified Extensible Firmware Interface (UEFI)** booting
- Working with GRUB, the bootloader, and the initrd system images
- Managing the boot sequence with systemd
- Intervening in the boot process to gain access to a system

It is very likely that you will not need to make many changes in the first two stages of the boot process, but these are the points that could be extremely helpful in cases of emergency, forensics, or major failures. That's why it's important to go through them carefully.

The third stage, the one managed by **systemd**, is where more actions and changes will be performed in order to manage which services are running, by default, in the system. We have already seen examples of most of the tasks to be performed in previous chapters; however, in this one, we will provide a comprehensive review.

Let's get started with stage one.

Understanding the boot process – BIOS and UEFI booting

Computers have hardware-embedded software controllers, also called **firmware**, that let you manage the very lowest layers of the hardware. This firmware performs the first recognition of what hardware is available in the system and what hardware features are enabled (such as **pre-boot network execution**, also called **PXE**).

In the architecture known as **PC** (short for **Personal Computer**), also referred to as x86, which Intel and IBM popularized, the embedded firmware is referred to as **Basic Input and Output System** (**BIOS**).

The BIOS boot process, with Linux, takes the following steps:

1. The machine is powered on, and the BIOS firmware is loaded.

2. The firmware initializes devices such as the keyboard, mouse, storage, and other peripherals.

3. The firmware reads the configuration, including the boot order, specifying which storage device is the one to continue the boot process with.

4. Once the storage device has been selected, BIOS will load the **Master Boot Record** (**MBR**) on it, which will enable the running of the **operating system loader**. In RHEL, the operating system loader is called **Grand Unified Bootloader** (**GRUB**).

5. GRUB loads the configuration and the **operating system kernel** and **initial RAM disk**, as specified in its configuration. In **Red Hat Enterprise Linux** (**RHEL**), the kernel is stored in a file called `vmlinuz`, and the initial boot image is in a file called `initrd`. All of the GRUB configuration and the `vmlinuz` and `initrd` files are stored in the `/boot` partition.

6. The initial boot image enables the loading of the first process of the system, also referred to as `init`, which, in RHEL 9, is **systemd**.

7. Finally, systemd loads the rest of the operating system.

For this process to happen, the disk has to have an MBR partition table, and the partition assigned to `/boot` has to be marked as bootable.

> **Tip**
> The MBR partition table format is very limited, allowing only four primary partitions and using extensions such as extended partitions to overcome this limit. It is not recommended that you use this type of partition unless it is necessary.

The UEFI boot process is very similar to the BIOS boot process. The main difference in the boot sequence is that UEFI can access and read disk partitions directly. The flow for it is as follows:

1. The machine is powered on, and the UEFI firmware is loaded.

2. The firmware initializes devices such as the keyboard, mouse, storage, and other peripherals.

3. The firmware reads the configuration, where it is specified which storage device and bootable partition to continue the boot process with (UEFI does not need an MBR to boot).

4. Once the storage device has been selected, the partitions on it are read from the **GUID Partition Table (GPT)**. The first partition with the VFAT format is accessed. Then, the EFI bootloader is loaded and run. The EFI bootloader in RHEL lives in the `/boot/efi` partition, and it continues to load GRUB.

5. Then, GRUB loads the **operating system kernel**, which, in RHEL, is stored in a file called `vmlinuz`, and the **initial boot image**, which is stored in a file called `initrd`. The GRUB configuration and the `vmlinuz` and `initrd` files are stored in the `/boot` partition.

6. The initial boot image enables the loading of the first process of the system, also referred to as `init`, which, in RHEL 9, is **systemd**.

7. Finally, systemd loads the rest of the operating system.

UEFI has several advantages over BIOS, enabling more complete pre-boot environments and other capabilities such as secure boot and support for GPT partitions that can go beyond the 2 TB limit that MBR partitions have.

The installer will take care of creating the boot and, if needed, UEFI partitions and binaries.

The part of pre-boot that needs to be known for the Red Hat Certified System Administrator certification is how to load the operating system loader from it. Through BIOS or UEFI, we can select from which storage device the operating system will load and move on to the next phase. Let's go to this next phase of the boot process.

Working with GRUB, the bootloader, and the initrd system images

Once the pre-boot execution has been completed, the system will be running the GRUB bootloader.

GRUB has the mission to load the main file of an operating system, the **kernel**, pass parameters and options to it, and load the **initial RAM disk**, also known as **initrd**.

GRUB can be installed using the `grub2-install` command. We will need to know which disk device will be used to boot, in this case, `/dev/vda`:

```
[root@rhel-instance ~]# grub2-install /dev/vda
Installing for i386-pc platform.
Installation finished. No error reported.
```

> **Important Note**
>
> You should point grub-install to the disk you will use to boot the system; this is the same one that you configured in the BIOS/UEFI to boot from.

This is intended to be used to manually rebuild a system or to fix a broken boot.

GRUB files are stored in /boot/grub2. The main configuration file is /boot/grub2/grub.cfg. However, if you take a close look at this file, you will see the following header:

```
[root@rhel-instance ~]# head -n 6 /boot/grub2/grub.cfg
#
# DO NOT EDIT THIS FILE
#
# It is automatically generated by grub2-mkconfig using
templates
# from /etc/grub.d and settings from /etc/default/grub
#
```

As you can see, this file is automatically generated and, therefore, not intended to be edited manually. How do we make changes to it then? There are two ways to do so:

- The first way is by following the instructions mentioned in the grub.cfg file. This means editing the /etc/default/grub file and/or the contents of the /etc/grub.d/ directory and then regenerating the GRUB configuration by running grub2-mkconfig.

- The second way is by using the grubby command-line tool.

> **Important Note**
>
> In RHEL, when there is a new version of the kernel, it is not updated, but a new kernel is installed alongside the previous one, adding a new entry in GRUB. In this way, there is an easy way to roll back to a previous working kernel if it's needed. During the installation, a new and updated initrd file is created for the new kernel.

Let's take a look at the current kernel configuration with grubby. The --default-kernel option will show which kernel file is loaded by default:

```
[root@rhel-instance ~]# grubby --default-kernel
/boot/vmlinuz-5.14.0-70.13.1.el9_0.x86_64
```

The --default-title option will show the name used during the boot:

```
[root@rhel-instance ~]# grubby --default-title
Red Hat Enterprise Linux (5.14.0-70.13.1.el9_0.x86_64) 9.0
(Plow)
```

We can see more information for the default kernel by using the --info option:

```
[root@rhel-instance ~]# grubby --info=/boot/ vmlinuz-5.14.0-
70.13.1.el9_0.x86_64
index=0
kernel="/boot/vmlinuz-5.14.0-70.13.1.el9_0.x86_64"
args="ro crashkernel=1G-4G:192M,4G-64G:256M,64G-:512M resume=/
dev/mapper/rhel-swap rd.lvm.lv=rhel/root rd.lvm.lv=rhel/swap"
root="/dev/mapper/rhel-root"
initrd="/boot/initramfs-5.14.0-70.13.1.el9_0.x86_64.img"
title="Red Hat Enterprise Linux (5.14.0-70.13.1.el9_0.x86_64)
9.0 (Plow)"
id="4353c8e5ca804a028005a3f1f8564cfc-5.14.0-70.13.1.el9_0.
x86_64"
```

We can see the following options passed to GRUB:

- index: This shows the index number of the entry.
- kernel: This is the file containing the kernel that will be loaded to run the core of the operating system.
- args: This refers to the arguments used by the kernel to boot the system.
- root: This is the partition, or logical volume, that will be assigned to the root / directory and mounted.
- initrd: This is the file containing the RAM disk to perform the initial part of the boot process.
- title: This is a descriptive title to be shown to the user during the boot process.
- id: This is the identifier of the boot entry.

> **Tip**
> You might want to run the grubby command to obtain the information for the kernel configured as default. To do so, you can do it by running the following command: grubby --info=$(grubby --default-kernel).

Let's make the boot process less verbose by adding the `quiet` and `rhbg` arguments that have been passed to the kernel:

```
[root@rhel-instance ~]# grubby --args="rhgb quiet"
--update-kernel=/boot/vmlinuz-5.14.0-70.13.1.el9_0.x86_64
[root@rhel-instance ~]# grubby \
--info=/boot/vmlinuz-5.14.0-70.13.1.el9_0.x86_64
index=0
kernel="/boot/vmlinuz-5.14.0-70.13.1.el9_0.x86_64"
args="ro crashkernel=1G-4G:192M,4G-64G:256M,64G-:512M resume=/
dev/mapper/rhel-swap rd.lvm.lv=rhel/root rd.lvm.lv=rhel/swap
rhgb quiet"
root="/dev/mapper/rhel-root"
initrd="/boot/initramfs-5.14.0-70.13.1.el9_0.x86_64.img"
title="Red Hat Enterprise Linux (5.14.0-70.13.1.el9_0.x86_64)
9.0 (Plow)"
id="4353c8e5ca804a028005a3f1f8564cfc-5.14.0-70.13.1.el9_0.
x86_64"
```

Let's test it by rebooting the machine with the `systemctl reboot` command. This is an example output:

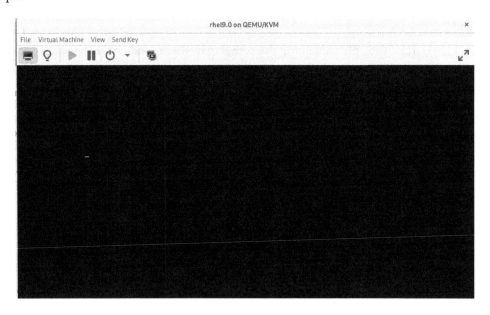

Figure 15.1 – A quiet boot

Tips

The `dmesg` command is a very important command when troubleshooting a system or looking for failures. All of the hardware details detected during the boot process can be seen in the output of this command.

In a normal boot, this might not be very useful as it goes too fast. However, if there are issues, it can help debug the situation from the console. To review these messages after booting, the `dmesg` command can be used:

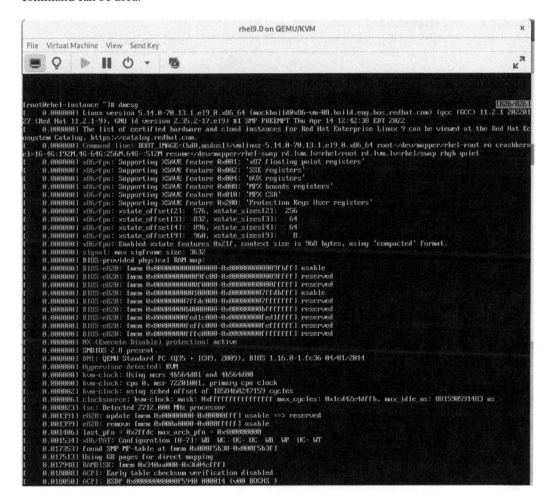

Figure 15.2 – Output of the dmesg command

We can remove an argument to the kernel by using the `--remove-args` option. Let's remove the `quiet` option:

```
[root@rhel-instance ~]# grubby --remove-args="quiet" --update-
kernel=/boot/vmlinuz-5.14.0-70.13.1.el9_0.x86_64
[root@rhel-instance ~]# grubby \
--info=/boot/vmlinuz-5.14.0-70.13.1.el9_0.x86_64
index=0
kernel="/boot/vmlinuz-5.14.0-70.13.1.el9_0.x86_64"
args="ro crashkernel=1G-4G:192M,4G-64G:256M,64G-:512M resume=/
dev/mapper/rhel-swap rd.lvm.lv=rhel/root rd.lvm.lv=rhel/swap
rhgb"
root="/dev/mapper/rhel-root"
initrd="/boot/initramfs-5.14.0-70.13.1.el9_0.x86_64.img"
title="Red Hat Enterprise Linux (5.14.0-70.13.1.el9_0.x86_64)
9.0 (Plow)"
id="4353c8e5ca804a028005a3f1f8564cfc-5.14.0-70.13.1.el9_0.
x86_64"
```

> **Important note**
> The `--info` and `--update-kernel` options accept the ALL option to review or perform actions on all the configured kernels.

If any administration task requires us to change kernel parameters, now we know how to do it. Let's move on to the next section of the boot process, `initrd`.

The **initrd** file, or **initial RAM disk**, contains a minimal system that is used to prepare the system to start. We found it in the previous configuration as `/boot/initramfs-5.14.0-70.13.1.el9_0.x86_64.img`. It can be regenerated using the `dracut` command. Let's see an example of how to rebuild the current `initrd` file:

```
[root@rhel-instance ~]# dracut --force --verbose
dracut: Executing: /usr/bin/dracut --force --verbose
dracut: dracut module 'busybox' will not be installed, because
command 'busybox' could not be found!
[omitted]
dracut: *** Including module: shutdown ***
dracut: *** Including modules done ***
dracut: *** Installing kernel module dependencies ***
dracut: *** Installing kernel module dependencies done ***
```

```
dracut: *** Resolving executable dependencies ***
dracut: *** Resolving executable dependencies done***
dracut: *** Hardlinking files ***
dracut: *** Hardlinking files done ***
dracut: *** Generating early-microcode cpio image ***
dracut: *** Constructing GenuineIntel.bin ****
dracut: *** Constructing GenuineIntel.bin ****
dracut: *** Store current command line parameters ***
dracut: *** Stripping files ***
dracut: *** Stripping files done ***
dracut: *** Creating image file '/boot/initramfs-5.14.0-
70.13.1.el9_0.x86_64.img' ***
dracut: *** Creating initramfs image file '/boot/initramfs-
5.14.0-70.13.1.el9_0.x86_64.img' done ***
```

In the preceding output, we can see what kernel modules and files, which are required for early access, are included in the `initrd` file. This step is useful for when our `initrd` file becomes corrupted and also for restoring a system from a backup, if done in different hardware, to include the proper storage drivers.

> **Tip**
> Check the manual page of `dracut` to learn more about the options to create `initrd` files. There's a Red Hat knowledge base article with instructions that be used to unpack `initrd`, along with an interesting exercise, at `https://access.redhat.com/solutions/24029`.

We have learned the very basics of the early stages of the boot process in order to be able to start troubleshooting boot issues, as required to become an RHCSA. This advanced topic could be covered in an entire book, but very little of it will be used in your daily tasks as a system administrator. That's why we only included the required aspects of it. We will include a specific use case in the *Intervening in the boot process to gain access to a system* section, and fix disk issues. Let's move on to the next topic on how services are managed in RHEL with **systemd**.

Managing the boot sequence with systemd

We have already learned how the firmware of the system will take care of pointing at a disk to run the operating system loader, which, in RHEL, is GRUB.

GRUB will load the kernel and initrd to prepare the system to start. Then, it's time to start the first process of the system, also referred to as process 1 or PID 1 (**PID** stands for **process identifier**). This process has to take care of loading all the required services in the system efficiently. In RHEL 9, the PID 1 is run by **systemd**.

In *Chapter 4*, *Tools for Regular Operations*, we described service and target management with systemd. In this chapter, let's review these interactions with the boot sequence.

The first two things related to the boot sequence that we can do with **systemd** are to reboot the system and to power it off. We will do this with the `systemctl` tool:

```
[root@rhel-instance ~]# systemctl reboot
```

We can see that the system will reboot. We can check how long the system has been running using the `uptime` command:

```
[root@rhel-instance ~]# uptime
 21:11:39 up 0 min,  1 user,  load average: 0,62, 0,13, 0,04
```

Now it's time to check `poweroff`. Before doing so, remember that after running this command, you will need to have a way to power the machine on again. Once we are aware of the process we are going to follow, let's run it:

```
[root@rhel-instance ~]# systemctl poweroff
```

Now I will power my machine *on* again.

There is a command that will stop the system without sending the signal to power the machine off, which is `systemctl halt`. The cases where this can be used are rare; however, it's good to know that it exists and what it does.

> **Important Note**
>
> The previously shown commands can be abbreviated to `reboot` and `poweroff`. If you check the file in `/usr/sbin/poweroff`, you will see that it is a symbolic link pointing to `systemctl`.

In *Chapter 4*, *Tools for Regular Operations*, we also reviewed how to set a default **systemd target** with `systemctl`. However, we can override the default configuration during boot time by passing the `systemd.unit` parameter to the kernel. We can do that using `grubby`:

```
[root@rhel-instance ~]# systemctl get-default
multi-user.target
[root@rhel-instance ~]# grubby --args="systemd.unit=emergency.
target" --update-kernel=/boot/vmlinuz-5.14.0-70.13.1.el9_0.
x86_64
[root@rhel-instance ~]# systemctl reboot
```

Now the system is rebooting. The `systemd.unit=emergency.target` parameter has been passed by **GRUB** to the **kernel**, and from the **kernel** to **systemd**, which, in turn, will ignore the default configuration and load the services required for the **emergency target**.

Now the system has been started in emergency mode and is waiting for the root password to give you control:

```
You are in emergency mode. After logging in, type "journalctl -xb" to view
system logs, "systemctl reboot" to reboot, "systemctl default" or "exit"
to boot into default mode.
Give root password for maintenance
(or press Control-D to continue): _
```

Figure 15.3 – The RHEL system booted in emergency mode

In emergency mode, there is no network configured and no other process running. You can make changes to the system knowing that no other user is accessing it. Also, only the / filesystem is mounted in read-only mode.

If a filesystem in the system is broken, this would be a good way to check it without any service accessing it. Let's try it with the command for checking the filesystem, which is called `fsck`:

```
[root@rhel-instance ~]# fsck /boot
fsck from util-linux 2.37.4
If you wish to check the consistency of an XFS filesystem or
repair a damaged filesystem, see xfs_repair(8).
```

The filesystem is OK. We could run `xfs_repair` on it, as it is an `xfs` filesystem (`fsck` detects the filesystem used), if it had issues that required a fix.

At this point, we might be thinking about how can we make changes to the root filesystem if it's already mounted as read-only at /? The process starts by remounting the / filesystem as read-write:

```
[root@rhel-instance ~]# mount -o remount -o rw /
```

Remember, you can access the manual pages for the command by running `man mount`. Now our root filesystem is mounted into / as read-write. We also need to mount /boot, so let's do it:

```
[root@rhel-instance ~]# mount /boot
```

With the /boot mounted, let's do some admin tasks, such as removing the arguments we have used in GRUB:

```
[root@rhel-instance ~]# grubby --remove-args="systemd.
unit=emergency.target" --update-kernel=/boot/vmlinuz-5.14.0-
70.13.1.el9_0.x86_64
[root@rhel-instance ~]# reboot
```

And we are back to a regular boot in the system. This might not be a practical way to go into emergency mode in Linux, but it shows you how to pass parameters to systemd at boot.

> **Tip**
>
> There is `rescue.target` that loads more services and makes the process somewhat easier. It does so by waiting for `sysinit.target` to complete, something that the emergency target does not do. A good exercise would be to repeat the previous sequence with `rescue.target`.

In the coming section, we will see how to make this change, alongside similar ones. Also, additionally we will see how to boot in a way in which no password is required.

Intervening in the boot process to gain access to a system

Sometimes, you need to intervene in a system that was handed over in which you do not have the password for the `root` user. This is an exercise that, although it sounds like an emergency situation, is more frequent than you would expect.

> **Important Note**
>
> The boot sequence must not have any disk encrypted for it to work, or you will need the password for the encrypted volumes.

The way to perform this procedure starts by stopping the boot process during the GRUB menu. This means we need to restart the system. Once the BIOS/UEFI checks have finished, the system will load GRUB. Then, we can stop the count by pressing the down or up arrow key while it's waiting for the selection of the kernel, as shown in the following screenshot:

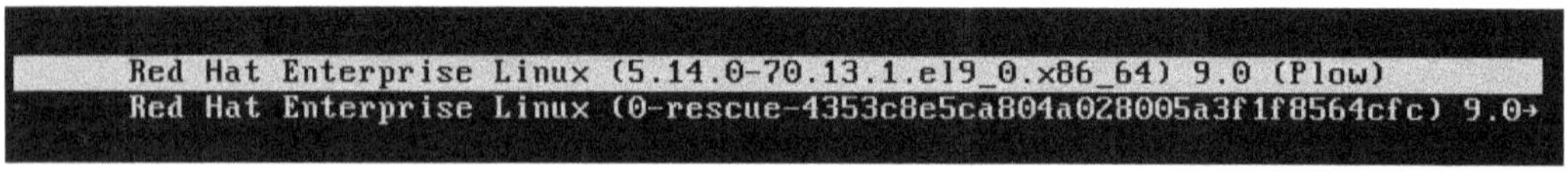

Figure 15.4 – The GRUB menu to select the kernel

We move back to the first entry. Then, we read the bottom of the screen where we find the instructions to edit the boot line:

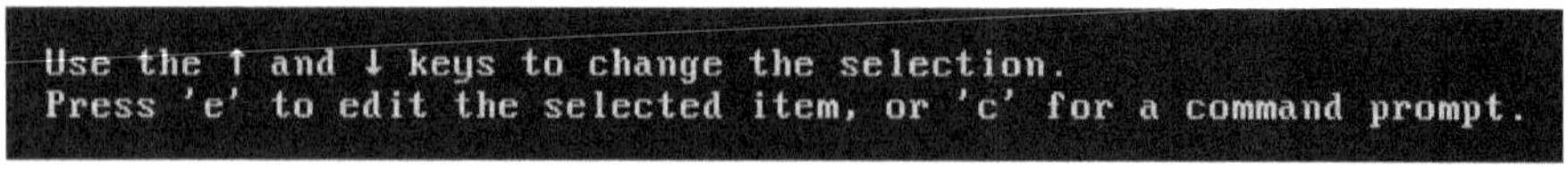

Figure 15.5 – The GRUB menu to select the kernel

If we press the *E* key, we will be able to edit the boot line selected in the menu. We will see the following five lines:

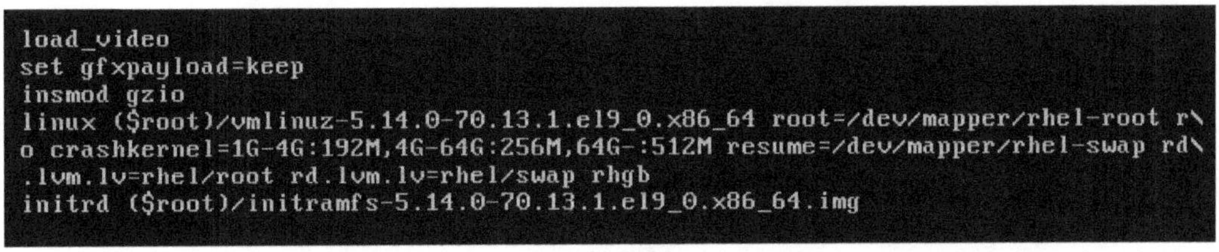

```
load_video
set gfxpayload=keep
insmod gzio
linux ($root)/vmlinuz-5.14.0-70.13.1.el9_0.x86_64 root=/dev/mapper/rhel-root r\
o crashkernel=1G-4G:192M,4G-64G:256M,64G-:512M resume=/dev/mapper/rhel-swap rd\
.lvm.lv=rhel/root rd.lvm.lv=rhel/swap rhgb
initrd ($root)/initramfs-5.14.0-70.13.1.el9_0.x86_64.img
```

Figure 15.6 – The GRUB menu to select the kernel

The first three lines with `load_video`, `set gfx_payload=keep`, and `insmod gzio` are setting options for GRUB. The next two options are the ones that are important. Let's review them:

- `linux`: This defines the kernel to be loaded and passes parameters to it.

- `initrd`: This defines where to load the `initrd` file and if there are any options for it.

> **Tip**
> Please note that the `linux` line is so long that it is wrapped, as we can see by the \ symbols, meaning that the line continues underneath.

Now we will go to the end of the `linux` line and add the `rd.break` option, as shown in the following screenshot:

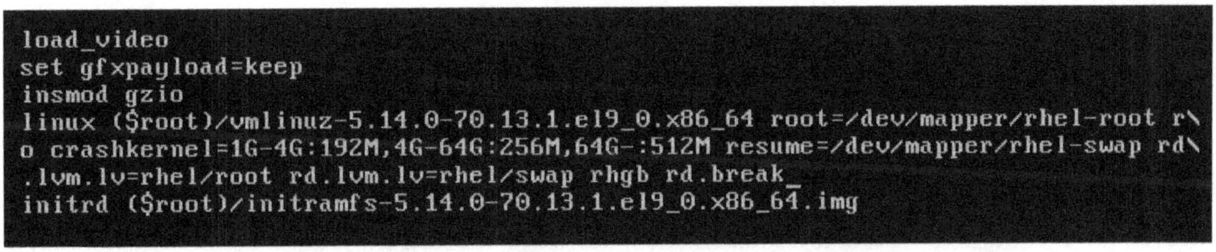

```
load_video
set gfxpayload=keep
insmod gzio
linux ($root)/vmlinuz-5.14.0-70.13.1.el9_0.x86_64 root=/dev/mapper/rhel-root r\
o crashkernel=1G-4G:192M,4G-64G:256M,64G-:512M resume=/dev/mapper/rhel-swap rd\
.lvm.lv=rhel/root rd.lvm.lv=rhel/swap rhgb rd.break_
initrd ($root)/initramfs-5.14.0-70.13.1.el9_0.x86_64.img
```

Figure 15.7 – The linux kernel line edited with the rd.break option

> **Tip**
> By default, the maintenance shell provided when using `rd.break` will ask you for the root password. If case you don't have the password, you can use `init=/bin/bash` instead of `rd.break` at the end of the `linux` line. Here, you will get a Bash shell to change the root's password.

To boot the edited line, we only need to press *Ctrl + X*. The `rd.break` option stops the boot process before `initrd` is loaded. The situation is now as follows:

- The root password is requested for maintenance.

- A single shell is loaded.

- The current root filesystem mounted on / is a minimal one with basic administration commands.

- The target root filesystem is mounted in /sysroot as read-only (instead of on /).

- No other filesystems are mounted.

- SELinux is not loaded.

The first thing we can do now is switch to the real, on-disk root filesystem with `chroot`:

```
switch_root:/# chroot /sysroot
sh-5.1#
```

Now our root filesystem is properly mounted, but it is read-only. Let's change that in the same way as we did in the previous section:

```
sh-5.1# mount -o remount -o rw /
```

Now we need to change the root user password with the `passwd` command:

```
sh-5.1# passwd
Changing password for user root
New password:
Retype new password:
passwd: all authentication tokens updated successfully
```

Now the password for the root user has been changed and the /etc/shadow file has been updated. However, it was modified without SELinux enabled; therefore, it could cause an issue in the next boot. To avoid that, there is a mechanism to fix the SELinux labels during the next boot. This mechanism consists of creating a hidden empty file, /.autorelabel, and then rebooting the system:

```
sh-5.1# touch /.autorelabel
```

Once the file has been created, it's time to reboot it to apply the SELinux changes. In this state, the machine might require you to force the power off and then power on. During the next boot, we will see how the system is applying the SELinux auto relabeling:

```
7.010960] [drm] number of cap sets: 0
7.014932] [drm] Initialized virtio_gpu 0.1.0 0 for virtio0 on minor 0
7.038929] Console: switching to colour frame buffer device 128x48
7.051130] virtio_gpu virtio0: [drm] fb0: virtio_gpu frame buffer device
7.140630] intel_pmc_core intel_pmc_core.0:  initialized
21.961638] selinux-autorelabel[657]: Relabeling / /boot /dev /dev/hugepages /dev/mqueue /dev/pts /dev/shm /run /sys /sys/fs/
group /sys/fs/pstore /sys/kernel/debug /sys/kernel/tracing
```

Figure 15.8 – SELinux autorelabel during boot

Now we can log in with the root user and its new password.

Summary

In this chapter, we reviewed the boot sequence. As you have seen, it is not long, but it is complex. It is also very important as no system can run if it cannot boot. We learned about the main differences between a BIOS-enabled system and a UEFI one, which enables some capabilities but also has its own requirements. Additionally, we learned about GRUB and its important role in the boot sequence, how to modify entries permanently with grubby, and how to make a one-time modification. Now, we understand the main files to boot, such as the kernel, vmlinuz, and the initial RAM disk, initrd.

This chapter also showed us how to start in emergency and rescue modes, as well as how to intervene in a system to reset the root password.

With these tools and procedures, we are now more prepared to handle any difficult situations in our systems. It's time to dive deeper and learn about kernel tuning and performance profiles in the next chapter.

16

Kernel Tuning and Managing Performance Profiles with tuned

As described occasionally in previous chapters, each system performance profile must be adapted to the expected usage of our system.

Kernel tuning plays a key role in this optimization, and in this chapter, we will be exploring this further in the following sections:

- Identifying processes, checking memory usage, and killing processes
- Adjusting kernel scheduling parameters to better manage processes
- Installing `tuned` and managing tuning profiles
- Creating a custom `tuned` profile
- Using the web console for observing performance metrics

By the end of this chapter, you will understand how kernel tuning is applied, how quick profiles can be used via `tuned` to suit general use cases for different system roles, and how to further extend those customizations for your servers.

Additionally, identifying processes that have become a resource hog and how to terminate them and/or prioritize them will be a useful way of getting a bit more juice out of our hardware when most needed.

Let's get hands-on and learn about these topics!

Technical requirements

You can continue the practice of using the **virtual machine** (**VM**) created at the beginning of this book in *Chapter 1*, *Getting RHEL Up and Running.* Any additional packages required for this chapter will be indicated alongside the text.

Identifying processes, checking memory usage, and killing processes

A process is a program that runs on our system—it might be a user logged in via **Secure Shell** (**SSH**) that has a bash terminal process running, or even the portion of the SSH daemon listening and replying to remote connections. Alternatively, it could be a program such as a mail client, a file manager, or more being executed.

Of course, processes take up resources in our system: memory, **Central Processing Unit** (**CPU**), disk, and more. Identifying or locating ones that might be misbehaving is a key task for system administrators.

Some of the basics were already covered in *Chapter 4*, *Tools for Regular Operations*, but it would be a good idea to have a refresher on these before continuing. However, here, we will be showing and using some of those tools in the context of performance tuning, such as the top command, which allows us to see processes and sort lists based on CPU usage, memory usage, and more. (Check the output of man top for a refresher on how to change the sorting criteria.)

One parameter to watch while checking system performance is the load average, which is a moving average made by the processes that are ready to run or are waiting for **input/output** (**I/O**) to complete. It's composed of three values—1, 5, and 15 minutes—and gives an idea of whether a load is increasing or decreasing. A rule of thumb is that if a load average is below 1, there is no resource saturation.

The load average is shown alongside many other tools, such as the aforementioned top , uptime or w commands.

If the system-load average is growing, the CPU or memory usage is spiking, and if some processes are listed there, they will be easier to locate in the top output. If the load average is also high and increasing, it might be possible that the I/O operations are increasing it. It is possible to install the iotop package, which provides the iotop command to monitor disk activity. When executed, it will show the processes in the system and the disk activity related to each one—that is, reads, writes, and swaps that might give us some more hints about where to look.

Once a process has been identified as taking too many resources, we can send a **signal** to control it.

A signal list can be obtained with the `kill -l` command, as illustrated in the following screenshot:

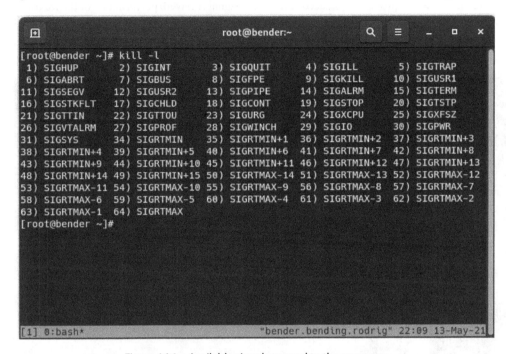

Figure 16.1 – Available signals to send to the processes

Note that each signal contains a number and a name—both can be used to send the signal to the process via its **process identifier (PID)**.

Let's review the most common ones, as follows:

Signal	ID	Usage
KILL	9	This is the most common signal, used to abruptly terminate a running process.
TERM	15	Notifies the process that it must exit so that the process can perform some cleanup operations before exiting.
USR1	10	Some programs use this signal to get notified about a configuration reload request without exiting the process itself.
HUP	1	Signal sent to processes when the controlling terminal or process ends.

Table 16.1 – Signal names, IDs, and their common usage

From the list shown in *Figure 16.1*, it's important to know that each signal has a **disposition**—that is, once a signal is sent, the process must, according to the signal received, perform one of the following actions: terminate, ignore the signal, perform a core dump, stop the process, or continue the process if it was stopped. The exact details about each signal can be checked at man 7 signal, as illustrated in the following screenshot:

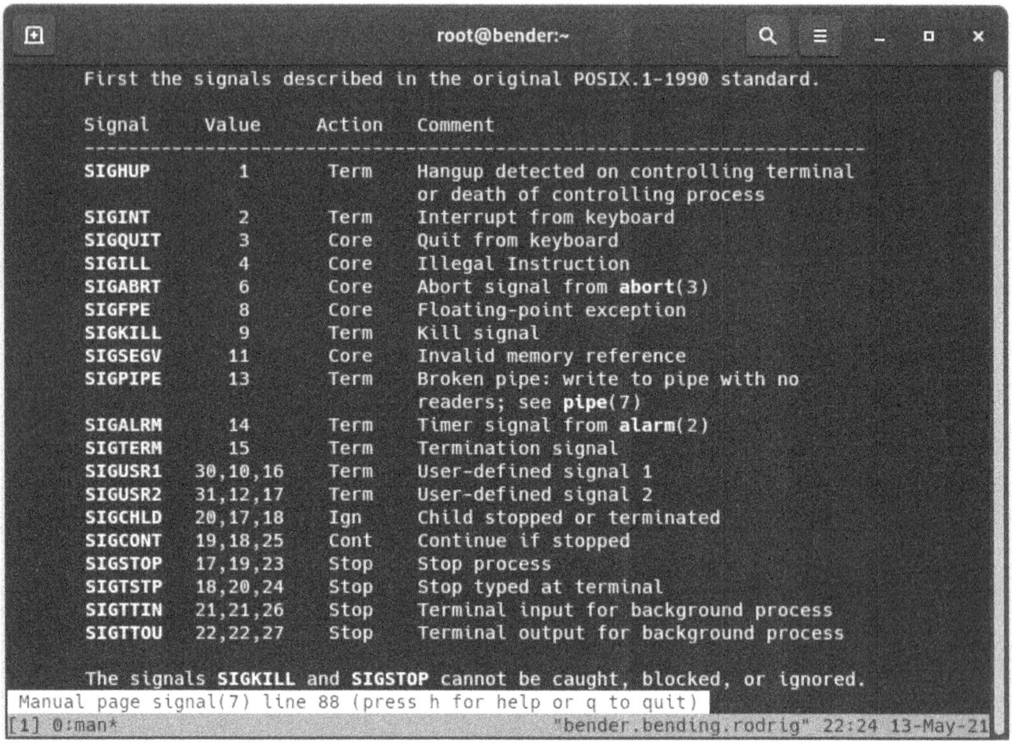

Figure 16.2 – The listing of signals, number equivalents, disposition (action), and behavior (man 7 signal)

One of the most typical usages when arriving at this point is to terminate processes that are misbehaving, so a combination of locating the process, obtaining the PID, and sending a signal to it is a very common task. In fact, it is so common that there are even tools that allow you to combine these stages in one command.

For example, we can compare ps aux|grep -i chrome|grep -v grep|awk '{print $2}'|xargs kill -9 with pkill -9 -f chrome. Both will perform the same action: search for processes named chrome, and send signal 9 (kill) to them.

Of course, even a user logging in is a process in the system (running SSH or the login shell, and so on); we can find the processes started by our target user via a similar construction (with ps, grep, and more) or with pgrep options such as pgrep -l -u user.

Bear in mind that, as the signals indicate, it's better to send a TERM signal to allow the process to run its internal cleanup steps before exiting, as directly killing them might result in leftovers in our system.

One interesting command that was widely used before terminal multiplexers such as tmux or screen became commonplace was nohup, which was prepended to commands that would last longer—for example, downloading a big file. This command captured the terminal hangout signal, allowing the executed process to continue execution, storing the output in a nohup.out file that could be checked later.

For example, to download the latest **Red Hat Enterprise Linux (RHEL) Image Standard Optical (ISO)** file from the customer portal, select one release—for example, 9.0. Then, once logged in at `https://access.redhat.com/downloads/content/479/ver=/rhel---9/9.0/x86_64/product-software`, we will select the binary ISO and right-click to copy the **Uniform Resource Locator (URL)** for the download.

> **Tip**
>
> The URLs obtained when copying from the customer portal are timebound, meaning they are only valid for a short period of time. Afterward, the download link is no longer valid and a new one should be obtained after refreshing the URL.

In a terminal, we will then execute the following command with the copied URL:

```
nohup wget URL_OBTAINED_FROM_CUSTOMER_PORTAL &
```

With the preceding command, nohup will not close the processes on terminal hang-up (disconnection), so wget will continue downloading the URL. The end ampersand symbol (&) detaches the execution from the active terminal, leaving it as a background job that we can check with the jobs command until it has finished.

If we forget to add the ampersand, the program will block our input, but we can press *Ctrl + Z* on the keyboard to stop the process. However, since we really want it to be continuing execution but in the background, we will execute bg, which will continue the execution of it.

If we want to bring back the program to receive our input and interact with it, we can move it to the foreground using the fg command.

If we press *Ctrl + C* instead, while the program has our input, it will receive a petition to interrupt and stop execution.

You can see that workflow in the following screenshot:

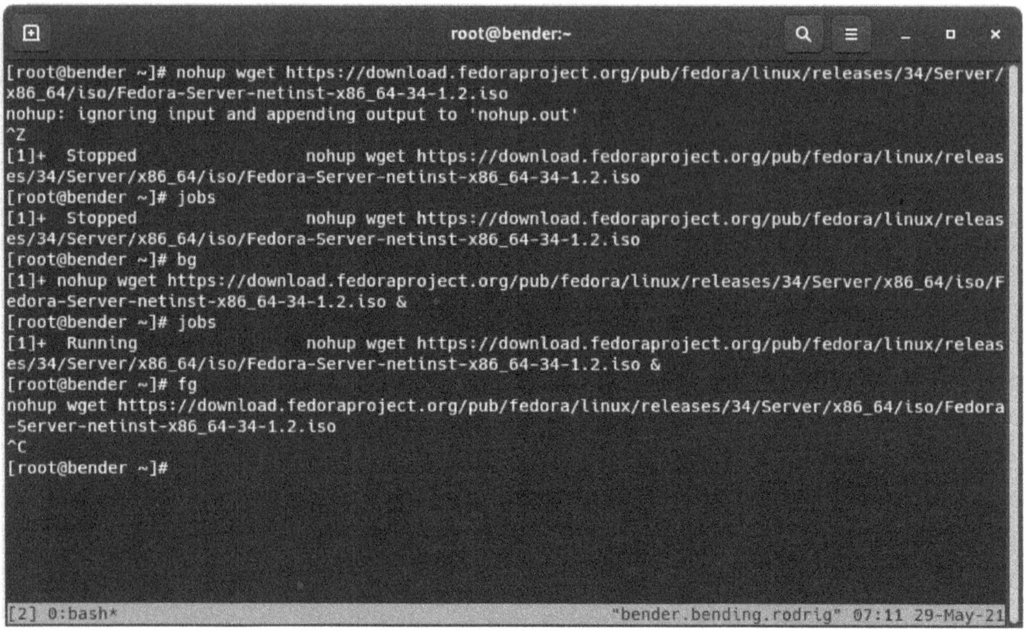
root@bender:~
[root@bender ~]# nohup wget https://download.fedoraproject.org/pub/fedora/linux/releases/34/Server/x86_64/iso/Fedora-Server-netinst-x86_64-34-1.2.iso
nohup: ignoring input and appending output to 'nohup.out'
^Z
[1]+ Stopped nohup wget https://download.fedoraproject.org/pub/fedora/linux/releases/34/Server/x86_64/iso/Fedora-Server-netinst-x86_64-34-1.2.iso
[root@bender ~]# jobs
[1]+ Stopped nohup wget https://download.fedoraproject.org/pub/fedora/linux/releases/34/Server/x86_64/iso/Fedora-Server-netinst-x86_64-34-1.2.iso
[root@bender ~]# bg
[1]+ nohup wget https://download.fedoraproject.org/pub/fedora/linux/releases/34/Server/x86_64/iso/Fedora-Server-netinst-x86_64-34-1.2.iso &
[root@bender ~]# jobs
[1]+ Running nohup wget https://download.fedoraproject.org/pub/fedora/linux/releases/34/Server/x86_64/iso/Fedora-Server-netinst-x86_64-34-1.2.iso &
[root@bender ~]# fg
nohup wget https://download.fedoraproject.org/pub/fedora/linux/releases/34/Server/x86_64/iso/Fedora-Server-netinst-x86_64-34-1.2.iso
^C
[root@bender ~]#
[2] 0:bash*
"bender.bending.rodrig" 07:11 29-May-21

Figure 16.3 – Suspending the process, resuming to the background, bringing it to the foreground, and aborting

In this case, we're downloading the Fedora 34 installation ISO (8 **gigabytes** or **GB**) using `nohup` and `wget`. Because we forgot to add the ampersand, we executed *Ctrl* + *Z* (which appears on the screen as `^Z`).

The job was reported as job `[1]` with a status of `Stopped` (which is also reported when executing `jobs`).

Then, we bring the job to the background execution with `bg`, and now `jobs` reports it as `Running`.

Afterward, we bring the job back to the foreground with `fg` and execute *Ctrl* + *C*, represented as `^C` on the screen, to finalize it.

This feature enables us to run multiple background commands—for example, we can copy a file in parallel to several hosts, as illustrated in the following screenshot:

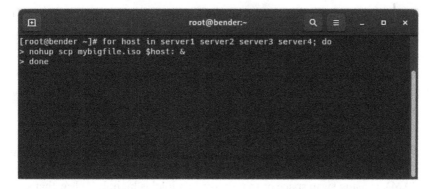

Figure 16.4 – Sample for the loop to copy a file to several servers with nohup

In this example, the copy operation performed over scp will be happening in parallel, and in the event of a disconnection from our terminal, the job will continue the execution and the output will be stored in nohup.out files in the folder that we were executing from.

> **Important Note**
>
> Processes launched with nohup will not be getting any additional input, so if the program asks for input, it will just stop execution. If the program asks for input, it's recommended that you use tmux instead, as it will still protect from terminal disconnection but also allow interaction with the launched program.

We will not always be willing to kill processes or stop or resume them; we might just want to deprioritize or prioritize them—for example, for long-running tasks that might not be critical.

Let's learn about this feature in the next section.

Adjusting kernel scheduling parameters to better manage processes

The Linux kernel is a highly configurable piece of software, so there's a whole world of tunables that can be used for adjusting its behavior: for processes, network cards, disk, memory, and more.

The most common tunables are the nice process value and the I/O priority, which regulate, respectively, the prioritization versus other processes of the CPU and I/O time.

For interacting with processes we're about to start, we can use the nice or ionice commands, prepending the command we want to execute with some parameters (remember to check the man contents for each one to get the full available range of options). Just remember that for nice, processes can go from –20 to +19, with 0 being the standard one, -20 the highest priority, and 19 the lowest priority (the higher the value, the nicer the process is).

Each process has a likelihood of getting kernel attention to run. By changing the priority via `nice` before execution or via `renice` once it's running, we can alter it a bit.

Let's think about a long-running process such as performing a backup—we want the task to succeed, so we will not be stopping or killing the process. However, at the same time, we don't want it to alter the production or level of service of our server. If we define the process with a `nice` value of 19, this means that any process in the system will get more priority—that is, our process will keep running but will not make our system busier.

This gets us into an interesting topic—many new users arriving in the Linux world, or administrators of other platforms, get a shock when they see that the system, with plenty of memory (**random-access memory** or **RAM**), is using swap space or that the system load is high. It is clear that some slight usage of swap space and having lots of free RAM just means that the kernel has optimized the usage by swapping out unused memory to disk. As long as the system doesn't feel sluggish, having a high load just means that the system has a long queue of processes to be executed; however, if the processes are *niced* to 19, for example, they are in the queue, as mentioned, any other process will go ahead of it in terms of priority.

When we're checking the system status with `top` or `ps`, we can also check for how long a process has been running, and that is also accounted for by the kernel. A new process just created that starts eating CPU and RAM has a higher chance of being killed by the kernel to ensure system operability (remember the **out-of-memory** (**OOM**) killer mentioned in *Chapter 4*, *Tools for Regular Operations*?).

For example, let's use `renice` on the process running our backup (containing the backup pattern in the process name to the lowest priority) with the following code:

```
pgrep -f backup | xargs renice -n 19
143405 (process ID) old priority 0, new priority 19
144389 (process ID) old priority 0, new priority 19
2924457 (process ID) old priority 0, new priority 19
3228039 (process ID) old priority 0, new priority 19
```

As you can see, `pgrep` has collected a list of PIDs, and that list has been piped as arguments for `renice` with a priority adjustment of 19, making processes nicer to others actually running in the system by giving others more chances to use CPU time.

Let's repeat the preceding example in our system by running a pi (π) calculation using `bc`, as illustrated in the `man` page for `bc`. First, we will time how long it takes for your system. Then, we will execute it via `renice`. So, let's get hands-on—first, let's time it, as follows:

```
time echo "scale=10000; 4*a(1)" | bc -l # Letter lowercase L
```

In my system, this was the result:

```
3.141592653...
...
375676
real 3m8,336s
user 3m6,875s
sys  0m0,032s
```

Now, let's run it with `renice`, as follows:

```
time echo "scale=10000; 4*a(1)" | bc -l &
pgrep -f bc |xargs renice -n 19 ; fg
```

In my system again, this was the result:

```
real 3m9,013s
user 3m7,273s
sys  0m0,043s
```

There's a slight difference of 1 second, but you can try running more processes to generate system activity in your environment to make it more visible and add more zeros to the scale to increase the time of execution. Similarly, `ionice` can adjust the priority of I/O operations that a process is causing (reads and writes)—for example, by repeating the action over the processes that take care of creating our backups, we could run the following command:

```
pgrep -f  backup|xargs ionice -c 3 -p
```

By default, it will not output information, but we can check the value via execution of the following command:

```
pgrep -f backup|xargs ionice -p
idle
idle
idle
idle
```

In this case, we've moved our backup processes so that I/O requests are handled when the system is idle.

The class, which we specified with the -c argument, can be one of the following:

- 0: None
- 1: Real-time
- 2: Best-effort
- 3: Idle

With -p, we specify the processes to act on.

Most of the settings that we can apply to our system came from specific ones, applied to each PID via the /proc/ virtual filesystem, such as adjusting the oom_adj file to reduce the value shown in the oom_score file. In the end, this determines whether the process should be higher on the list when OOM has to kill some process to try saving the system from catastrophe.

Of course, there are system-level settings such as /proc/sys/vm/panic_on_oom that can tune how the system has to react (panic or not) if OOM has to be invoked.

Additionally, the disks have a setting to define the scheduler being used—for example, for a disk named sda, it can be checked via cat /sys/block/sda/queue/scheduler.

The scheduler used for a disk has different approaches and depends on the kernel version—for example, in RHEL 7, it used to be noop, deadline, or cfq. However, in RHEL 8 and RHEL 9, those were removed, so we have md-deadline, bfq, kyber, and none.

This is such a big and complex topic that there is even a specific manual for it at https://access.redhat.com/documentation/en-us/red_hat_enterprise_linux/9/html-single/monitoring_and_managing_system_status_and_performance/index. So if you're interested in going deeper, have a look at it.

I hope to have achieved two things here, as follows:

- Making clear that the system has a lot of options for tuning and that it has its own documentation for it, even a Red Hat Certified Architect exam at https://www.redhat.com/en/services/training/rh442-red-hat-enterprise-performance-tuning.
- It's not an easy task—one idea has been reinforced several times in this book: test everything using your system's workload, as results might vary from one system to another.

Fortunately, there's no need to feel afraid about system tuning—it's something we can become more proficient in with experience at all levels (including knowledge, hardware, workloads, and more). But on the other hand, systems also include some easier ways to perform quick adjustments that will fit many scenarios, as we will see in the next section.

Installing tuned and managing tuning profiles

Hopefully, after a bit of scaremongering happening in the previous section, you already have a mindset prepared for an easier path.

Just in case, ensure the `tuned` package is installed, or install it with `dnf -y install tuned`. The package provides a *tuned* service that must be enabled and started for operation. As a refresher, we can achieve this by running the following command:

```
systemctl enable tuned
systemctl start tuned
```

Now we're ready to interact and get more information about this service, which announces itself at `dnf info tuned` as a daemon that tunes the system dynamically based on observation and is currently acting on an Ethernet network and hard disks.

Interaction with the daemon is performed via the `tuned-adm` command. For illustration, in the following screenshot, we're showing the command-line options that are available and a list of profiles:

```
[root@bender ~]# tuned-adm
usage: tuned-adm [-h] [--version] [--debug] [--async] [--timeout TIMEOUT]
                 [--loglevel LOGLEVEL]
                 {list,active,off,profile,profile_info,recommend,verify,auto_profile,profile_mode}
                 ...
[root@bender ~]# tuned-adm list
Available profiles:
- accelerator-performance      - Throughput performance based tuning with disabled higher latency STO
P states
- balanced                     - General non-specialized tuned profile
- desktop                      - Optimize for the desktop use-case
- hpc-compute                  - Optimize for HPC compute workloads
- intel-sst                    - Configure for Intel Speed Select Base Frequency
- latency-performance          - Optimize for deterministic performance at the cost of increased powe
r consumption
- network-latency              - Optimize for deterministic performance at the cost of increased powe
r consumption, focused on low latency network performance
- network-throughput           - Optimize for streaming network throughput, generally only necessary
on older CPUs or 40G+ networks
- optimize-serial-console      - Optimize for serial console use.
- powersave                    - Optimize for low power consumption
- throughput-performance       - Broadly applicable tuning that provides excellent performance across
a variety of common server workloads
- virtual-guest                - Optimize for running inside a virtual guest
- virtual-host                 - Optimize for running KVM guests
Current active profile: virtual-guest
[root@bender ~]#
```

Figure 16.5 – The tuned-adm command-line options and profiles

As you can see, there are some options for listing, disabling, and grabbing information about a profile, getting recommendations about which profile to use, verifying that settings have not been altered, automatically selecting a profile, and more.

One thing to bear in mind is that newer versions of the `tuned` package might bring additional profiles or configurations (stored in the `/usr/lib/tuned/` folder hierarchy), so the output might differ in your system.

Let's review some of the most common ones in the following table:

Profile	Use case
`virtual-host`	Optimizes server usage as host for VMs
`virtual-guest`	Optimizes the server as a VM running on top of `virtual-host`
`powersaving`	Tries to reduce power consumption as much as possible
`balanced`	Balances between power-saving and performance
`desktop`	Based on balanced, better responsiveness for applications

Profile	Use case
`throughput-performance`	Maximum throughput
`latency-performance`	Reduces latency and provides performance
`network-throughput`	Based on throughput performance, but with additional network tuning

Table 16.2 – The tuned-adm profiles and use cases

As mentioned earlier, each configuration is always a trade-off—more power consumption is required when increasing performance; otherwise, improving throughput might also increase latency.

Let's enable the `latency-performance` profile for our system. To do so, we will execute the following command:

```
tuned-adm profile latency-performance
```

We can verify that it has been activated with `tuned-adm active`, where we can see it shows `latency-performance`, as shown in the following screenshot:

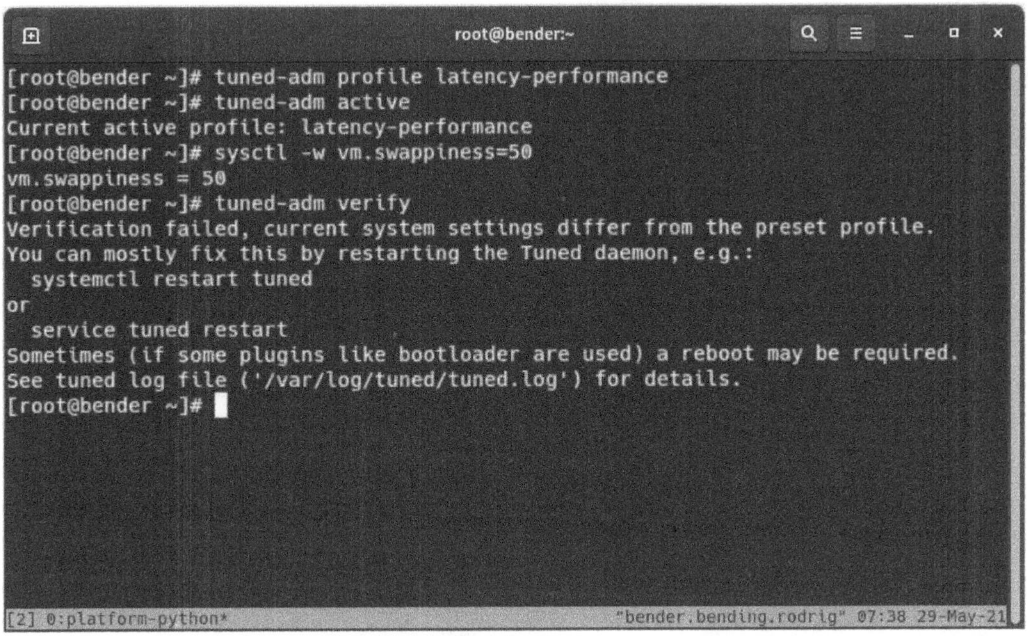
root@bender:~
[root@bender ~]# tuned-adm profile latency-performance
[root@bender ~]# tuned-adm active
Current active profile: latency-performance
[root@bender ~]# sysctl -w vm.swappiness=50
vm.swappiness = 50
[root@bender ~]# tuned-adm verify
Verification failed, current system settings differ from the preset profile.
You can mostly fix this by restarting the Tuned daemon, e.g.:
systemctl restart tuned
or
service tuned restart
Sometimes (if some plugins like bootloader are used) a reboot may be required.
See tuned log file ('/var/log/tuned/tuned.log') for details.
[root@bender ~]#
[2] 0:platform-python*
"bender.bending.rodrig" 07:38 29-May-21

Figure 16.6 – The tuned-adm profile activation and verification

Additionally, we modified the system with `sysctl -w vm.swappiness=69` (on purpose) to demonstrate the `tuned-adm verify` operation, as it reported that some settings changed from the ones defined in the profile.

> **Important Note**
>
> As of writing, dynamic tuning is disabled by default—to enable or to check the current status, check that `dynamic_tuning=1` appears in the `/etc/tuned/tuned-main.conf` file. It is disabled in the performance profiles as, by default, it tries to balance power consumption and system performance, which is the opposite of what performance profiles try to do.

Additionally, bear in mind that the **Cockpit** interface introduced in this book also features a way to change the performance profile, as shown in the following screenshot. Once you have clicked on the **Performance profile** link from the main Cockpit page, you will see this dialog open:

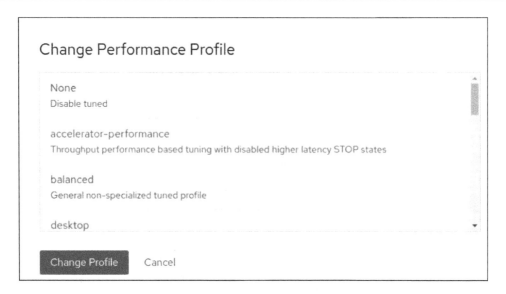

Figure 16.7 – Changing the tuned profile within the Cockpit web interface

In the next section, we will examine how tuned profiles work under the hood and how to create a custom one.

Creating a custom tuned profile

Once we've commented on the different tuned profiles, we can ask the following questions: *How do they work? How do we create one?*

For example, let's examine `latency-performance` in the next lines of code, by checking the `/usr/lib/tuned/latency-performance/tuned.conf` file.

In general, the syntax of the file is described in the man `tuned.conf` page, but the file, as you will be able to examine, is an **initialization** (**ini**) file—that is, a file organized in categories, expressed between brackets and pairs of keys and values assigned by the equals (=) sign.

The main section defines a summary of the profile if it inherits from another profile via `include`, and the additional sections depend on the plugins installed.

To learn about the available plugins, the documentation included on the man page (man `tuned.conf`) instructs us to execute `rpm -ql tuned | grep 'plugins/plugin_.*.py$'`, which provides an output similar to this:

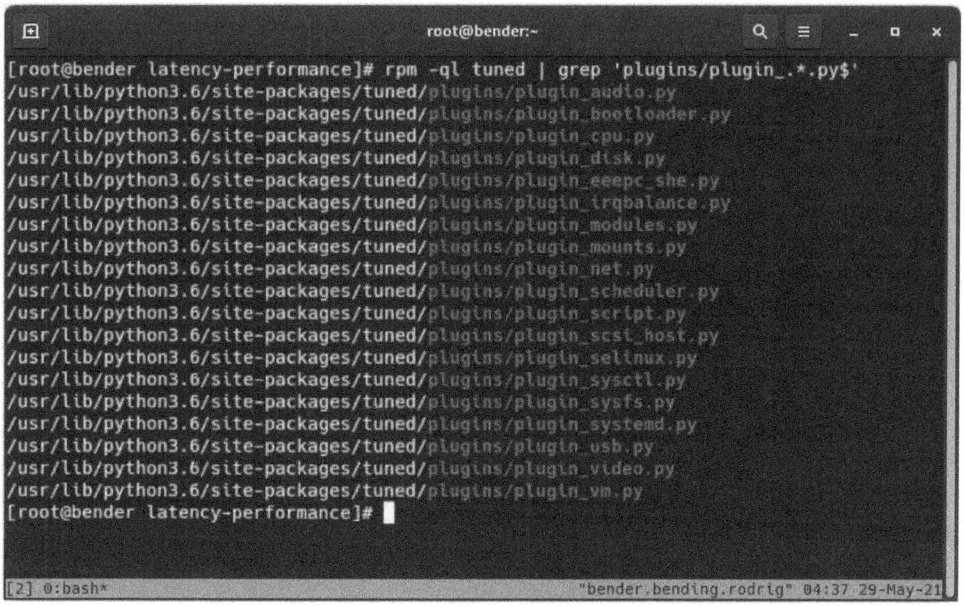

```
[root@bender latency-performance]# rpm -ql tuned | grep 'plugins/plugin_.*.py$'
/usr/lib/python3.6/site-packages/tuned/plugins/plugin_audio.py
/usr/lib/python3.6/site-packages/tuned/plugins/plugin_bootloader.py
/usr/lib/python3.6/site-packages/tuned/plugins/plugin_cpu.py
/usr/lib/python3.6/site-packages/tuned/plugins/plugin_disk.py
/usr/lib/python3.6/site-packages/tuned/plugins/plugin_eeepc_she.py
/usr/lib/python3.6/site-packages/tuned/plugins/plugin_irqbalance.py
/usr/lib/python3.6/site-packages/tuned/plugins/plugin_modules.py
/usr/lib/python3.6/site-packages/tuned/plugins/plugin_mounts.py
/usr/lib/python3.6/site-packages/tuned/plugins/plugin_net.py
/usr/lib/python3.6/site-packages/tuned/plugins/plugin_scheduler.py
/usr/lib/python3.6/site-packages/tuned/plugins/plugin_script.py
/usr/lib/python3.6/site-packages/tuned/plugins/plugin_scsi_host.py
/usr/lib/python3.6/site-packages/tuned/plugins/plugin_selinux.py
/usr/lib/python3.6/site-packages/tuned/plugins/plugin_sysctl.py
/usr/lib/python3.6/site-packages/tuned/plugins/plugin_sysfs.py
/usr/lib/python3.6/site-packages/tuned/plugins/plugin_systemd.py
/usr/lib/python3.6/site-packages/tuned/plugins/plugin_usb.py
/usr/lib/python3.6/site-packages/tuned/plugins/plugin_video.py
/usr/lib/python3.6/site-packages/tuned/plugins/plugin_vm.py
[root@bender latency-performance]#
```

Figure 16.8 – The available tuned plugins in our system

> **Important Note**
> If two or more plugins try to act over the same devices, the replace=1 setting will mark the difference between running all of them or only the latest one.

Coming back to the latency-performance profile, this has three sections: main, cpu, and sysctl.

For the CPU, it sets the performance governor, which we can check—if supported via cat /sys/devices/system/cpu/*/cpufreq/scaling_governor—for each CPU that is available in our system. Bear in mind that in some systems, the path might differ or might not even exist, and we can check the available ones via the execution of cpupower frequency-info –governors, with powersave and performance being the most common ones.

The name of the section for each plugin might be arbitrary if we specify the type keyword to indicate which plugin to use, and we can use some devices to act on via the devices keyword, allowing—for example—the definition of several disk sections with different settings based on the disk being configured. For example, we might want some settings for the system disk—let's say sda—that are different for the disk, so we use the data backups at sdb, as illustrated here:

```
[main_disk]
type=disk
```

```
devices=sda
readahead=>4096
[data_disk]
type=disk
devices=!sda
spindown=1
```

In the preceding example, the disk named sda gets configured with readahead (which reads sectors ahead of the current utilization to have the data cached before actually being requested to access it), and we're telling the system to use spindown on data disks that might only be used at backup time, thus reducing noise and power consumption when not in use.

Another interesting plugin is sysctl, used by several of the profiles, which defines settings in the same way as the sysctl command. Because of this, the possibilities are huge: defining the **Transmission Control Protocol** (TCP) window sizes for tuning networking, virtual memory management, transparent huge pages, and more.

> **Tip**
>
> It is hard to start from scratch with any performance tuning, and as tuned allows us to inherit settings from a parent, it makes sense to find which one of the available profiles is the closest to what we want to achieve, check what is being configured in it, and—of course—compare it with the others (as you can see, there are also examples for other plugins) and apply it to our custom profile.

To get an idea about how the defined system profiles touch a system, my RHEL 9 system shows the following output for cat /usr/lib/tuned/*/tuned.conf|grep -v ^#|grep '^\ ['|sort -u:

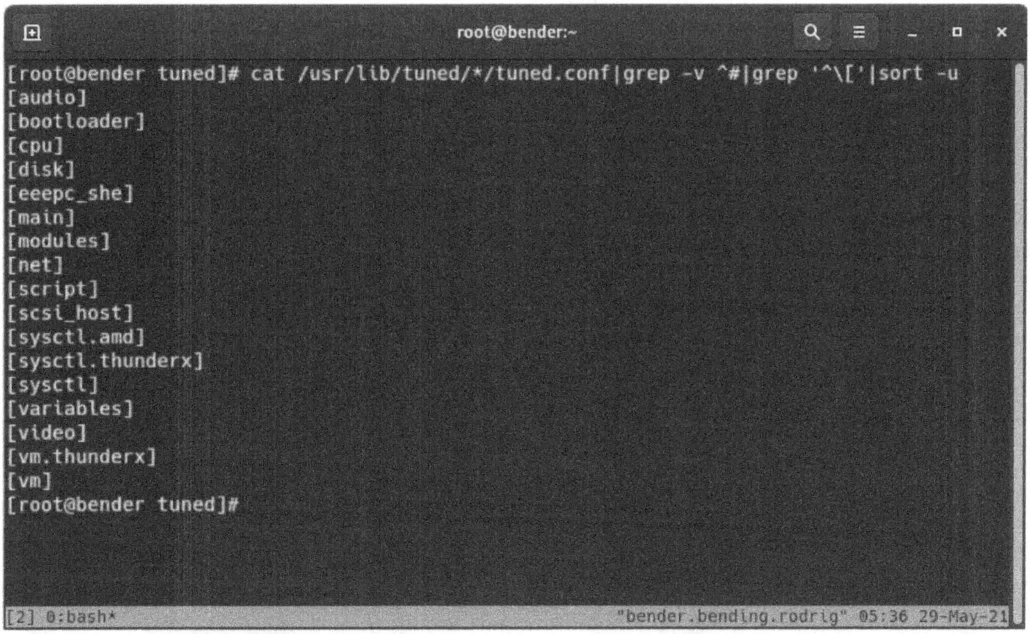

Figure 16.9 – Sections in the system-supplied profiles

So, as you can see, they touch a lot of areas. I would like to highlight the `script` section, which defines a shell script to execute used by the `powersave` profile, and the `variables` section, used by `throughput-performance` to define regular expressions for later matching and applying settings based on the CPU.

Once we're ready, we will create a new folder at `/etc/tuned/newprofile`. A `tuned.conf` file must be created, containing the main section with the summary and additional sections for the plugins we want to use.

When creating a new profile, it might be easier if we copy the profile we're interested in from the `/usr/lib/tuned/$profilename/` into our `/etc/tuned/newprofile/` folder and start the customization from there.

Once it's ready, we can enable the profile with `tuned-adm profile newprofile`, as we introduced earlier in this chapter.

You can find more information about the profiles available in the official documentation at `https://access.redhat.com/documentation/en-us/red_hat_enterprise_linux/9/html-single/monitoring_and_managing_system_status_and_performance/index#tuned-profiles-distributed-with-rhel_getting-started-with-tuned`.

With this, we've set up our own custom profile for tuning our performance settings.

In the next section, let's explore how to graphically check the performance of our system.

Using the web console for observing performance metrics

Having adjusted the tuned profile or even having a custom one defined, as we covered earlier in this chapter, requires a confirmation that our changes are there and improving the performance... and a very good fit for this is using the web console for obtaining this data.

Just in case, make sure that you enable the `cockpit` service by executing `systemctl enable --now cockpit.socket`. Once enabled, you can connect the IP address of the host at port `9090` with your browser to get access to it.

In my case, the IP of the host is `192.168.2.60` and the final URL is `https://192.168.2.60:9090`, which, when accessed, shows this warning in the browser:

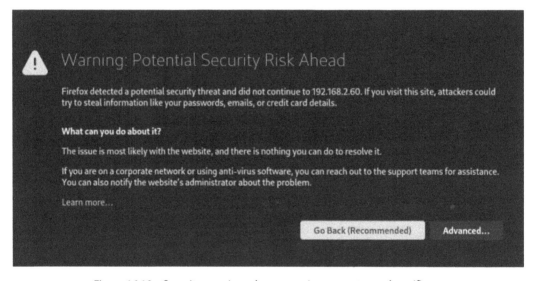

Figure 16.10 – Security warning when accessing a non-trusted certificate

When clicking on **Advanced**, we get the following dialog:

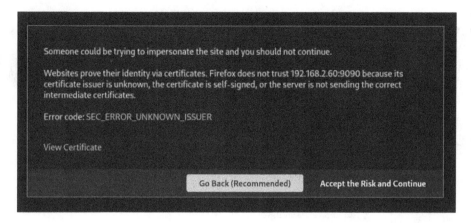

Figure 16.11 – The advanced non-trusted SSL certificate dialog in the browser

Let's click on **Accept the Risk and Continue** so that we get to the login screen of Cockpit:

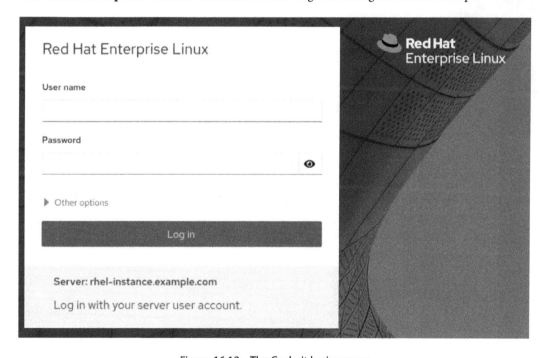

Figure 16.12 – The Cockpit login screen

After entering your user credentials (in my case, I used the *root* user), we get to the dashboard screen, as follows:

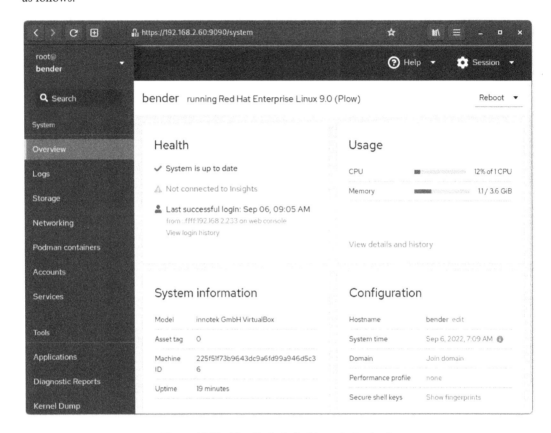

Figure 16.13 – The Cockpit dashboard after login

When we click on the **View details and history** option under **Usage**, we get to the following screen:

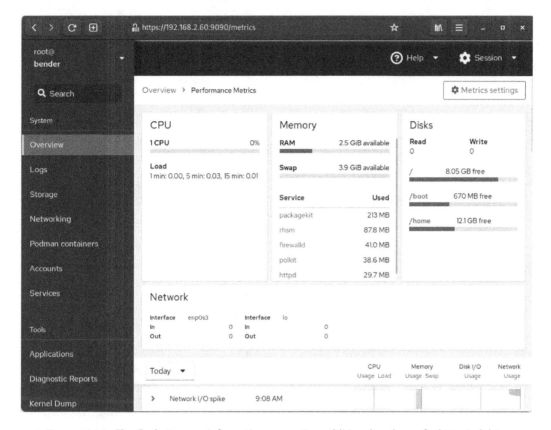

Figure 16.14 – The Cockpit usage information requesting additional packages for historical data

As we can see, Cockpit informs us that we're missing the `cockpit-pcp` package for grabbing metrics history:

Figure 16.15 – Additional software installation dialog in Cockpit

Bear in mind that if the package was missing, once enabled, it will take a while to start reporting metrics and events. But once this has been done, you can see a new section similar to this one:

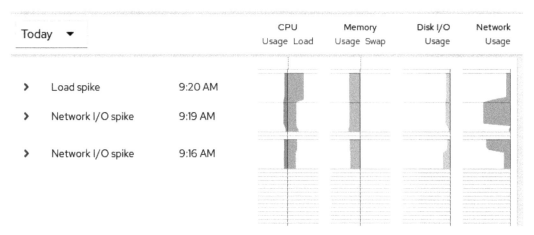

Figure 16.16 – Cockpit performance information once pmlogger
has been running for a while in our system

As you can see, we can quickly evaluate the data obtained and the events recorded to match against known bottlenecks with our system. Using this data, we can check whether the performed changes in the tuned profiles of other settings have improved the overall system experience.

Note that Cockpit uses **Performance Copilot** (**pcp**) in the background for collecting information about our system and showing it graphically. Remember that this is only a graphical interface and that other tools might be available in our system for this task.

With these guidelines, we come to the end of this chapter.

Summary

In this chapter, we learned about identifying processes, checking their resource consumption, and how to send signals to them.

In terms of the signals, we learned that some of them have some additional behavior, such as terminating processes nicely or abruptly, simply sending a notification that some programs understand as reload configuration without restarting, and more.

Also, related to processes, we learned how to adjust their priority compared to other processes in terms of CPU and I/O so that we can adjust long-running processes or disk-intensive ones to not affect other running services.

Finally, we introduced the `tuned` daemon, which includes several general use-case profiles that we can use directly in our system, allowing `tuned` to apply some dynamic tuning. Alternatively, we can fine-tune the profiles by creating one of our own to increase system performance or optimize power usage. We also learned how to graphically monitor the performance of our system via the `cockpit` web console that provides additional tools and allows us to evaluate the outcome of the tunings performed.

In the next chapter, we will learn about how to work with containers, registries, and other components so that applications can run as provided by the vendor while being isolated from the server running them.

17

Managing Containers with Podman, Buildah, and Skopeo

In this chapter, we'll learn how to use **Podman** and Red Hat **Universal Base Image**, also called **UBI**. Together, Podman and UBI provide users with the software they need to run, build, and share enterprise-quality containers on **Red Hat Enterprise Linux** (RHEL).

In recent years, being able to understand and use containers has become a key requirement for Red Hat system administrators. In this chapter, we'll review the basics of containers, how containers work, and the standard tasks for managing containers.

You will also learn how to run containers with simple commands, build enterprise-quality container images, and deploy them on a production system. Finally, you will learn when to use more advanced tools such as **Buildah** and **Skopeo**.

In this chapter, we will cover the following topics:

- Introduction to containers
- Running a container using Podman and UBI
- When to use Buildah and Skopeo

Technical requirements

In this chapter, we will review the basic usage of Podman, Buildah, and Skopeo, as well as how to build and run containers using Red Hat UBI.

We will create and run containers on the local RHEL 9 system, which we deployed in *Chapter 1, Getting RHEL Up and Running*. You will need to have the `container-tools` package from the **Application Stream** repository installed.

Introduction to containers

Containers provide users with a new way to run software on Linux systems. Containers provide all the dependencies related to a given piece of software, in a consistent and redistributable manner. Containers were first made popular by Docker, Google, and Red Hat, and many others joined Docker to create a set of open standards called the **Open Container Initiative** (**OCI**). The popularity of the OCI standards has facilitated a large ecosystem of tools where users don't have to worry about compatibility between popular container images, registries, and tools. Containers have become standardized in recent years and most major tools follow three standards governed by the OCI, outlined here:

- **Image specification**: Governs how container images are saved on disk

- **Runtime specification**: Specifies how containers are started by communicating with the operating system (in particular, the Linux kernel)

- **Distribution specification**: Governs how images are pushed and pulled from registry servers

You can learn more about this at `https://opencontainers.org/`.

All container tools (Docker, Podman, Kubernetes, and so on) need an operating system to run the container, and each operating system can choose different sets of technology to secure containers, so long as they comply with the OCI standards. RHEL uses the following operating system capabilities to securely store and run containers:

- **Namespaces**: These are a technology in the Linux kernel that helps isolate processes from one another. Namespaces prevent containerized processes from having visibility of other processes on the host operating system (including other containers). Namespaces are what make a container feel like a **virtual machine** (**VM**).

- **Control groups** (**Cgroups**): These limit the number of **central processing units** (**CPUs**), memory, disk **input/output** (**I/O**), and/or network I/O available to a given process/container. This prevents the *noisy neighbor* problem.

- **Security-Enhanced Linux** (**SELinux**): As described in *Chapter 10*, *Keeping Your System Hardened with SELinux*, this provides an extra layer of operating system security that can limit the damage caused by security exploits. SELinux is nearly transparent when used in conjunction with containers and provides mitigation of security breakouts, even when there are vulnerabilities in tools such as Podman, Docker, or runC.

Many systems administrators use VMs to isolate applications and their dependencies (libraries and so on). Containers provide the same level of isolation but reduce the overhead of virtualization. Since containers are simple processes, they do not need a **virtual CPU** (**vCPU**) with all of the overhead of translation. Containers are also smaller than VMs, which simplifies management and automation. This is particularly useful for **continuous integration/continuous delivery** (**CI/CD**).

RHEL provides users with container tools and images that are compatible with all OCI standards. This means they work in a way that is very familiar to anyone who has used Docker. For those unfamiliar with these tools and images, the following concepts are important:

- **Layers**: Container images are constructed as a set of layers. New containers are created by adding new layers (even to delete things) that reuse existing lower layers. The ability to use existing prepackaged containers is convenient for developers who simply want to make changes to their applications and test them in a reproducible way.

- **Distribution and deployments**: Since containers provide all the dependencies coupled with an application, they are easy to deploy and redistribute. Combining them with container registries makes it easy to share container images, and collaboration, deployments, and rollbacks are much quicker and easier.

The container tools provided with RHEL make it easy to deploy containers at a small scale, even for production workloads. But to manage containers at scale and with reliability, container orchestration such as Kubernetes is a much better fit. Red Hat, following the lessons learned from building Linux distributions, has created a Kubernetes distribution called **OpenShift**. If you need to deploy containers at scale, we recommend you take a look at this platform. The container tools and images provided in RHEL, and introduced in this chapter, will provide a solid foundation for deploying to Kubernetes/OpenShift if and when you are ready for it. The tools introduced in this chapter have been built in a way that will prepare your applications to be deployed in Kubernetes when you are ready for it.

Installing container tools

To manage pods, containers, and container images, we can use Podman; to create OCI images, we can use Buildah; and to inspect container images and repositories on registries, we can use Skopeo. So, let's install them:

```
[root@rhel-instance ~]# dnf install podman buildah skopeo -y
[omitted]
```

Let's take a look at the main tools that we have installed, as follows:

- podman: This command is used to run containers. You may use it in every case in which you find the use of the docker command in the examples you discover on the internet. We will use this command in this chapter to run our containers.

- buildah: This is a tool that's used to create container images. It uses the same Dockerfile definitions as Docker but without the need for a daemon.

- skopeo: This tool is used to introspect containers and check the different layers so that we can review whether they contain any non-compliant issues.

You now have a machine installed with all of the tools you will need to build, run, and manage containers on an RHEL 9 system.

Running a container using Podman and UBI

Now that you have the container tools from the Application Stream repository installed, let's run a simple container based on Red Hat UBI that contains a set of official container images and extra software based on RHEL. To run a UBI image, it only takes a single command, as illustrated in the following code snippet:

```
[root@rhel-instance ~]# podman run -it registry.access.redhat.
com/ubi9/ubi bash
Trying to pull registry.access.redhat.com/ubi9:latest...
Getting image source signatures
Checking if image destination supports signatures
Copying blob bf30f05a2532 done
Copying blob c6e5292cfd5f done
Copying config 168c58a383 done
Writing manifest to image destination
Storing signatures
[root@e38453f5e055 /]#
```

Check the hostname; it has changed from `rhel-instance` to `e38453f5e055`, which means we are inside the container.

> **Tip**
>
> These tutorials run commands as root, but one of the benefits of Podman is that it can run containers as a regular user without special permissions or a running daemon in the system.

You now have a fully isolated environment to execute whatever you want. You can run any commands you'd like in this container. It's isolated from the host and from other containers that might be running, and you can even install software on it.

> **Note**
>
> Red Hat UBI is based on software and packages from RHEL. This is the official image to use with RHEL and provides a rock solid, enterprise-ready base for your containers. UBI will be used throughout this chapter.

Running a one-off container such as this is useful for testing new configuration changes and new pieces of software without interfering with the software directly on the host.

> **Note**
>
> Red Hat UBI contains a minimal set of tools for this image to be used as an enterprise-ready base. If you want to run common commands that have not been installed in the system such as ps, you need to install them first using dnf, like a normal system.

Let's install the package that contains the ps command:

```
[root@e38453f5e055 /]# dnf install procps-ng
Updating Subscription Management repositories.
Unable to read consumer identity
Subscription Manager is operating in container mode.
 [output omitted]
 [output omitted]
Installed:
 procps-ng-3.3.174.el9.x86_64
 Complete!
```

Let's take a look at the processes running in the container, as follows:

```
[root@e38453f5e055 /]# ps -efa
UID          PID    PPID  C STIME TTY          TIME CMD
root           1       0  0 21:13 pts/0    00:00:00 bash
root          39       1  0 21:17 pts/0    00:00:00 ps -efa
```

As you can see, the only processes that are running are the shell we are using and the command we have just run. It is a completely isolated environment.

Now, exit the container by running the following command:

```
[root@e38453f5e055 /]# exit
[root@rhel-instance ~]#
```

Now that we have a working set of container tools and a UBI container image cached locally, we're going to move on to some more basic commands.

Basic container management – pull, run, stop, and remove

In this section, we'll run some basic commands to get familiar with using containers. First, let's pull some more images, as follows:

```
[root@rhel-instance ~]# podman pull registry.access.redhat.com/
ubi9/ubi-minimal
```

```
...
[root@rhel-instance ~]# podman pull registry.access.redhat.com/
ubi9/ubi-micro
...
[root@rhel-instance ~]# podman pull registry.access.redhat.com/
ubi9/ubi-init
...
```

We now have several different images cached locally. Let's take a look at these here:

```
[root@rhel-instance ~]# podman images
REPOSITORY                                    TAG          IMAGE
ID       CREATED       SIZE
registry.access.redhat.com/ubi9/
ubi-micro     latest        8bbc013a1407   13 days ago  26.2 MB
registry.access.redhat.com/ubi9/
ubi-init      latest        a991566cfe2c   13 days ago  244 MB
registry.access.redhat.com/
ubi9              latest        168c58a38365   2 weeks ago  228 MB
registry.access.redhat.com/ubi9/
ubi-minimal   latest        4a8128b051b8   2 weeks ago  129 MB
```

Notice that we have four images cached locally. Red Hat UBI comes in multiple flavors, as outlined here:

- **UBI Standard** (ubi9/ubi): An RHEL-based container base image with **YellowDog Updater Modified** (YUM)/**Dandified YUM** (DNF) in the image. It can be used in a similar way to any other Linux base image. This image is targeted at 80% of people's use cases and can easily be consumed from within a Dockerfile or Containerfile. The trade-off with this image is that it is larger than some of the other images.

- **UBI Minimal** (ubi9/ubi-minimal): This base image minimizes the size by using a small package manager called microdnf that is written in C instead of Python, such as for the standard YUM/DNF. This C implementation makes it smaller and pulls fewer dependencies into a container image. This base image can be used in any Dockerfile or Containerfile simply by using the microdnf command instead of yum. This image saves about 129 MB in memory.

- **UBI Micro** (ubi9/ubi-micro): This base image is built without a package manager. It cannot be used with a standard Dockerfile or Containerfile. Instead, users add software to this image using the Buildah tool on the container host. This image is the smallest base image provided in RHEL.

- **UBI Init** (`ubi9/ubi-init`): Based on the RHEL standard image, this image also supports the use of `systemd` in the container. This makes it easy to install a few pieces of software, start them with `systemd`, and treat the container in a similar way to a VM. This image is best for users who don't mind slightly larger images and just want ease of use.

Now that you understand the basics of the four types of base images, let's start a container in the background so that we can inspect it while it's running. Start it in the background with the following command:

```
[root@rhel-instance ~]# podman run -itd --name background ubi9
bash
b75aeb784f71d990f4a6413acc1e09046ad966fce6df5e7232517f8ad642
d4ae
```

Notice that when we start the container, the shell returns to normal and we can't type commands in the container. Our Terminal doesn't enter a shell in the container. The -d option specified that the container should run in the background. This is how most server-based software, such as web servers, runs on a Linux system.

We can still connect our shell to a container running in the background if we need to troubleshoot one, but we have to determine which container we want to connect to. To do this, list all of the containers that are running with the following command:

```
[root@rhel-instance ~]# podman ps
CONTAINER ID   IMAGE                                       COMMAND
CREATED         STATUS                 PORTS         NAMES
b75aeb784f71   registry.access.redhat.com/
ubi9:latest    bash          18 seconds ago  Up 18 seconds
ago              background
```

We could reference the container using the `CONTAINER ID` value, but we have started the container with the name `background` to make it easier to reference. We can enter the container and see what is going on inside it with the `exec` subcommand, as follows:

```
[root@rhel-instance ~]# podman exec -it background bash
[root@b75aeb784f71 /]#
```

After you type a few commands, exit the container by running the following command:

```
[root@b75aeb784f71 /]# exit
```

Now, let's stop the containerized process by running the following command:

```
[root@rhel-instance /]# podman stop background
```

Check it's stopped by running the following command:

```
[root@rhel-instance ~]# podman ps -a
CONTAINER ID   IMAGE                                     COMMAND
CREATED          STATUS                         PORTS      NAMES
b75aeb784f71   registry.access.redhat.com/
ubi9:latest   bash           2 minutes ago  Exited (137) 33 seconds
ago              background
```

Notice that the state is Exited. This means the process has been stopped and is no longer in memory, but the storage is still available on disk. The container can be restarted, or we can delete it permanently with the following command:

```
[root@rhel-instance ~]# podman rm background
b75aeb784f71d990f4a6413acc1e09046ad966fce6df5e7232517f8ad642
d4ae
```

This deleted the storage, which means the container is gone forever; you can remove all the stopped containers as a cleanup step. Verify that no containers are left by running the following command:

```
[root@rhel-instance ~]# podman ps -a
CONTAINER ID   IMAGE                                     COMMAND
CREATED              STATUS          PORTS     NAMES
```

This section taught you some basic commands. Now, let's move on to attaching storage.

Attaching persistent storage to a container

Remember that the storage in a container is ephemeral. Once the podman rm command is executed, the storage is deleted. If you have data that you need to save after the container has been removed, you need to use a volume. To run a container with a volume, execute the following command:

```
[root@rhel-instance ~]# podman run -it --rm -v /mnt:/mnt:Z
--name data ubi9 bash
[root@228b44c14e2f /]#
```

The preceding command has mounted /mnt into the container, and the Z option has told it to appropriately change the SELinux labels so that data can be written to it. The --rm option ensures that the container is removed as soon as you exit the shell. Now, you can save data on this volume, and it won't be removed when you exit the container. Add some data by running the following command:

```
[root@12ad2c1fcdc2 /]# touch /mnt/test.txt
[root@12ad2c1fcdc2 /]# exit
```

```
exit
[root@rhel-instance ~]#
```

Now, inspect the test file you created by running the following command:

```
[root@rhel-instance ~]# ls /mnt/
test.txt
```

Notice that the file is still on the system, even though the container has been removed and its internal storage has been deleted.

Deploying a container on a production system with systemd

Since Podman is not a daemon, it relies on systemd to start a container when the system boots. Podman makes it easy to start a container with systemd by creating a systemd **unit file** for you. The process of running a container with systemd looks like this:

1. Run a container with Podman exactly how you want it to run in production.
2. Export a systemd unit file.
3. Configure systemd to use this unit file.

First, let's run an example container, as follows:

```
[root@rhel-instance ~]# podman run -itd --name systemd-test
ubi9 bash
ba3a30f45379b7cb5006844ebbd4e626962055aad8784b8288b199261d
f92b1f
```

Now, let's export the systemd unit file that we'll use to start this container, as follows:

```
[root@rhel-instance ~]# podman generate systemd --name --new
systemd-test > /etc/systemd/system/podman-test.service
```

Enable and start the service by running the following command:

```
[root@rhel-instance ~]# systemctl daemon-reload
[root@rhel-instance ~]# systemctl enable --now podman-test
Created symlink /etc/systemd/system/default.target.wants/
podman-test.service → /etc/systemd/system/podman-test.service.
```

Test that the container is running by executing the following command:

```
[root@rhel-instance ~]# systemctl status podman-test
 podman-test.service - Podman container-systemd-test.service
Loaded: loaded (/etc/systemd/system/podman-test.service;
enabled; vendor preset: disabled)
Active: active (running) since Thu 2022-08-17 21:29:30 EDT;
13min ago
[output omitted]
...
```

Now, check that the container is running by using the podman command, as follows:

```
[root@rhel-instance ~]# podman ps
CONTAINER ID   IMAGE                              COMMAND
CREATED                STATUS            PORTS   NAMES
12f8c495f20c  registry.access.redhat.com/
ubi9:latest   bash      About a minute ago  Up About a minute
ago           systemd-test
```

This container will now start every time the system boots; even if you kill the container with Podman, systemd will always make sure this container is running. Podman and systemd make it easy to run containers in production. Now, let's stop the container with systemctl and disable it, as follows:

```
[root@rhel-instance ~]# systemctl stop podman-test
[root@rhel-instance ~]# systemctl disable podman-test
```

Building a container image using a Dockerfile or Containerfile

Now that we know how to run containers, let's learn how to build container images. Container images are commonly built with a file that serves as a blueprint for how to build them the same way every time. A **Dockerfile** or **Containerfile** contains all of the information necessary to build container images. It makes it easy to script how a container will get built. A Containerfile is just like a Dockerfile, but its name attempts to make it more agnostic and not tied to the Docker tooling. Either type of file can be used with the container tools that come with RHEL. Start by creating a file called Containerfile inside a test folder with the following content:

```
FROM registry.access.redhat.com/ubi9/ubi
RUN yum update -y
```

This simple Containerfile pulls the UBI standard base image and applies all of the latest updates to it. Now, let's build a container image by running the following command:

```
[root@rhel-instance ~]# mkdir test
[root@rhel-instance ~]# mv Containerfile test/
[root@rhel-instance ~]# podman build -t test-build ./test
STEP 1: FROM registry.access.redhat.com/ubi9/ubi
STEP 2: RUN yum update -y
... [output omitted] ...
```

You now have a new image called test-build that has a new layer containing all of the updated packages from the Red Hat UBI repositories, as illustrated in the following code snippet:

```
[root@rhel-instance ~]# podman images
REPOSITORY                                      TAG       IMAGE
ID        CREATED        SIZE
localhost/test-
build                           latest   98abd9c1754a 9 minutes
ago   260 MB
... [output omitted] ...
```

The workflow for building images from a Dockerfile or Containerfile is nearly identical to how Docker was in RHEL 7 or any other operating system. This makes it easy for system administrators and developers to move to Podman.

Configuring Podman to search registry servers

Container registries are like file servers for container images. They allow users to build and share container images, resulting in better collaboration. Often, it's useful to pull container images from public registry servers that are located on the internet, but in many instances, corporations have private registries that are not public. Podman makes it easy to search multiple registries, including private registries, on your company's network.

Podman comes with a configuration file that allows users and administrators to select which registries are searched by default. This makes it easy for users to find the container images that administrators want them to find.

A set of default registries to search for are defined in /etc/containers/registries.conf. Let's take a quick look at this file by filtering all the comments in it, as follows:

```
[root@rhel-instance ~]# cat /etc/containers/registries.conf |
grep -v ^#
```

```
unqualified-search-registries = ["registry.access.redhat.com",
"registry.redhat.io", "docker.io"]
short-name-mode = "enforcing"
```

As you can see, we have the `unqualified-search-registries` variable for secure registries, which includes the two main Red Hat registries, `registry.access.redhat.com` and `registry.redhat.io`, as well as the `docker.io` Docker registry. All of these registries are secured with **Transport Layer Security (TLS)** certificates, but Podman can also be configured to pull images without encryption using the `registries.insecure` section.

Separately from TLS, all images provided by Red Hat are signed and provide a signature store that can be used to verify them. This is not configured by default and is beyond the scope of this chapter.

The `short-name-mode` option supports three modes to control the behavior of short-name resolution:

- `enforcing`: Use `enforcing` if only one `unqualified-search-registries` is set. In this case, if there is more than one registry, and the user program is running in a Terminal (for example, `stdout` and `stdin` are a **teletypewriter (TTY)**), then it prompts the user to choose one of the specified search registries. In another case, if the program is not running in a Terminal, the ambiguity cannot be resolved and it will lead to an error.

- `permissive`: This behaves the same way as `enforcing` but does not give an error if the program is not running in a Terminal. Instead, it falls back to using all `unqualified-search-registries`.

- `disabled`: This will use all `unqualified-search-registries` without asking.

By default, `short-name-mode` is configured with `enforcing`.

To verify that Podman is using and searching the proper registries, run the following command:

```
[root@rhel-instance ~]# podman info | grep registries -A 4
registries:
  search:
  - registry.access.redhat.com
  - registry.redhat.io
  - docker.io
```

> **Tip**
>
> If you want to publish your own images, you can do this in the service that Red Hat offers to do so: `https://quay.io`. You can also configure `registries.conf` to search `quay.io` for images you store there.

Summary of Podman options

Let's review the options that we used with Podman in this chapter:

Command	Usage
`podman run`	Runs a container with a specified image name
`podman run -it ubi8 bash`	Interactively runs the `bash` command in a new container created using a `ubi8` image
`podman pull`	Retrieves an image from the registry to be used at a later time
`podman ps`	Lists running containers
`podman stop`	Stops a running container
`podman rm`	Removes the instance of a container (not its image)
`podman exec`	Executes a command in a running container
`podman generate`	Helps generate configuration files such as `systemd` units
`podman info`	Shows information on Podman configuration
`podman build`	Builds a container image using the specifications coming in a Containerfile or Dockerfile
`podman images`	Lists the images available in the local system

Table 17.1 – List of Podman options

As you can see, Podman includes options for managing the full container life cycle. Most Podman commands are compatible with `docker`. Podman even provides a package (`podman-docker`) that supplies an alias from `podman` to `docker` so that users can continue to type a command they are familiar with. While Podman and Docker feel quite similar to use, Podman can be run as a regular user and does not require a daemon to be continuously running. Let's move on to the next section to explore some advanced use cases.

When to use Buildah and Skopeo

Podman is a general-purpose container tool and should solve 95% of a user's needs. Podman leverages Buildah and Skopeo as libraries and pulls these tools together under one interface. That said, there are edge cases where a user may want to leverage Buildah or Skopeo separately. We will explore two such use cases here.

Building container images with Buildah

Building from a Dockerfile or Containerfile is quite easy, but it does come with some trade-offs. For example, Buildah is good in the following situations:

- When you need granular control over committing image layers. This can be necessary when you want two or three commands to run and then commit a single layer.

- When you have difficult-to-install software – for example, some third-party software comes with standardized installers that don't understand they are being run in a Dockerfile. Many of these `install.sh` installers assume they have access to the entire filesystem.

- When a container image does not provide a package manager. UBI Micro builds very small images because it does not have a Linux package manager installed, nor any of the dependencies of a package manager.

For this example, let's build on top of UBI Micro to demonstrate why Buildah is such a great tool. First, create a new container to work with, as follows:

```
[root@rhel-instance ~]# buildah from registry.access.redhat.
com/ubi9/ubi-micro
ubi-micro-working-container
```

The preceding command created a reference to a new container called `ubi-micro-working-container`. Once Buildah creates this reference, you can build upon it. To make this easier, let's start over and save the reference in a shell variable, as follows:

```
[root@rhel-instance ~]# microcontainer=$(buildah from registry.
access.redhat.com/ubi9/ubi-micro)
```

Now, you can mount the new container as a volume. This lets you modify the container image by changing the files in a directory. Run the following command to do this:

```
[root@rhel-instance ~]# micromount=$(buildah mount
$microcontainer)
```

Once the container storage has been mounted, you can modify it in any way you like. These changes will eventually be saved as a new layer in the container image. This is where you could run an installer (`install.sh`), but in the following example, we will use the package manager on the host to install packages in UBI Micro:

```
[root@rhel-instance ~]# yum install \
    --installroot $micromount --releasever 9  --setopt install_
weak_deps=false  --nodocs -y  httpd
... [output omitted] ...
```

```
[root@rhel-instance ~]# yum clean all \
    --installroot $micromount
... [output omitted] ...
```

When the package installation completes, we will unmount the storage and commit the new image layer as a new container image called ubi-micro-httpd, as illustrated in the following code snippet:

```
[root@rhel-instance ~]# buildah umount $microcontainer
232e2cdde4171c465f63b013214e5f1a161a1e62969e265c92487c92ce1f
73b9
[root@rhel-instance ~]# buildah commit $microcontainer
ubi-micro-httpd
Getting image source signatures Copying blob 5f70bf18a086
skipped: already exists
Copying blob a5198d17d541 skipped: already exists
Copying blob 80a8684cd0a5 done
Copying config c857c77f14 done
Writing manifest to image destination
Storing signatures
c857c77f14674d79114010d80031b6fc5ae309f524d0ef10bf2f1d242e3
41202
```

You now have a new container image with httpd installed, built on UBI Micro. Only a minimal set of dependencies have been pulled in. Look at how small the image is:

```
[root@rhel-instance ~]# podman images
REPOSITORY                                  TAG        IMAGE
ID        CREATED        SIZE
localhost/ubi-micro-
httpd                                       latest     c857c7
7f1467 About a minute ago  26.3 MB
```

Buildah is a wonderful tool that gives you a lot of control over how builds are done. Now, we will move on to Skopeo.

Inspecting a remote container with Skopeo

Skopeo is specifically designed and built to work on remote container repositories. With the following command, you can easily remotely inspect the available tags for an image:

```
[root@rhel-instance ~]# skopeo inspect docker://registry.
access.redhat.com/ubi9/ubi
```

```
{
    "Name": "registry.access.redhat.com/ubi9/ubi",
    "Digest":
"sha256:aee6d39282dabc3374a01d4a81f97c6827cbcdcf155cadb5a42
966134205b05d",
    "RepoTags": [
        "9.0.0-1468",
        "9.0.0-1571",
        "9.0.0-1468.1655190709",
... [output omitted] ...
```

Remote inspection is useful for determining whether you want to pull an image and, if so, with which tag. Skopeo can also be used to copy between two remote registry servers without caching a copy in the local storage. For more information, see the `skopeo` man pages.

Summary

In this chapter, we reviewed the basics of how to run, build, and share containers on RHEL 9. With this, you are prepared to create containers, run them, manage them, and even use `systemd` to ensure they're always running in a production environment.

You are now ready to leverage the functionality and ease of deployment that containers provide. While a deep dive into all of the intricacies of migrating software into containers is outside the scope of this book, containers simplify the packaging and delivery of applications ready to be executed with all of their dependencies.

Containers are now a strong focus within the IT industry. Containers alone simplify how applications are packaged and delivered, but orchestration platforms such as OpenShift (based on Kubernetes) make it easier to deploy, upgrade, and manage containerized applications at scale.

Congratulations – you have made it to the end of this chapter! Now, it's time to move on to the next chapter and take a self-assessment to ensure you've absorbed the material; this will help you practice your skills. There are two more chapters to go.

Part 4 – Practical Exercises

This part includes practical exercises to review what was learned in the previous parts. It includes an intermediate practice and a more advanced one, allowing you to assess your progress.

The following chapters are included in this part:

- *Chapter 18, Practice Exercises – 1*
- *Chapter 19, Practice Exercises – 2*

Practice Exercises – 1

In this practice exercise, we will run a set of steps to check the knowledge you have acquired throughout this book. As opposed to previous chapters, not all steps will be indicated, and it's therefore left to your discretion to perform the steps required to accomplish your desired goals. It is recommended to avoid referencing past chapters for guidance. Instead, try to use your memory or the tools available in the system. This exercise, if performed correctly, will train you effectively for official exams.

It is strongly advised to start this exercise with a clock to keep track of time.

Technical requirements

All the practice exercises in this chapter require the use of a virtual machine (VM), running **Red Hat Enterprise Linux 9** (**RHEL 9**) installed with the base installation. Additionally, new virtual drives will be required for storage operations.

For the exercises, it is assumed you possess the following:

- RHEL 9 installed with the base operating system using the **Minimal Install** software selection.
- Access to the Red Hat Customer Portal, with an active subscription.
- The VM must be expendable. This is because actions performed on it during the exercise might render it unusable and require it to be reinstalled.

Tips for the exercise

This is a list of general recommendations for any test, and most fall under the category of common sense, but it's always important to have them in our mind before performing any such test:

- Read all the questions before starting the official exam or any test.
- Specific words have specific meanings that give hints about the requirements or ways to accomplish the goals. This is why reading everything first might give you multiple perspectives on how to complete the test.

- Make yourself comfortable. Install your favorite editor and run `updatedb` to have a fresh database of packages and installed files ready for use. Define your keyboard layout. Install `tmux` and learn how to use it so that you can open new tabs and name them without requiring extra windows.

- Locate dependencies between requests, as some goals depend on others for completion. Find those dependencies to see how you can locate the solution without later having to come back and redo some steps as a result of choosing the wrong path.

- Use a timer. This is important to get an idea of which exercises will take you more time to complete, in order to see the areas that you need to improve in.

- Don't remember specific command lines. Learn how to use the documentation available in the system via `man`, `/usr/share/docs`, or arguments such as `--help` for the commands required.

- Ensure that changes persist and are still active after a reboot. Some changes might be active while you run them, but those must be persisted. Examples may include firewall rules, services to start at boot, and so on.

- Remember to use `dnf whatprovides /path/COMMAND` to find packages providing a file you might be missing.

- Check the following link: `https://www.redhat.com/en/services/training/ex200-red-hat-certified-system-administrator-rhcsa-exam?=Objectives`. This will provide you with the official *EX200* exam objectives.

Practice exercise 1

> **Important Note**
> The following exercise has, by design, been created so that there will be no highlights on commands, packages, and so on. Remember what you've learned so far in order to detect the keywords to see what needs to be done.

Don't jump into the walk-through too early. Try to remember what was covered.

Exercises

1. Configure the time zone to GMT.
2. Allow passwordless login for the `root` user using SSH.
3. Create a user (named `user`) that can connect to the machine without a password.
4. The `user` user should change their password every week, with 2 days' warning and 1 day of usage once expired.

5. The `root` user must be able to SSH as `user` without a password so that nobody can connect remotely as `root` using a password.

6. The `user` user should be able to become the `root` user without a password and also execute commands without a password.

7. When a user tries to log in over SSH, display a legal message about not allowing unauthorized access to this system.

8. SSH must listen on port `22222` instead of the default one (`22`).

9. Create a group named `devel`.

10. Make `user` a member of `devel`.

11. Store user membership in a file named `userids` in the home folder for `user`.

12. The `user` user and `root` user should be able to connect to `localhost` via SSH without specifying the port, and default to compression for the connection.

13. Find all man page names in the system, and put the names into a file named `manpages.txt`.

14. Print usernames for users without logins permitted to the system. For each username, print the **user ID (UID)** and groups for that user.

15. Monitor available system resources every 5 minutes. Do not use cron. Store as `/root/resources.log`.

16. Add a per-minute job to report the available percentage of free disk space, and store it in `/root/freespace.log` so that it shows both the filesystem and free space.

17. Configure the system to only leave 3 days of logs.

18. Configure the log rotation for `/root/freespace.log` and `/root/resources.log`.

19. Configure the time synchronization against `pool.ntp.org`, using Fast Sync.

20. Provide **Network Time Protocol (NTP)** server services for the `172.22.0.1/24` subnet.

21. Configure system stats for collection every minute.

22. Configure the password length for users in the system to be 12 characters long.

23. Create a bot user named `privacy` that will keep its files only visible to itself by default.

24. Create a folder named `/shared` that can be accessed by all users and that defaults new files and directories to still be accessible to users of the `devel` group.

25. Configure a network connection with IPv4 and IPv6 addresses named `mynic`, using the following data:

    ```
    Ip6: 2001:db8:0:1::c000:207/64 g
    gateway 2001:db8:0:1::1
    Ipv4 192.0.1.3/24
    gateway 192.0.1.1
    ```

26. Allow the host to use the hostname google to reach www.google.com, and the hostname redhat to reach www.redhat.com.

27. Report the files modified from those that the vendor distributed, and store them in /root/altered.txt.

28. Make our system installation media packages available via HTTP under the /mirror path for other systems to use as a mirror, configuring the repository in our system. Remove the kernel packages from that mirror so that other systems (even ours) can't find new kernels. Prevent the glibc packages from being installed from this repository without removing them.

29. As user, make a copy of the /root folder in the /home/user/root/ folder and keep it in sync every day, synchronizing additions and deletions.

30. Check that our system conforms to the **Payment Card Industry Data Security Standard (PCI-DSS)** standard.

31. Add a second hard drive of 30 GB to the system. However, use only 15 GB to move the mirror to it, making it available at boot using compression and deduplication. Make it available under /mirror/mirror.

32. Create a second copy of the mirror under /mirror/mytailormirror, removing all packages starting with the letter k*.

33. Create a new volume in the remaining space of the added hard drive (15 GB), and use it to extend the root filesystem.

34. Create a boot entry that allows you to boot into emergency mode in order to change the root password.

35. Create a custom tuning profile that defines the readahead to be 4096 for the first drive and 1024 for the second drive. This profile should also crash the system should an **out-of-memory (OOM)** event occur.

36. Disable and remove the installed httpd package. Then, set up the httpd server using the registry.redhat.io/rhel9/httpd-24 image.

For this section, we'll copy each item in the list of goals, and then provide an explanation below it, using proper syntax highlighting and explanations.

Exercise 1 solution

1. Configuring the time zone to GMT

We can check the current system date by executing the date command. At the very last part of the line that is subsequently printed, the time zone will be shown. In order to configure it, we can use the timedatectl command, or alter the /etc/localtime symbolic link.

So, to achieve this goal, we can use one of the following:

- `timedatectl set-timezone GMT`
- `rm -fv /etc/localtime; ln -s /usr/share/zoneinfo/GMT /etc/localtime`

Now, `date` should report the proper time zone.

2. Allowing passwordless login to the root user using SSH

Doing this will require the following:

- SSH must be installed and available (that means installed and started).
- The `root` user should have an SSH key generated and added to the list of authorized keys.

First, let's tackle this with SSH, as seen in the following code snippet:

```
dnf -y install openssh-server; systemctl enable sshd; systemctl start sshd
```

Now, let's generate an SSH key by pressing *Enter* to accept all defaults, like so:

```
ssh-keygen
```

Now, let's add the generated key (`/root/.ssh/id_rsa`) to the authorized keys, like so:

```
cd; cd .ssh; cat id_rsa.pub >> authorized_keys; chmod 600 authorized_keys
```

To validate this, we can execute `ssh localhost date`, after which we will be able to get our current system's date and time without providing a password.

3. Creating a user named user that can connect to the machine without a password

This requires creating a user and an SSH key that is added in a similar fashion to the `root` user. The next options will also be relevant to a user, but for the purposes of this demonstration, we will tackle them as separate tasks:

```
useradd user
su - user
```

Now, let's generate an SSH key by pressing *Enter* to accept all defaults, like so:

```
ssh-keygen
```

Now, let's add the generated key (/root/.ssh/id_rsa) to the authorized keys, as follows:

```
cd; cd .ssh; cat id_rsa.pub >> authorized_keys; chmod 600
authorized_keys
```

To validate this, we can execute ssh localhost date, and we will be able to get the current system date and time without providing a password.

Then, use logout to return to our root user.

4. The user user should change their password every week, with 2 days' warning and 1 day of usage once expired

This requires us to tune the user restrictions, as follows:

```
chage -W 2 user
chage -I 1 user
chage -M 7 user
```

5. The root user must be able to SSH as user without a password so that nobody can connect remotely as the root user using a password

This requires two steps. The first is to enable user with the root user's authorized key, and then tune the sshd daemon, as follows:

```
cat /root/.ssh/id_rsa.pub >> ~user/.ssh/authorized_keys
```

Edit the /etc/ssh/sshd_config file and add or replace the PermitRootLogin line so that it looks like this:

```
PermitRootLogin prohibit-password
```

Save and then restart the sshd daemon, like so:

```
systemctl restart sshd
```

6. The user user should be able to become root and also execute commands without a password

This means configuring the /etc/sudoers file by adding the following line:

```
user ALL=(ALL) NOPASSWD:ALL
```

7. When a user tries to log in over SSH, display a legal message about not allowing unauthorized access to this system

Create a file—for example, /etc/ssh/banner—with the message to display: for example, "Get out of here".

Modify /etc/ssh/sshd_config and set the Banner /etc/ssh/banner line, then restart the sshd daemon with systemctl restart sshd.

8. SSH must listen on port 22222 instead of the default one

This is a tricky one. The first step is to alter /etc/ssh/sshd_config and define port 22222. Once this is done, restart sshd with the following command:

```
systemctl restart sshd
```

This, of course, will fail... why?

The firewall must be configured, like this:

```
firewall-cmd --add-port=22222/tcp --permanent
firewall-cmd --add-port=22222/tcp
```

SELinux must then be configured, like so:

```
semanage port -a -t ssh_port_t -p tcp 22222
```

Now, the sshd daemon can be restarted by running the following command:

```
systemctl restart sshd
```

9. Creating a group named devel

Use the following command:

```
groupadd devel
```

10. Making user a member of devel

Use the following command:

```
usermod -G devel user
```

11. Storing user membership in a file called userids in the home folder for user

Use the following command:

```
id user > ~user/userids
```

12. The user user and root user should be able to connect to the localhost via SSH without specifying the port, and default to compression for the connection

We altered the default SSH port to be 22222.

Create a file named .ssh/config for both user and root, with the following contents:

```
Host localhost
Port 22222
    Compression yes
```

13. Finding all man page names in the system, and putting the names into a file named manpages.txt

Man pages are stored in /usr/share/man. Therefore, use the following command:

```
find  /usr/share/man/ -type f > manpages.txt
```

14. Printing usernames for users without a login so that they can be permitted access to the system, and printing the UID and groups for each user

The following command first builds a list of users in the system with the nologin shell:

```
for user in $(cat /etc/passwd| grep nologin|cut -d ":" -f 1)
do
echo "$user -- $(grep $user /etc/group|cut -d ":" -f 1|xargs)"
done
```

From that list, check the membership in the /etc/group file, leaving only the group name and using xargs to concatenate them into a string to be printed.

The previous example makes use of for loops and inline execution of commands, via $ ().

15. Monitoring available system resources every 5 minutes without using cron, and storing them as /root/resources.log

The ideal way to monitor something would be cron, but as we're told not to use it, this only leaves us with systemd timers. (You can check the files tested via the following link: https://github. com/PacktPublishing/Red-Hat-Enterprise-Linux-RHEL-9-Administration/ tree/main/chapter-18-exercise1.)

Create a /etc/systemd/system/monitorresources.service file with the following contents:

```
[Unit]
Description=Monitor system resources
[Service]
Type=oneshot
ExecStart=/root/myresources.sh
```

Create a /etc/systemd/system/monitorresources.timer file with the following contents:

```
[Unit]
Description=Monitor system resources
[Timer]
OnCalendar=*-*-* *:0,5,10,15,20,25,30,35,40,45,50,55:00
Persistent=true
[Install]
WantedBy=timers.target
```

Create a /root/myresources.sh file with the following contents:

```
#!/bin/bash
df > /root/resources.log
```

Enable the new timer, as follows:

```
systemctl daemon-reload
systemctl enable  --now monitorresources.timer
```

Does it work? If not, `journalctl -f` will give some details. SELinux prevents us from executing a `root` file, so let's convert it into a binary type and mark it as executable, as shown in the following snippet:

```
chcon -t bin_t /root/myresources.sh
chmod +x /root/myresources.sh
```

16. Adding a per-minute job to report the available percentage of free disk space and storing it in /root/freespace.log so that it shows the filesystem and free space

`df` reports used disk space and available space, so we need to do some math.

This will report the mounted location, size, used space, and available space, with `;` as a separator. Refer to the following example:

```
df|awk '{print $6";"$2";"$3";"$4}'
```

`bash` allows us to do some math operations, but these lack fractional parts. Luckily, we can do a trick—we will be looping over it, as follows:

```
for each in $(df|awk '{print $6";"$2";"$3";"$4}'|grep -v
"Mounted")
do
    FREE=$(echo $each|cut -d ";" -f 4)
    TOTAL=$(echo $each|cut -d ";" -f 2)
    echo "$each has $((FREE*100/TOTAL)) free"
done
```

The `for` loop will check all the available data, grab some specific fields, separate them with `;`, and then run the loop for each line, stored in the `$each` variable.

We cut the output and then get the fourth field. This is the available space.

We cut the output and then we get the second field. This is the total number of blocks.

As `bash` can do integer divisions, we can multiply by 100 and then divide to get the percentage and add a string as part of the output.

Alternatively (but not as illustrative), we could have discounted to 100 the percentage used already given by `df` and saved some steps of the calculation.

We also need to store the output in a file. To do this, we can either wrap the whole loop in a redirection or add it in the `echo` line so that it appends to a file.

And we also need to do it via cron—we show the full solution next.

Create a /root/myfreespace.sh script with the following contents:

```
for each in $(df|awk '{print $6";"$2";"$3";"$4}'|grep -v
"Mounted")
do
    FREE=$(echo $each|cut -d ";" -f 4)
    TOTAL=$(echo $each|cut -d ";" -f 2)
    echo "$each has $((FREE*100/TOTAL)) free"
done
```

Then, use chmod 755 /root/myfreespace.sh to make it executable.

Run crontab -e to edit root's crontab, and add the following line:

```
*/1 * * * * /root/myfreespace.sh >> /root/freespace.log
```

17. Configuring the system to only leave 3 days of logs

This can be done by editing /etc/logrorate.conf, with the following settings:

```
daily
rotate 3
```

Remove other occurrences of weekly, monthly, and so on, to leave only the one we want.

18. Configuring the log rotation for /root/freespace.log and /root/resources.log

Create a /etc/logrotate.d/rotateroot file, with the following contents:

```
/root/freespace.log {
    missingok
    notifempty
    sharedscripts
    copytruncate
}
/root/resources.log {
    missingok
    notifempty
    sharedscripts
```

```
    copytruncate
}
```

19. Configuring the time synchronization against pool.ntp.org with fast sync

Edit /etc/chrony.conf and add the following line:

```
pool pool.ntp.org iburst
```

Then, run the following command:

```
systemctl restart chronyd
```

20. Providing NTP server services for the 172.22.0.1/24 subnet

Edit /etc/chrony.conf by adding the following line:

```
Allow 172.22.0.1/24
```

Then, run the following command:

```
systemctl restart chronyd
```

21. Configuring system stats for collection every minute

Run the following command:

```
dnf -y install sysstat
```

We now need to modify /usr/lib/systemd/system/sysstat-collect.timer. Let's do this by creating an override, as follows:

```
cp /usr/lib/systemd/system/sysstat-collect.timer /etc/systemd/system/
```

Edit /etc/systemd/system/sysstat-collect.timer by replacing the OnCalendar value so that it looks like this:

```
OnCalendar=*:00/1
```

Then, reload the units with the following command:

```
systemctl daemon-reload
```

22. Configuring the password length for users in the system to be 12 characters long

Edit /etc/security/pwquality.conf with the following line:

```
minlen =  12
```

23. Creating a bot user named privacy that keeps its files only visible to itself by default

To do this, run the following:

```
adduser privacy
su - privacy
echo "umask 0077" >> .bashrc
```

This resolution uses umask to remove permissions from others on all newly created files.

24. Creating a folder named /shared that can be accessed by all users and defaults new files and directories to still be accessible to users of the devel group

To do this, run the following:

```
mkdir /shared
chown root:devel /shared
chmod 777 /shared
chmod +s /shared
```

25. Configuring a network connection with IPv4 and IPv6 addressing named mynic, using the following data: 2001:db8:0:1::c000:207/64 g gateway 2001:db8:0:1::1 ipv4 192.0.1.3/24 gateway 192.0.1.1

See the following snippet for how to accomplish this:

```
nmcli con add con-name mynic type ethernet ifname eth0 ipv6.
address 2001:db8:0:1::c000:207/64 ipv6.gateway 2001:db8:0:1::1
ipv4.address 192.0.1.3/24 ipv4.gateway 192.0.1.1
```

26. Allowing the host to use the hostname google to reach www. google.com, and the hostname redhat to reach www.redhat.com

Run and record the IPs obtained, as shown here:

```
ping www.google.com
ping www.redhat.com
```

Note down the IPs obtained previously.

Edit /etc/hosts by adding the following:

```
IPFORGOOGLE google
IPFORREDHAT redhat
```

Then, save and exit.

27. Reporting the files modified from those that the vendor distributed, and storing them in /root/altered.txt

See the following snippet for how to accomplish this:

```
rpm  -Va > /root/altered.txt
```

28. Making our system installation media packages available via HTTP under the /mirror path for other systems to use as a mirror, and configuring the repository in our system. Removing the kernel packages from that mirror so that other systems (even ours) can't find new kernels. Ignoring the glibc packages from this repository to be installed without removing them

This is a complex one, so let's examine it step by step.

Install http and enable it using the following code:

```
dnf -y install httpd
firewall-cmd  --add-service=http -permanent
firewall-cmd  --add-service=http
systemctl start httpd
systemctl enable httpd
```

Create a folder under /mirror, then copy the source media packages and make them available over http, like so:

```
mkdir /mirror /var/www/html/mirror
mount /dev/cdrom /mnt
rsync -avr -progress /mnt/ /mirror/
mount -o bind /mirror /var/www/html/mirror
chcon  -R -t httpd_sys_content_t /var/www/html/mirror/
```

Remove the kernel packages, as follows:

```
find /mirror -name kernel* -exec rm '{}' \;
```

Create repository file metadata by using the following commands:

```
dnf -y install createrepo
cd /mirror
createrepo .
```

Create a repository file using the repository we created, and set it up on the system, ignoring the glibc* packages from it.

Edit /etc/yum.repos.d/mymirror.repo by adding the following contents:

```
[mymirror]
name=My RHEL9 Mirror
baseurl=http://localhost/mirror/
enabled=1
gpgcheck=0
exclude=glibc*
```

29. As user, make a copy of the /root folder in the /home/user/root/ folder and keep it in sync every day, synchronizing additions and deletions

See the following snippet for how to accomplish this:

```
su - user
crontab -e
```

Edit crontab and add the following line:

```
@daily rsync  -avr --progress --delete root@localhost:/root/ /
home/user/root/
```

30. Checking whether our system conforms to the PCI-DSS standard

See the following snippet for how to accomplish this:

```
dnf -y install openscap  scap-security-guide openscap-utils
oscap xccdf eval --report pci-dss-report.html --profile pci-dss
/usr/share/xml/scap/ssg/content/ssg-rhel9-ds.xml
```

31. Adding a second hard drive of 30 GB to the system, but using only 15 GB to move the mirror to it, making it available at boot using compression and deduplication, and available under /mirror/mirror

Compression and deduplication in this sentence mean **Virtual Data Optimizer** (**VDO**). We need to move the mirror we currently have to it and make the old mirror, we had, go there instead.

If we have the installation media, we can choose to copy it over and repeat the kernel removal or transfer. To do so, let's first create a VDO volume in a partition in our new hard drive (sdb), like so:

```
fdisk /dev/sdb
n <enter>
p <enter>
1 <enter>
<enter>
+15G <enter>
w <enter>
q <enter>
```

This will create a partition of 15 GB from the start. Let's create a VDO volume on it, by using the following command:

```
dnf -y install vdo kmod-kvdo
pvcreate /dev/sdb1
vgcreate myvdo /dev/sdb1
lvcreate --type vdo --name myvdo vdo  -L 15G
```

```
mkfs.xfs /dev/myvdo/vdo
# Let's umount cdrom if it was still mounted
umount /mnt
# Mount vdo under /mnt and copy files over
mount /dev/myvdo/vdo/mnt
rsync -avr -progress /mirror/ /mnt/mirror/
# Delete the original mirror once copy has finished
rm -Rfv /mirror
umount /mnt
mount /dev/myvdo/vdomyvol /mirror
```

At this point, the old mirror was copied into a `mirror` folder on the VDO volume. This is mounted under `/mirror`, hence it has the original mirror under `/mirror/mirror` as requested. We might need to perform the following actions:

- Bind the `/mirror` mount to `/var/www/html/mirror/` to make the file available.
- Restore SELinux context to allow the `httpd` daemon to access files in `/var/www/html/mirror/`.

Adjust the repository file we created to point to the new path.

32. Creating a second copy of the mirror under /mirror/ mytailormirror and removing all packages starting with k*

See the following snippet for how to accomplish this:

```
rsync -avr -progress /mirror/mirror/ /mirror/mytailormirror/
find /mirror/mytailormirror/ -name "k*" -type f -exec rm '{}'
\;
cd /mirror/mytailormirror/
createrepo .
```

33. Creating a new volume in the remaining space (15 GB) of the hard drive and using it to extend the root filesystem

See the following snippet for how to accomplish this:

```
fdisk /dev/sdb
n <enter>
p <enter>
```

```
<enter>
<enter>
w <enter>
q <enter>
pvcreate /dev/sdb2
# run vgscan to find out the volume name to use (avoid myvdo as
is the VDO from above)
vgextend $MYROOTVG /dev/sdb2
# run lvscan to find out the LV storing the root filesystem and
pvscan to find the maximum available space
lvresize -L +15G /dev/rhel/root
```

34. Creating a boot entry that allows us to boot into emergency mode in order to change the root password

See the following snippet for how to accomplish this:

```
grubby --args="systemd.unit=emergency.target" --update-kernel=/
boot/vmlinuz-$(uname -r)
```

35. Creating a custom tuning profile that defines the readahead to be 4096 for the first drive and 1024 for the second drive – this profile should also crash the system should an OOM event occur

Refer to the following command:

```
dnf -y install tuned
mkdir -p /etc/tuned/myprofile
```

Edit the /etc/tuned/myprofile/tuned.conf file by adding the following contents:

```
[main]
summary=My custom tuned profile
[sysctl]
vm.panic_on_oom=1
[main_disk]
type=disk
devices=sda
readahead=>4096
[data_disk]
```

```
type=disk
devices=!sda
readahead=>1024
```

36. Disabling and removing the installed httpd package, and setting up the httpd server using the registry.redhat.io/rhel9/httpd-24 image

See the following snippet for how to accomplish this:

```
dnf remove -y httpd
dnf -y install podman
podman login registry.redhat.io # provide RHN credentials
podman pull registry.redhat.io/rhel9/httpd-24
mkdir /var/www #only if it doesn't exist
podman run -d --name httpd -p 80:8080 -v /var/www:/var/www:Z
registry.redhat.io/rhel9/httpd-24
```

19
Practice Exercise – 2

In this second practice exercise chapter, we will run a set of exercises to check the knowledge you've acquired throughout this book. In contrast with this book's chapters, not all the steps will be specified; it's left up to your discretion to perform the steps required to accomplish the necessary goals. It is recommended that you avoid checking back on the chapters for guidance and instead try to use your memory or the tools available in the system. This experience will be a key factor when you take on the official exams.

It is strongly advised that you time yourself during this exercise so that you know how long it took for you to complete it.

Technical requirements

All the practical exercises in this chapter require a **virtual machine** (**VM**) running **Red Hat Enterprise Linux 9** (**RHEL 9**) to be installed with the base installation. Additionally, new virtual drives will be required for storage operations.

The exercises assume that you have the following:

- RHEL 9 installed with a base operating system **Minimal Install** software selection.

- Access to the Red Hat Customer Portal with an active subscription.

- The VM must be expendable; that is, actions you perform on it might render it unusable, so it will have to be reinstalled.

Tips for the exercise

This is a list of general recommendations for any test, most of which are common sense, but it's always interesting to keep them in mind:

- Read the questions in their entirety before starting the exam.

- Specific words have specific meanings that give hints about the requirements or ways to accomplish the exercise. That's why, again, reading everything first might add or remove possibilities.

- Make yourself comfortable: install your favorite editor, run `updatedb` so that you have a fresh database of packages and files ready for you, and define your keyboard layout. Install and learn the basics of how to use `tmux` so that you can open new tabs and name them without requiring extra windows.

- Locate dependencies between requests. Some goals depend on others for completion, so find those dependencies to see how you can build up the solution without having to go back and redo some steps because of taking the wrong path.

- Use a timer. It's important to get an idea of which exercises took more time to complete so that you can find areas to improve upon.

- Don't remember specific commands. Instead, learn how to use the documentation available in the system by using `man`, `/usr/share/docs` arguments such as `--help` for the commands, and so on.

- Ensure that changes persist and are still active after a reboot. Some changes might be active while you run them, but those must be persisted: firewall rules, services to start at boot, and so on.

- Remember that you can use `dnf whatprovides "*/COMMAND"` to find packages regarding a file you might be missing.

- Check `https://www.redhat.com/en/services/training/ex200-red-hat-certified-system-administrator-rhcsa-exam?=Objectives` for the official *EX200* exam objectives.

Practice exercise 2

> **Important note**
> By design, in the following exercise, commands, packages, and so on will not be highlighted. Remember what you've learned so far to detect the keywords to see what needs to be done.

Don't jump into the solution too early; try to think about it and remember what was covered.

Exercises

1. Download the necessary file from this book's GitHub repository at `https://raw.githubusercontent.com/PacktPublishing/Red-Hat-Enterprise-Linux-RHEL-9-Administration/main/chapter-19-exercise2/users.txt`.

2. Use the `users.txt` file to generate users in the system in an automated way using the values provided, in the following order: `username`, `placeholder`, `uid`, `gid`, `name`, `home`, `shell`.

3. Create a group named `users` and add that group as the primary group to all users, leaving their own groups, named after each user, as secondary groups.

4. Change the home folders for the users so that they are group-owned.

5. Set up an HTTP server and enable a web page for each user, with a small introduction for each web page that is different between users.

6. Allow all users in the `users` group to become `root` without a password.

7. Create SSH keys for each user and add each key to `root` and the other users so that each user can SSH like the other users; that is, without a password.

8. Disable password access to the system with SSH.

9. Set each user with a different password using `/dev/random` and store the password in the `users.txt` file in the second field of the file.

10. If the number of letters in the username is a multiple of 2, add that fact to each user description web page.

11. Create a container that runs the `yq` Python package as the entry point.

12. Configure password aging for users that are not a multiple of 2 so that they're expiring.

13. Configure a daily compressed log rotation for a month of logs using date-named files.

14. Save all logs generated in the day in `/root/errors.log`.

15. Install all available updates for system libraries.

16. Repair the broken `rpm` binary using a previously downloaded package available in the `/root` folder.

17. Make all processes that are executed by the user `doe` run with a low priority and the ones from `john` run with a higher priority (+/- 5).

18. Make the system run with the highest throughput and performance.

19. Change the system network interface so that it uses an IP address that's higher than the one it was using. Add another IPv6 address to the same interface.

20. Create and add `/opt/mysystem/bin/` to the system `PATH` variable for all users.

21. Create a firewall zone, assign it to an interface, and make it the default zone.

22. Add a repository hosted at `https://myserver.com/repo/` with a **GNU Privacy Guard** (**GPG**) key from `https://myserver.com/mygpg.key` to the system since our server might be down. Configure it so that it can be skipped if it's unavailable.

Exercise 2 resolution

In this section, we'll copy each item from the list of goals, and explain them while using proper syntax highlighting.

1. Downloading the necessary file from this book's GitHub repository at https://raw.githubusercontent. com/PacktPublishing/Red-Hat-Enterprise-Linux-RHEL-9-Administration/main/chapter-19-exercise2/users.txt

Here's how to do this:

```
wget https://raw.githubusercontent.com/PacktPublishing/
Red-Hat-Enterprise-Linux-RHEL-9-Administration/main/chapter-19-
exercise2/users.txt
```

2. Using the users.txt file to generate users in the system in an automated way using the values provided, in the following order: username, placeholder, uid, gid, name, home, shell

First, let's examine the users.txt file with the following code:

```
cat users.txt
user;x;1000;1000;myuser1;/home/user1; /bin/false
john ;x ;1001 ;1001; John; /home/john ;/bin/false
doe ;x ;1002 ;1002; Doe; /home/doe ; /bin/sh
athena ;x ;1011 ;1011; Athena Jones; /home/ajones ; /bin/sh
pilgrim ;x ;2012 ;2012; Scott Pilgrim; /home/spilgrim ; /bin/sh
laverne; x ; 2020;2020; LaVerne;/home/LaVerne;/bin/bash
```

As described in the request, the fields in that file are username, placeholder, uid, gid, name, home, and shell. The placeholder is not asked to create a user as it's usually the password so that we can work with the other data while ignoring that.

As we can also see, each field is separated by at least a ; symbol, but some have extra spaces before or after them. Since we also have surnames, we can't just remove all spaces; we need to do this before and after the actual text we want.

We need to use cut with the ; field separator, but first, we need to read the file line by line.

We can achieve this with bash's built-in read function:

```
cat users.txt|while read -r line; do echo ${line};done
```

Using this as a base, we can start building up everything we're going to need to create users. Let's start by working on the individual steps and then build up the full command line.

We have lots of lines, so for each one, we need to define the fields and remove end/start spaces:

```
NEWUSERNAME=$(echo ${line}|cut -d ";" -f 1)
NEWUID=$(echo ${line}|cut -d ";" -f 3)
NEWGID=$(echo ${line}|cut -d ";" -f 4)
NEWNAME=$(echo ${line}|cut -d ";" -f 5)
NEWSHELL=$(echo ${line}|cut -d ";" -f 6)
```

In the preceding examples, we're echoing each line and cutting the field specified with -f using the ; field delimiter. This allows us to select exactly the field containing the data we're looking for. To make this easier, we can store each in a variable so that we can reuse that snippet of code and still have a clear understanding of what each script will be doing.

The preceding code will work, *but* it will fail with the spaces, so we need to extend them to just capture the actual text without the spaces. Let's use xargs for this:

```
NEWUSERNAME=$(echo ${line}|cut -d ";" -f 1|xargs)
NEWUID=$(echo ${line}|cut -d ";" -f 3|xargs)
NEWGID=$(echo ${line}|cut -d ";" -f 4|xargs)
NEWNAME=$(echo ${line}|cut -d ";" -f 5|xargs)
NEWHOME=$(echo ${line}|cut -d ";" -f 6|xargs)
NEWSHELL=$(echo ${line}|cut -d ";" -f 7|xargs)
```

Groups need to be created; for that, we are going to use NEWUSERNAME as the group name and NEWGID as the **group ID (GID)**:

```
groupadd -g "${NEWGID}" "${NEWUSERNAME}"
```

The next step is to build the command line for adding a user:

```
useradd --d "${NEWHOME}" --m --s "${NEWSHELL}" --u "${NEWUID}"
--g "${NEWGID}" --c "${NEWNAME}" "${NEWUSERNAME}"
```

Now that everything's ready, let's build the solution:

```
cat users.txt| while read -r line ; do
NEWUSERNAME=$(echo ${line}|cut -d ";" -f 1|xargs)
NEWUID=$(echo ${line}|cut -d ";" -f 3|xargs)
NEWGID=$(echo ${line}|cut -d ";" -f 4|xargs)
NEWNAME=$(echo ${line}|cut -d ";" -f 5|xargs)
NEWHOME=$(echo ${line}|cut -d ";" -f 6|xargs)
NEWSHELL=$(echo ${line}|cut -d ";" -f 7|xargs)
```

```
groupadd -g "${NEWGID}" "${NEWUSERNAME}"
useradd -d "${NEWHOME}" -m -s "${NEWSHELL}" -u "${NEWUID}" -g
"${NEWGID}" -c "${NEWNAME}" "${NEWUSERNAME}"
done
```

> **Note**
>
> Depending on the system that you are using to run this code it could happen that the **user IDs (UIDs)** or GIDs already exist and users or groups are not created. In that case, you can change the number in the second or third column of the affected user or group in the users.txt file or just ignore the error and keep doing the following exercises.

3. Creating a group named myusers and adding that group as the primary group to all users, leaving their own groups, named after each user, as secondary groups

In this case, we need to create a group called users and then we will loop over the users once the new group has been created, and then modify the users group to get the myusers group and add their own as secondary groups:

```
groupadd myusers
cat users.txt| while read -r line ; do
NEWUSERNAME=$(echo ${line}|cut -d ";" -f 1|xargs)
usermod -g myusers -G "${NEWUSERNAME}" "${NEWUSERNAME}"
done
```

4. Changing the home folders for the users so that they are group-owned

Here's how to do this:

```
cat users.txt| while read -r line ; do
NEWUSERNAME=$(echo ${line}|cut -d ";" -f 1|xargs)
NEWHOME=$(echo ${line}|cut -d ";" -f 6|xargs)
chown -R ${NEWUSERNAME}:myusers ${NEWHOME}/
done
```

5. Setting up an HTTP server and enabling a web page for each user, with a small introduction for each that is different between users

Here's how to do this:

```
dnf -y install httpd
firewall-cmd --add-service=http
firewall-cmd --runtime-to-permanent
setsebool httpd_enable_homedirs on

cat users.txt| while read -r line ; do
NEWUSERNAME=$(echo ${line}|cut -d ";" -f 1|xargs)
NEWUID=$(echo ${line}|cut -d ";" -f 3|xargs)
NEWGID=$(echo ${line}|cut -d ";" -f 4|xargs)
NEWNAME=$(echo ${line}|cut -d ";" -f 5|xargs)
NEWHOME=$(echo ${line}|cut -d ";" -f 6|xargs)
NEWSHELL=$(echo ${line}|cut -d ";" -f 7|xargs)
mkdir -p "${NEWHOME}"/public_html/
chown "${NEWUSERNAME}:${NEWUSERNAME}" "${NEWHOME}"/public_html/
chmod 755 "${NEWHOME}"/public_html/
chmod 711 "${NEWHOME}"
echo "Hello, my name is ${NEWNAME} and I'm a user of this
system" > ${NEWHOME}/public_html/index.html
done
```

Finally, we'll need to enable homedirs by editing /etc/httpd/conf.d/userdir.conf and modifying UserDir so that it becomes Userdir public_html:

```
cd /etc/httpd/conf.d
sed -i 's/UserDir disabled/UserDir enabled/' userdir.conf
sed -i 's/#UserDir public_html/UserDir public_html/' userdir.
conf
systemctl restart httpd
curl http://localhost/~john/
Hello, my name is John and I'm a user of this system
```

6. Allowing all users in the users group to become root without a password

This can be done in several ways, but since all the users are in the users group, we can add that group, like so:

```
echo "%users ALL=(ALL) NOPASSWD: ALL" >> /etc/sudoers
```

7. Creating SSH keys for each user and adding each key to root and the other users so that each user can SSH like the other users; that is, without a password

First, let's create keys for each user and add the keys to root:

```
mkdir /root/.ssh
touch /root/.ssh/authorized_keys
cat users.txt| while read -r line ; do
NEWHOME=$(echo ${line}|cut -d ";" -f 6|xargs)
mkdir -p ${NEWHOME}/.ssh/
ssh-keygen -N '' -f ${NEWHOME}/.ssh/id_rsa
cat ${NEWHOME}/.ssh/id_dsa.pub >> /root/.ssh/authorized_keys
done
```

Now, let's copy the authorized keys for each user:

```
cat users.txt| while read -r line ; do
NEWUSERNAME=$(echo ${line}|cut -d ";" -f 1|xargs)
NEWHOME=$(echo ${line}|cut -d ";" -f 6|xargs)
cp /root/.ssh/authorized_keys ${NEWHOME}/.ssh/
chown -R ${NEWUSERNAME}:myusers ${NEWHOME}/.ssh/
done
```

Validate that users can SSH just like any other user:

```
USERS=$(cat users.txt|cut -d ";" -f1|xargs)
for user in ${USERS};
do
for userloop in ${USERS};
do
su -c "ssh ${user}@localhost" ${userloop}
```

```
done
done
```

The preceding command should work for all the users because we copied authorized_keys, right? This isn't the case as some users have their shell disabled. You will need to exit the SSH by writing the exit command after every user is running SSH with success.

8. Disabling password access to the system with SSH

Edit /etc/ssh/sshd_config and replace any value of PasswordAuthentication with no.

Then, restart sshd, like so:

```
systemctl restart sshd
```

9. Setting each user with a different password using /dev/random and storing the password in the users.txt file in the second field of the file

From /dev/random, we can get random data but it's binary, so it's probably not valid if we want to use it for logging in later. We can use a hash function over the data we've received and use that as the password, as follows:

```
MYPASS=$(dd if=/dev/urandom count=1024 2>&1|md5sum|awk '{print
$1}')
```

This will be the password, without the need for it to be encrypted.

With usermod, we can define a password from its encrypted seed, so we will be combining both.

Additionally, we're told to store the generated password in users.text, so we will need to edit the file.

But there's a problem: editing a specific field in the .txt file might not be easy, but we can just rewrite it completely:

```
cat users.txt| while read -r line ; do
MYPASS=$(dd if=/dev/random count=12>&1|md5sum|awk '{print $1}')
NEWUSERNAME=$(echo ${line}|cut -d ";" -f 1|xargs)
NEWUID=$(echo ${line}|cut -d ";" -f 3|xargs)
NEWGID=$(echo ${line}|cut -d ";" -f 4|xargs)
NEWNAME=$(echo ${line}|cut -d ";" -f 5|xargs)
NEWHOME=$(echo ${line}|cut -d ";" -f 6|xargs)
NEWSHELL=$(echo ${line}|cut -d ";" -f 7|xargs)
```

```
echo -e
"${NEWUSERNAME};${MYPASS};${NEWUID};${NEWGID};${NEWNAME};
${NEWHOME};${NEWSHELL}" >> newusers.txt
echo ${MYPASS} | passwd ${NEWUSERNAME} --stdin
done
cp users.txt users.txt.bkp
cp newusers.txt users.txt
```

In this way, we've rewritten the users.txt file to a new file by adding all the fields we had and have overwritten users.txt with our new copy.

The last command in the loop reads the password from the variable and feeds it to the passwd file, which will encrypt and store it while reading it from stdin.

10. If the number of letters in the username is a multiple of 2, adding that fact to each user description web page

Here's how to do this:

```
cat users.txt| while read -r line ; do
NEWUSERNAME=$(echo ${line}|cut -d ";" -f 1|xargs)
NEWHOME=$(echo ${line}|cut -d ";" -f 6|xargs)
LETTERSINNAME=$(( $(echo ${NEWUSERNAME}|wc -m) - 1 ))
if [ "$(({$LETTERSINNAME} % 2 ))" == "0" ]; then
echo "My name is multiple of 2" >> ${NEWHOME}/public_html/
index.htm
fi
done
```

In this example, we repeat the same field calculation, but we add the wc command to get the number of characters and remove one to adjust it to the number of letters.

In the comparison, we evaluate the remainder when dividing by 2 so that when there's no remainder, this means that our number of letters is a multiple of 2.

11. Creating a container that runs the yq Python package as the entry point

When we read "Python package", we should think about PIP. PIP is not recommended to be used on systems directly as it might alter the system-provided Python libraries, and it's better to use a virtual environment for it. Alternatively, you can use a container that will keep it isolated.

As described in *Chapter 17, Managing Containers with Podman, Buildah, and Skopeo*, the easiest way to do this is by creating a file that defines the container creation steps.

For containers, it will be also required to install the `podman` package and the `container-tools` modules if you don't have them in your system.

As this file is a Python package, we require a container that already has Python in it; for example, `https://catalog.redhat.com/software/containers/ubi9/python-39/61a61032bfd4a5234d59629e`.

So, let's create `ContainerFile` with the following contents (available at `https://github.com/PacktPublishing/Red-Hat-Enterprise-Linux-RHEL-9-Administration/blob/main/chapter-19-exercise2/ContainerFile`):

```
FROM registry.access.redhat.com/ubi9/python-39
MAINTAINER RHEL9 Student <student@redhat.com>
LABEL name="yq image" \
maintainer="student _AT_ redhat.com" \
vendor="Risu" \
version="1.0.0" \
release="1" \
summary="yq execution container" \
description="Runs yq"
ENV USER_NAME=risu \
USER_UID=10001 \
LC_ALL=en_US.utf8
RUN pip3 install --upgrade pip --no-cache-dir && \
pip3 install --upgrade yq --no-cache-dir
USER 10001
VOLUME /data
ENTRYPOINT ["/opt/app-root/bin/yq"]
CMD ["-h"]
```

When combined with `podman build -t yq -f ContainerFile`, it will pull the `ubi9` image with Python so that we can just run the `pip3 install` command to install yq, which will be then assigned as our entry point.

For example, if we define an invalid entry point (because we might not know where the program is installed), we can use `podman run -it --entrypoint /bin/bash <podmanid>`. We can get the `podman` ID by running `podman images` and checking the generation date for each of the available pods in our system.

The created container can be tested with `podman run -it <podmanid>`, where it will output information about what the `yq` command does.

Note that `yq`, as expressed in its repository at `https://github.com/kislyuk/yq`, requires that we have installed the `jq` command, but we left it out on purpose to demonstrate how to create a container.

12. Configuring password aging for users that are not a multiple of 2 so that they're expiring

Here's how to do this:

```
cat users.txt| while read -r line ; do
NEWUSERNAME=$(echo ${line}|cut -d ";" -f 1|xargs)
NEWHOME=$(echo ${line}|cut -d ";" -f 6|xargs)
LETTERSINNAME=$(( $(echo ${NEWUSERNAME}|wc -m) - 1 ))
if [ "$((${LETTERSINNAME} % 2 ))" != "0" ]; then
chage -M 30 ${NEWUSERNAME}
fi
done
```

Here, we've reused the loop from *question 10*, but inverted the conditional. Since there are no requirements regarding the kind of password aging we can use, we just need to define the maximum number of days before a password change is required to be 30 days.

13. Configuring a daily compressed log rotation for a month of logs using date-named files

First, we need to make sure that `logrotate` is installed. We can do this by running the following command:

```
dnf -y install logrotate
```

Once installed, edit the `/etc/logrotate.conf` file so that it contains the following:

```
rotate 30
daily
compress
dateext
```

We need to ensure that no other period is defined (monthly, weekly, and so on).

14. Saving all logs generated in the day in /root/errors.log

This has a trick to it: some programs will log to the journal, while some of them will log to `*.log` files.

The date for today can be obtained with `+%Y-%m-%d`, as follows. This format, which uses the year-month-day format, is commonly used in program logs:

```
grep "$(date '+%Y-%m-%d')" -Ri /var/log/*.log|grep -i error
>  /root/errors.log
journalctl --since "$(date '+%Y-%m-%d')" >> /root/errors.log
```

By doing this, we combine both outputs. We could, of course, try to sort the entries by date so that they correlate, but bear in mind that the first `grep` command does a recursive search, so the filename is being prepended, making it harder to sort.

15. Installing all available updates for system libraries

Usually, the system libraries contain the `lib` substring in them, so the update should be a matter of running the following command:

```
dnf upgrade *lib*
```

As it will ask for confirmation, review the listed packages to make sure that no errors occurred.

16. Repairing the broken rpm binary using a previously downloaded package available in the /root folder

This is a tricky but useful knowledge check.

First, let's make sure that the `rpm` package is available by running the following command:

```
dnf download rpm
```

Verify that the file exists with the following command:

```
ls -l rpm*.rpm
```

Check the file to make sure we have a way to go back in case we break it beyond repair:

```
rpm -qip rpm*.rpm
```

Now, let's look at the destructive action that will help us validate we are solving the issue:

```
rm -fv /usr/bin/rpm
```

From here, it's like *look ma, no hands...* no RPM is available to install the rpm*.rpm package, but we still need to install it to fix the issue.

rpm packages are compressed cpio archives, so what we can do is use the following command:

```
rpm2cpio rpm*.rpm |cpio -idv
```

This will extract the compressed rpm contents (without the need to run a script).

Move the uncompressed rpm file back into /usr/bin, like so:

```
mv usr/bin/rpm /usr/bin/rpm
```

Verify the installation and operation of rpm with the following command:

```
rpm -V rpm
```

It will complain, saying at least that the date has changed. However, it may have also updated the sizes and md5sum if the downloaded file was newer.

Move the system to a sane state by reinstalling the rpm package, like so:

```
rpm -i rpm*.rpm
```

This will make the system complain because the package was already installed (it will state that it will overwrite rpm, rpm2archive, rpm2cpio, rpmdb, rpmkeys, and more).

If the rpm version differs, we can just upgrade it with the following command:

```
rpm -Uvh rpm*.rpm
```

Then, we can verify this with the following command:

```
rpm -V rpm
```

Nothing should be reported as changed regarding what the database contains. If we get a timestamp modification or if we cannot upgrade, we can run the installation with the --force argument to tell rpm that it's OK to continue and overwrite the files.

Alternatively, once the rpm binary has been restored with cpio, we can use the following command:

```
dnf -y reinstall rpm
```

Another approach for this could have been to scp the rpm binary from a similar system or to use rescue media.

17. Making all processes that are executed by the user doe run with a low priority and the ones from john run with a higher priority (+/- 5)

We have no way of making this a default, but we can combine a cron job to do so.

Execute -e crontab as root to edit the root's crontab and set up a job that runs every minute, like so:

```
*/1 * * * *  pgrep -u doe |xargs renice +5
*/1 * * * * pgrep -u john|xargs renice -5
```

This will use pgrep for all **process IDs (PIDs)** for john and doe and feed them via xargs to the renice process.

Alternatively, we could use something like the following:

```
renice +5 $(pgrep -u doe)
```

This can be used as an alternative to the xargs command.

18. Making the system run with the highest throughput and performance

tuned is a system daemon we can install to automatically apply some well-known parameters to our system, which will become the base for our specific optimizations later. Here's how we can install it:

```
dnf -y install tuned
systemctl enable tuned
systemctl start tuned
tuned-adm profile throughput-performance
```

19. Changing the system network interface so that it uses an IP address that's higher than the one it was using and adding another IPv6 address to the same interface

Using nmcli, check the current system IP's address, like so:

```
nmcli con show
```

This should be the output:

```
[root@rhel-instance ~]# nmcli con show
NAME     UUID                                     TYPE       DEVICE
enp1s0   ed51018f-0061-3546-bc68-fdba76516d60     ethernet   enp1s0
[root@rhel-instance ~]#
```

Figure 19.1 – Output of nmcli con show

With this, we can find which system interface is being used and connected. Let's say it's enp1s0, which is connected on the connection named enp1s0.

Let's use nmcli con show "enp1s0"|grep address to find the current addresses.

If our address is—for example—10.0.0.6, we can use the following code:

```
nmcli con mod "Wired Connection" ipv4.addresses 10.0.0.7
nmcli con mod "Wired Connection" ipv6.addresses
2001:db8:0:1::c000:207
```

Verify this with the following command:

```
nmcli con show "Wired Connection"|grep address
```

20. Creating and adding /opt/mysystem/bin/ to the system PATH variable for all users

Edit the /etc/profile.d/mysystempath.sh file and place the following contents in it:

```
export PATH=${PATH}:/opt/mysystem/bin
```

To validate this, add the +x attribute to the file and create a folder with the following commands:

```
chmod +x /etc/profile.d/mysystempath.sh
mkdir -p /opt/mysystem/bin
```

Relogging with the user should show the new path when executing the following command:

```
echo ${PATH}
```

21. Creating a firewall zone, assigning it to an interface, and making it the default zone

This is a tricky question. In this book, we've explained how to query zones and how to change the default one, and even shown screenshots of `cockpit` for managing the firewall, so now that you're an experienced user, this shouldn't be hard.

The first thing you need to do when you don't know how to do something is check the manual page. Here's the command to run in this instance:

```
man firewall-cmd
```

This doesn't show a lot of interesting information. However, toward the end of the man pages, there's a section called *SEE ALSO*, where we can find out about `firewalld.zones(5)`. This means that we can check *Section 5* of the manual for `firewalld.zones`.

We don't usually specify the section as there might not be a lot of duplicates, so we can just run the following command:

```
man firewalld.zones
```

This instructs us to check the default ones in `/usr/lib/firewalld/zones` and `/etc/firewalld/zones`, so let's do that:

```
cp /usr/lib/firewalld/zones/public.xml /etc/firewalld/zones/
dazone.xml
```

Now, let's edit the new copied file, called `/etc/firewalld/zones/dazone.xml`, and change its name from `Public` to `dazone`. Then, we need to reload the firewall:

```
firewall-cmd --reload
```

Let's validate that the new zone is there with the following command:

```
firewall-cmd --get-zones
```

Let's make it the default zone:

```
firewall-cmd --set-default-zone=dazone
```

Now, add the default interface (`ens3`):

```
firewall-cmd --add-interface=ens3 --zone=dazone
```

It will fail. This is expected since ens3 has already been assigned to a zone (public). So, let's use the following commands:

```
firewall-cmd --remove-interface=ens3 --zone=public
firewall-cmd --add-interface=ens3 --zone=dazone
```

As you can see, even without prior knowledge about creating new zones, we've been able to use our system knowledge about finding information to accomplish this goal.

22. Adding a repository hosted at https://myserver.com/repo/ with the GPG key from https://myserver.com/mygpg.key to the system since our server might be down and configuring it so that it can be skipped if it's unavailable

If we don't remember the syntax for a repository, we can use one of the examples available on the system. To do this, go to /etc/yum.repos.d/, list the available files and pick one to create a myserver.repo file with the following contents:

```
[myserver]
name=My server repository
baseurl=https://myserver.com/repo/
enabled=1
gpgcheck=1
gpgkey=https://myserver.com/mygpg.key
```

How do we skip it if it's unavailable? Let's check the man page for yum. Again, not much information is provided here, but in the *SEE ALSO* section, man dnf.conf is specified. This lists a Boolean that might help us, so let's add this to our repository file:

```
skip_if_unavailable=1
```

With that, we've completed our objectives.

Index

`Packt.com`

Subscribe to our online digital library for full access to over 7,000 books and videos, as well as industry leading tools to help you plan your personal development and advance your career. For more information, please visit our website.

Why subscribe?

- Spend less time learning and more time coding with practical eBooks and Videos from over 4,000 industry professionals

- Improve your learning with Skill Plans built especially for you

- Get a free eBook or video every month

- Fully searchable for easy access to vital information

- Copy and paste, print, and bookmark content

Did you know that Packt offers eBook versions of every book published, with PDF and ePub files available? You can upgrade to the eBook version at `packt.com` and as a print book customer, you are entitled to a discount on the eBook copy. Get in touch with us at `customercare@packtpub.com` for more details.

At `www.packt.com`, you can also read a collection of free technical articles, sign up for a range of free newsletters, and receive exclusive discounts and offers on Packt books and eBooks.

Other Books You May Enjoy

If you enjoyed this book, you may be interested in these other books by Packt:

Linux Kernel Debugging

Kaiwan Billimoria

ISBN: 978-1-80107-503-9

- Explore instrumentation-based printk along with the powerful dynamic debug framework
- Use static and dynamic Kprobes to trap into kernel/module functions
- Catch kernel memory defects with KASAN, UBSAN, SLUB debug, and kmemleak
- Interpret an Oops in depth and precisely identify it's source location
- Understand data races and use KCSAN to catch evasive concurrency defects
- Leverage Ftrace and trace-cmd to trace the kernel flow in great detail
- Write a custom kernel panic handler and detect kernel lockups and hangs
- Use KGDB to single-step and debug kernel/module source code

Windows and Linux Penetration Testing from Scratch - Second Edition

Phil Bramwell

ISBN: 978-1-80181-512-3

- Get to know advanced pen testing techniques with Kali Linux
- Gain an understanding of Kali Linux tools and methods from behind the scenes
- Get to grips with the exploitation of Windows and Linux clients and servers
- Understand advanced Windows concepts and protection and bypass them with
- Kali and living-off-the-land methods
- Get the hang of sophisticated attack frameworks such as Metasploit and Empire
- Become adept in generating and analyzing shellcode
- Build and tweak attack scripts and modules

Packt is searching for authors like you

If you're interested in becoming an author for Packt, please visit `authors.packtpub.com` and apply today. We have worked with thousands of developers and tech professionals, just like you, to help them share their insight with the global tech community. You can make a general application, apply for a specific hot topic that we are recruiting an author for, or submit your own idea.

Share your thoughts

Now you've finished *Red Hat Enterprise Linux 9 Administration*, we'd love to hear your thoughts! Scan the QR code below to go straight to the Amazon review page for this book and share your feedback or leave a review on the site that you purchased it from.

`https://packt.link/r/1803248807`

Your review is important to us and the tech community and will help us make sure we're delivering excellent quality content.

Download a free PDF copy of this book

Thanks for purchasing this book!

Do you like to read on the go but are unable to carry your print books everywhere? Is your eBook purchase not compatible with the device of your choice?

Don't worry, now with every Packt book you get a DRM-free PDF version of that book at no cost.

Read anywhere, any place, on any device. Search, copy, and paste code from your favorite technical books directly into your application.

The perks don't stop there, you can get exclusive access to discounts, newsletters, and great free content in your inbox daily

Follow these simple steps to get the benefits:

1. Scan the QR code or visit the link below

https://packt.link/free-ebook/9781803248806

2. Submit your proof of purchase
3. That's it! We'll send your free PDF and other benefits to your email directly

www.ingramcontent.com/pod-product-compliance
Lightning Source LLC
Chambersburg PA
CBHW081452050326
40690CB00015B/2769